FOOD ACTIVISM TODAY

Food Activism Today

Sustainability, Climate Change, and Social Justice

Donald M. Nonini and Dorothy C. Holland

NEW YORK UNIVERSITY PRESS

New York

NEW YORK UNIVERSITY PRESS
New York
www.nyupress.org

Library of Congress Cataloging-in-Publication Data
Names: Nonini, Donald Macon, author. | Holland, Dorothy C., author.
Title: Food activism today : sustainability, climate change, and social justice /
Donald M. Nonini and Dorothy C. Holland.
Description: New York, NY : New York University Press, [2024] | Series: Social
transformations in American anthropology | Includes bibliographical references and index.
Identifiers: LCCN 2023020453 | ISBN 9781479810970 (hardback ; alk. paper) |
ISBN 9781479810987 (paperback ; alk. paper) | ISBN 9781479810994 (ebook other) |
ISBN 9781479811267 (ebook)
Subjects: LCSH: Local foods—United States. | Food supply—Environmental aspects—
United States. | Food supply—Social aspects—United States.
Classification: LCC HD9005 .N66 2024 | DDC 381/.410973—dc23/eng/20230526

LC record available at https://lccn.loc.gov/2023020453 This book is printed on acid-free
paper, and its binding materials are chosen for strength and durability. We strive to use
environmentally responsible suppliers and materials to the greatest extent possible in publishing our books.

Manufactured in the United States of America

10 9 8 7 6 5 4 3 2 1

Also available as an ebook

CONTENTS

PREFACE AND A NOTE ABOUT VOICE

There is a predicament, a seemingly unsolvable dilemma, at the heart of this book's argument. This predicament is the universal incapacity of local food activists to reconcile their commitment to sustainable local food and farming with the imperative to provide the large number of poor and working-class people in the United States with access to the adequate fresh, nutritious, and culturally appropriate food they need to sustain themselves on terms that preserve their autonomy and dignity. This predicament was repeatedly brought up to us by our consultants, but no one said it more clearly and poignantly than Jay Thomas of Durham, who observed, "It's a foundational problem because from a farmer's perspective, working in sustainable agriculture, they need to be selling their vegetables for more than they are [able] in order to make ends meet. And from a justice perspective, underprivileged folks need to be able to buy the food from those farmers for less than they're selling it for now." This predicament is the conundrum that this book is devoted to unpacking—a challenge that only a combined approach drawing on ethnography and the political economy of class and race in the contemporary United States is up to meeting.

A word about voice: Throughout this book, the first author (Don Nonini) takes full responsibility for writing all chapters of this book except one (chapter 10, "Assessing the Transformative Significance of Food Activism"), which was originally written by Dottie (Dorothy) Holland as part of a coauthored journal article, a part that she intended to transform into this chapter. She was unable to do so before she died in April 2019. Don did so with minimal changes and takes responsibility for including it in its edited form as a chapter in this book.

Don Nonini and Dottie Holland began this project by engaging in many hours of close collaboration in early coteaching and research design from 2005 to 2008. We coapplied for the successful National Science Foundation grant that funded this research project. Moreover, from the period of in-site ethnographic data collection in 2009–2011 up to Dottie's death in April 2019, she worked as a full partner with Don in overseeing our research team's collection of data and in joint discussions with our research team members and then with Don in interpreting and analyzing our ethnographic data in the years 2011–2018. She played an equal role in the design of the book's argument and actively reviewed and edited drafts of chapters written by Don until her medical condition made it impossible for her to continue in early 2019. The argument of the book has shifted since then, but the basic framework of the book remains as a joint construction.

It is as tribute to Dottie's role in the fashioning of this book that the surviving author has used the first-person plural "we" in its chapters. I (i.e., Don Nonini) have tried to avoid other, rhetorical uses of "we." I have made one exception to writing as "we": in the second half of the conclusion, I use first-person singular voice to present my political, economic, and ecological analytical projections into the future, because I do not know whether Dottie would have agreed with them.

That said, Don as first and surviving author takes sole responsibility for any errors made.

Introduction

Living in One World, Imagining Another

Anybody who starts farming doesn't sit down and go, "Wow, I'm gonna get rich doing this." . . . The market will only bear what the market will bear. I can't charge $10 a pound for my pork sausage and expect to get it. . . . So, anyway, we're doing this not because it's an altruistic endeavor but because it's the right thing to do, and we do need to make [it] economically sustainable. Farming is the one profession that we cannot live without. And it's a noble profession, and I want to be part of encouraging people to start farming and to get back to farming because we are losing small farms and farmers at a ridiculously fast rate. And I don't know about you, but when I'm old, I don't want to have to rely on other countries for my food. That's the place we will be if we don't grow farmers, if we don't protect spaces to farm in, if we don't protect genetic diversity in our plants and animals.
—Sonya James, Felicity Farms, Lincoln County, North Carolina

Local Food Activism and the Local Food Movement in the United States and North Carolina

In 2011, when we met Sonya James, a small-scale livestock farmer and restaurant owner in the Western Piedmont region of North Carolina, she and her partner, Jessie Newman, were in their seventh year of raising one hundred Tamworth heritage hogs and a few chickens and ducks on their ten-acre farm, while cooking and baking in a restaurant they owned near Charlotte that offered meals prepared from locally grown livestock and produce, even as they operated a mobile local food stand in the Charlotte area. We were drawn to interviewing Sonya and Jessie not only for their personal investment in local, small-scale and nonindustrial farming but also because they were celebrated leaders in the Charlotte local food movement known for their commitment to artisanal livestock rearing and local food cuisine and for supporting other local farmers, restaurant owners, and others working together to develop the local food economy of Charlotte, over and against the dominant, globally sourced industrial food sector.

Despite their incessant labor on their farm and in their restaurant, Sonya and Jessie undertook a livelihood that was economically precarious at best. They depended almost entirely on their own labor on their farm and the labor of nearby small farmers at a time when farms producing industrial food in the US hired scores or hundreds of farm laborers. Profit margins were small to nonexistent, as Sonya's interview attested. Much depended on the operation of their restaurant and on maintaining its reputation for providing fresh, delicious, and humanely grown meat within the highly competitive regional Charlotte upscale restaurant economy. But the odds against long-term success were high.

Sonya and Jessie were devoted to growing small numbers of heritage hogs that provided high-quality and delicious pork grown under free-range conditions for their own restaurant and other restaurants, but they were subject to cutthroat competition from industrial pork wholesalers like Sysco and Smithfield, as some Charlotte restauranteurs purchased these wholesalers' pork while falsely advertising it as artisanal and locally grown by Felicity Farms. These wholesalers could easily undercut Sonya and Jessie's prices, because unlike Sonya and Jessie, wholesalers purchased their pork from large-scale industrial processors whose costs were underwritten by government subsidies for the soybean- and corn-based feeds that millions of hogs consumed in massive amounts and because industrial hog slaughter and processing were undertaken by automated machinery operated by unorganized, hyperexploited, and largely immigrant workers.

Relative to fifty years previously, there were few small-scale farmers growing local food with whom Sonya and Jessie could cooperate, since farmland in their county, located on the edges of Charlotte's suburban expansion, was difficult to find, extraordinarily costly, and highly taxed given the suburban housing developments that hemmed in small farms on all sides in the region. This made it almost impossible for younger people seeking to become small farmers to find farmland. Small farming was thus already in decline when we met Sonya and Jessie in 2011.

It was only due to Sonya and Jessie's long hours spent farming, cooking, baking, and overseeing their restaurant staff that they were able to earn a "living wage," that is, a livelihood, at all. Integral to their labor was their combining meat from their own farm's heritage livestock with the produce grown by other local farmers in the meals their restaurant offered, as well as their continuous efforts to promote local farming and nonindustrial food growing, which attracted upscale Charlotte consumers to the restaurant. Yet the meal prices that Sonya and Jessie had to ask for to stay in business precluded most Charlotte residents, and certainly those belonging to its poor population, from ever being able to patronize their restaurant.

As an unhappy coda, in September 2020, at the height of the COVID-19 pandemic, with a population facing lockdown, and after more than a decade in operation, Sonya and Jessie closed their restaurant. Jessie, its chef, wrote, "Due to covid restriction, the state of our economy and all this uncertainty, we are closing our doors for good. Anyone who truly knows me, knows that my craft here at [the restaurant] was my passion. I came . . . [here] to show how amazing it can be, to buy products from my neighbor farmers and that freshness and quality is utmost important" (Harvest Moon Grille 2020). As of the time of this writing, not only had Sonya and Jessie's restaurant not reopened, but their ten-acre farm in Lincoln County appeared no longer to be in operation as a working farm.

Sonya and Jessie's story of barely getting by, hanging on for years, and the ultimate failure of their small enterprise illustrates the precarious and challenging situation faced by the local food sector and those, like Sonya and Jessie, who are devoted to it. Their story is broadly representative of so many people who work on behalf of sustainable, local farming and the food it produces in the Charlotte region and throughout the United States. So, too, does their story, amplified many times over, have implications for the lack of access to fresh and nutritious food that poor populations in Charlotte and elsewhere in the US experience and for efforts by activists to address the problem. We began our research on food activism in four sites in North Carolina in 2011 with the conviction that it was imperative for us to investigate how food activists like Sonya and Jessie in North Carolina were seeking to make local farming and food more sustainable and working to make locally grown food more accessible to the poor population, as an index to the situation of food activists and the local food sector in the US more broadly.

Why Study Local Food Activism?

There has been increasing public interest in the United States around sustainable agriculture, "eating local," and "eating organic." In many parts of the United States and beyond, there is heightened interest in "local food" and in the widely circulating ideas that food produced on small farms in the vicinity of where it is eaten is not only better tasting than food produced by the globally sourced agroindustrial food system but also ethically superior on the grounds of its health, environmental, and community benefits. This ethic and the desires that inform it are epitomized in the slogan, "know your local farmer!" At the same time, many other people have become concerned to move beyond the deep inequalities of access to such food that have become entrenched by the class and racial injustices of US society in order to increase access to this food by its poor and working-class population.

Local food and farming intuitively are greatly attractive to large numbers of people. Reasons range from the pleasures of eating fresh and nutritious food grown by one's neighbors and people living nearby to the gratifications that people experience eating together, to the joys people attain when they master new skills and arts of gardening, farming, or livestock raising, to the clear importance that people attach to teaching their children where food comes from and how to grow it, and to the ethical energies that animate sharing fresh and nutritious food with neighbors and community residents who are hungry.

The US population is interested in local, sustainably grown food and farming for another reason than just its compelling physical, social, and aesthetic attractions: the politics of economic security. There is a poorly acknowledged but widely held impulse among the US population toward economic democracy defined by food self-sufficiency. What would be more despotic than a population's not being allowed to gain access to food or grow its own food or even being prevented from learning how to grow it? This is a popular impulse that has been activated for many as they have experienced and witnessed the precarious situations many people face in the insecurity of the times in which we live. It is an impulse toward what food activists call "community food security," and we say more about it shortly.

All these aspects of growing, distributing, and sharing locally grown foods make it abundantly clear that learning about the meanings of food and farming in local settings has many attractions. Those who are most deeply involved in constructing those meanings—food activists and other actors in local food systems—are excellent subjects from whom to learn these meanings.

The widespread popularity of locally grown, sustainable food can be witnessed in the proliferation of related media, including books on nutritional food and cookbooks, *Iron Chef* and similar programs on television's Food Channel, and speakers tours by celebrity authors, organic farmers, and chefs, such as Michael Pollan of the *New York Times*; Joel Salatin, the author and proprietor of Polyface Farms; the chef Alice Waters of Chez Panisse; and Will Allen, an African American urban gardener-organizer from Milwaukee and MacArthur Award winner. Increased public demand for "organic," "fresh," "natural," and "local" food has led to the rise of national corporations such as Whole Foods to service upper-middle-class consumers, while Walmart has become the largest supplier of US Department of Agriculture (USDA)–certified "organic" produce (sourced, however, globally) for lower-income consumers. Meanwhile, consumers in droves have fled supermarket center aisles stocked with packaged, manufactured foodstuffs to prowl the peripheral aisles for "fresh" produce, fruits, and bakery goods (Taparia and Koch 2015). Sales of such produce by these and other retailers are reinforced by corporate campaigns that brand companies' "organic" and "sustainable" products, including not only produce but also livestock, fish, bleach, and paper towels.

A burgeoning scholarly literature in the new field of critical food and agriculture studies has sought to document these changes while linking them to the emergence of specific movements around food (Allen 2004; Allen et al. 2003; Beus and Dunlap 1990; Dahlberg 1991; Guthman 2008a; Guthman and DuPuis 2006; Hassanein 2003; Vallianatos et al. 2004; Wekerle 2004).[1] North Carolina cities and towns are on the tour circuits of the celebrities mentioned earlier. They have increasing numbers of residents seeking organic, fresh, natural, and local produce and meats; and the state's cities and towns have been the sites of rapidly growing numbers of new farmers markets, urban gardens, periurban small farms, and "community supported agriculture" arrangements.

Since the 1980s, another scholarly approach has focused on the emergence of the charitable or emergency food sector, its ongoing reorganization, its projects, and its values (e.g., Poppendieck 1998, 2014; Riches 2018). These are often manifested in neighborhood- and school-based "food drives" and omnipresent church food pantries and kitchens that distribute charity food to "poor neighbors" in the state's cities and towns (Berner and O'Brien 2004).

Over the same period since the 1980s, tens of thousands of people have become urban gardeners, while urban farms yielding produce, fruit, nuts, eggs, and so on for sale have proliferated. Thousands of gardens and farms have emerged on rooftops, in backyards, and on derelict house lots. Tens of thousands of small-scale urban and periurban farmers, making use of their own and farmworkers' physical labor, low levels of fossil fuels, and their deep knowledge of local soils, crops, and growing conditions, have turned to sell their foodstuffs in hundreds of new farmers markets established throughout the US, with the numbers of farmers markets nationally having increased 450 percent between 1994 and 2012 and totaling eight thousand by 2016 (Robinson and Farmer 2017, xv).

The economic value of produce sold through these farmers markets has continually increased. "Community supported agriculture" (CSA) businesses have proliferated, with twelve thousand active nationally in 2016 (Robinson and Farmer 2017, xvi) and thousands of new subscribers joining them as a way of gaining access to the fresh locally grown foods they provide, while providing a steady income to the farmers who grow the food. In 2015, almost 115,000 farmers engaged in direct sales to customers through farmers markets, on-farm stores, roadside/tailgate stands, CSAs, and pick-your-own sites (National Agricultural Statistics Service 2015). Other thousands sold their produce to institutions like schools and hospitals.

Older and a few newer institutions associated with locally grown direct-sales farms have become increasingly visible in urban and rural

landscapes: food consumer cooperatives, culinary schools, produce ag-
gregation hubs, and incubator farms providing hands-on experience to
aspiring young farmers, as well as hundreds of gourmet restaurants on
cities' streets featuring "fresh organic produce" and "artisanal meats"
at the top of their menus. In cities, scaled-up coalitions of local food-
system stakeholders have come together at regional scales to cooperate
in "food policy councils" to plan and coordinate policies to encourage
local farming and fresh and nutritious local food consumption.

These features of "the local food system" have not come into existence
over the past three to four decades due only to the hard work and ingenuity of local small-scale farmers and their farmworkers, restaurant
chefs and proprietors, and other food workers and artisans (cheese makers, charcutiers, beekeepers, etc.). Thousands of other people who were
not farmers or in the food industry have contributed tens of thousands
of their labor hours, have engaged in numerous economic transactions
and interpersonal exchanges over local food, and have built or helped
build new enterprises (farmers markets, food co-ops, etc.) in order to
improve their local food systems in cities, towns, and rural areas across
the US. Many thousands of other people have also worked thousands of
hours conscientiously committed to improving their local food systems
as gardeners, as volunteers or interns (e.g., on CSAs), as organizers of
meetings to set up county land trusts (to hold farmland in perpetuity),
as leaders of autumn farm tours, and in many other capacities. We call
all those people who have also supported and promoted changes in the
ways food is grown, prepared, distributed, and consumed in the United
States "food activists." This book deals with their experiences in four locales in the state of North Carolina. We are convinced that what we have
found there can provide transposable crucial insights into food activism
elsewhere in the United States.

In this book, we focus on food activists who aim to improve their
local food system. By "improving" a local food system, in this book,
we mean two things: making it more robust and enduring—in a word
"sustainable"—or making it more "just" so that social and economic inequalities do not threaten access to foods sufficient in quantity or culturally appropriate quality for the people in need of it.[2] We refer to those
who undertake the effort toward more sustainable local food systems as
"sustainable food and farming activists." Almost all of the activists with

whom we spoke were *not* local farmers or farmworkers—if only for the straightforward reason that local farmers and farmworkers were either too busy with their on-farm labors or, in the most frequent situation when their on-farm incomes were insufficient to live from, engaged in other part-time or seasonal employment to get by (Paul 2017). It is crucial to acknowledge the activism of nonfarmers because without them, local farmers and other food workers alone would have been completely unable to carry out the achievements we describe in this book.

Other activists have sought to make our local food system more socially just. From the early 1980s onward, a new sector in the US food system has grown rapidly, aiming to remediate chronic food insecurity or hunger among an increasingly large population of poor people in the US: the "emergency food" or "surplus food" sector. Consisting in part of government benefits, including transfer payments (i.e., the Supplemental Nutrition Assistance Program [SNAP] and the Special Supplemental Nutrition Program for Women, Infants, and Children [WIC] vouchers) and surplus food provided to food insecure populations (USDA's Emergency Food Assistance Program [TEFAP]), the emergency food sector has also come to include new charitable organizations distributing food, like food banks, food pantries, and community kitchens operated by churches, nongovernment organizations, and public-private partnerships. We particularly noted the increasingly dominant role of these organizations that provided charitable food to food-insecure people as a routinized mechanism of "emergency" food provision to millions of people.

During our research, we began seeing these organizations operating together as a highly institutionalized and hierarchical system. We began to discover that there were reasons why large numbers of food-insecure people appeared to be unwilling to act as recipients of charitable foods or were sometimes ignored or harassed by government agencies and charitable food outlets and had serious unmet needs. Responding to their needs, there were activists outside the large-scale charitable sector or working on its edges who labored many hours to create community-wide food resources that would lead to enduring food security for the large number of poor people increasingly at risk of hunger. Rather than focus on meeting only recurring individual needs, they sought collective solutions—ones that food-insecure people could play a role in creating. It was they who not only passed out produce to needy neighbors or parishioners but also started and kept up community gardens; they used their knowledge to teach local poor youths to "grow their own" by providing them with small plots of land, veggie starts, compost, and the like; they volunteered to work in food pantries or church kitchens; they worked with poor applicants to help them secure SNAP benefits, but they also might help a neighbor start up a poultry microenterprise to provide eggs for their families or to sell a few for pocket money. They were what we call "community food security activists."

It is important to be clear about one thing: many of these activists were themselves poor or had modest incomes, and many of them were people of color. If they were "do-gooders," they did good in their own neighborhoods, for their neighbors. We refer to all those people who worked in these ways, either within the charitable food sector or outside of it in order to provide more adequate, nourishing, reliable, and culturally appropriate local food to food-insecure people, as "food security activists" or an exceptional few as "community food security activists"—we clarify the differences between the two shortly. It was vital to us, for reasons we give later, that these activists sought to relate to food-insecure persons in ways that were just and conducive to the latter's welfare and dignity irrespective of their class, racial, or gender identities.

A crucial distinction in this book is between charity-based food security activists, who worked for charitable food organizations (food pantries, etc.) and those few who were "community food security activists." The latter work to ensure the food security of the people whose communities (locally defined) they belong to in ways that seek to maximize the power of those residing in the community to grow, prepare, and consume the foods they deem culturally appropriate over time. This is consistent with the definition by the national Community Food Security Coalition (CFSC) of "community food security" as "a condition in which all community residents obtain a safe, culturally acceptable, nutritionally adequate diet through a sustainable food system that maximizes community self-reliance and social justice" (Food Security 2022).

Between 1994 and the CFSC's closing in 2012, most community food security activists tended to ally their orientations with the CFSC, which many joined with their organizations—by 2010, there were more than three hundred organizations that were CFSC members, but only three were in North Carolina (authors' personal observation). Community food security activists may work *with* other activists in the charitable food sector to provide members of their communities with food assistance, but they are primarily accountable to the people whose food needs they serve, not to food charities.

Like sustainable farming activists, thousands of food security activists (both charity-based and community food activists) across the US have labored many hours to improve their local food economy by seeking to transform it materially (e.g., starting a church garden to grow food for elders), spent many hours in meetings and discussions to build such projects, and occasionally were able to bring new food-based enterprises into existence (e.g., a mobile food market in a neighborhood in an urban "food desert").[3] They have sought to make food access for all who were in need of it as equitable, affordable, and culturally appropriate as possible. While we found a flourishing charitable food sector in North Carolina, we found far fewer community food security activists compared to other states. Eventually, we discovered a number of exceptional activists who "crossed over"—they were simultaneously sustainable farming/food *and* community food security activists. For reasons that will become clear, these were extraordinary practitioners of social change.

Theoretically, our research from the outset focused on the "moral logics" of both kinds of food activists vis-à-vis the moral logic of the neoliberal corporate industrial food sector, that is, on their discourses and embodied practices about what the relationships between people and food *ought* to be as different from that sector's. Empirically, our initial reason for focusing on both kinds of activists was that we had hypothesized that certain food movements that had emerged outside North Carolina, in particular on the US West Coast and in the Midwest, showed sustainable farming and community food security activists either working together or even being the same people active in a convergent movement—and we saw this as a national trend that would be evident if we undertook ethnographic research in North Carolina. For example, the activist-scholar Patricia Allen argued in her classic book *Together at the Table*, "Local food systems may solve many of the problems that concern alternative agrifood system advocates. They are considered to have environmental benefits, such as reducing energy use; and social benefits such as creating new opportunities for solving problems of hunger and homelessness; and economic benefits such as improving opportunities for employment. . . . Thus they tie together the priorities of the sustainable agriculture and community food security movements" (2004, 66).

Our hypothesis proved to be mistaken. Unlike North Carolina, urban areas on the West Coast and in the Midwest provided a fertile ground for many such crossover activists to emerge (Allen 2004). We believe that these areas had large working-class populations drawn from racial or ethnic groups long resident in these regions with substantial ties to rural-based farmers', environmental justice, farmworkers', or displaced migrants' movements (Mares and Peña 2011; Morales 2011). These cultural connections made ties between the two kinds of activists within these groups feasible and often allowed them to be able to work across urban and rural divides.

In contrast, North Carolina's larger urban areas have become "landscapes of consumption" whose economic fortunes were tied to the state's flourishing (if now somewhat waning) globalized high-tech and services economy, while most of its rural areas have become depressed "landscapes of production"—due to extensive deindustrialization and agricultural decline arising from food imports, resulting in few jobs and high unemployment (Holland et al. 2007, 18–31; Nonini et al. 2015). The highly unequal economic urban-rural divide in North Carolina has been exacerbated by two simultaneous demographic shifts. While large numbers of highly educated, mostly white professionals, technicians, and businesspeople have *moved into* its cities from out of state, rural areas in North Carolina have long experienced people *moving out of* them, including ambitious whites and many African Americans fleeing racial discrimination. The result has been that rural areas have been left with majorities of relatively poor, downwardly mobile whites and politically subordinate minorities of very poor African Americans, American Indians, and Latinx people.

Taken together, these factors in North Carolina have not been conducive to positive relationships between rural-oriented sustainable food activists on the one hand and urban-based community food security activists, or for that matter charity-based food security activists, on the other. Such cooperation has arisen more readily elsewhere, where cultural and economic divisions do not coincide so acutely with the urban-rural geographic divide. The legacies of racial divisions and class exploitation and neglect still lay heavily on North Carolina's landscapes.

Our Research Project: Research on Food and Farming for All (ROFFA)

From the summer of 2010 through spring of 2012, we and several colleagues undertook a collaborative ethnographic project to study the "local food movement" of North Carolina in four different sites. To capture the diversity of the state's local food systems, we selected two rural sites, one in the mountains of western North Carolina and the other in the impoverished northeastern region of the state, and two urban sites, one the metropolitan region of a large city in the western Piedmont and the other a smaller regional city in the state's central Piedmont. Members of our research group interviewed leaders, activists, and volunteers working in nonprofit organizations committed to sustainable agriculture and to food access for food-insecure populations. They met with farmers, permaculturists, small-scale food-business owners and vendors, farmers market managers, community gardeners, clergy, food-bank managers, and food-pantry and community-kitchen operators who provided "emergency food" to hungry people, with Agricultural Extension agents, nutritionists, and urban planners, and with many others. Members of our research group visited farms, community gardens, and farmers markets; attended and participated in meetings held by nonprofit organizations, church groups, gardening groups, farmers market boards, food policy councils, nutritional educators, regional farming organizations, and national "community food security" organizations; and observed and participated in work in community gardens and related sites. As we noted earlier, almost all the farmers we met were not food activists, and therefore we did not (and indeed could not) systematically interview them.

Our researchers found that sustainable farming/food activists often manifested very different perspectives from food security activists, except when the two "crossed over" and were the same people and sustainable farming activists became *community* food security activists or at least when both kinds of activists worked closely together— extraordinary situations whose importance we came to learn much from. The practices and projects, the labors, transactions, and enterprises of these activists are set out in this book. By investigating both kinds of food activists together and examining their interrelations, we believe that the research findings we report in this book are unique.

However, as crucially important as these desirable qualities of growing, eating, sharing, and promoting locally grown foods are, we discovered in the course of our research that reporting about them was not the only compelling reason to learn more about what local food activists in North Carolina were up to and why. To understand where we are heading as we examine together sustainable farming/food activists and food security activists (both community- and charity-oriented), it is necessary to reimagine the world humans live in at a planetary scale—the scale much written about in other contexts but also related to the themes of this book.

Small-scale local farming and the local food sector that are exemplified in Sonya and Jessie's story have not been isolated from the broader environmental, political, and economic trends that have shaped US society over the past four decades. These shifts represent convergent cultural/political, environmental, and climate crises. This goes far to explain why, despite recent growth, local food economies have not expanded much beyond their niche-market status as specialized services offered to a well-to-do, largely white, demographic. Meanwhile, at the other (lopsidedly large) end of the income and wealth distribution, these trends currently leave a large number of consumers unable to afford locally grown, fresh foods, while a growing number of poor people remain food insecure.

We start with the shift toward neoliberalism.

What Is Neoliberalism?

This book deals with local food systems in the United States and specifically in North Carolina by examining the experiences and practices of food activists who are seeking to shift local food economies from industrial agriculture toward more environmentally sustainable local farming and food production and to provide the food so produced to food-insecure people living in their locales. (However, few studies as yet [but see Gray 2014] have focused on the wages and working conditions of farmworkers—a lacuna we return to in this book's conclusion.)

The environment, the climate, species diversity, the human population, farming, food, and hunger are not only biomaterial processes applicable to life on Earth and refractory to human intervention but are also culturally mediated and socially constructed phenomena (Biersack 2006). Our capacity as humans to confront these phenomena necessarily requires that we see them *as connected to* human social activity and cultural productions, although their material dimensions *cannot be reduced* to their social and cultural dimensions. As humans conceive and experience these material phenomena, they draw on prevailing cultural ideas and visions about them, on social interactions to cope with them, on culturally defined individual plans to adapt to them, on socially enjoined collective strategies to mitigate them, and on cultural justifications to preserve or transform them. Humans produce these social and cultural phenomena every day as they labor on, interact with, and transform nature and the human place in it as they constantly remake it (Smith 2008). As such, humans enter into relations with other relevant social realities and cultural processes that they themselves have previously created.

One such cultural process and social reality is something that we, joining the analyses of many other scholars, call "neoliberalism." Neoliberalism goes by a variety of other academic and popular names: "market fundamentalism," "economism," "economic determinism," or the theory of the "free-market economy" or of the "private-enterprise system." Neoliberalism is simultaneously a discourse of theories arising from the Western discipline of economics, a doctrine about how humans should optimally organize their economic lives to suit market realities (competition, profits, etc.), a class-based ideology that valorizes those who have wealth over those do not, and a rhetoric of terms used to persuade or coerce—or to justify coercing—people who are unwilling to live and behave in accordance with its prescriptions. Most important, neoliberalism defined in all these ways has become the dominant form of governance that since the 1980s has increasingly come to structure the everyday lives of more than three hundred million people in the United States and of hundreds of millions of people elsewhere on Earth.

Neoliberalism is no longer an academic or abstract matter but one of everyday politics. Its political ascendance in the United States and elsewhere in the advanced industrialized countries since the 1980s and its role in promoting economic globalization mean that it has had and continues to have a tragic effect on how humans have responded to the processes of climate change and environmental degradation that now threaten the continued survival of humans and the nonhuman species on which humans depend for their sustenance. Neoliberalism, a reflection ultimately of the underlying system of capitalism that it promotes, has rendered the megaprocesses of climate change and ecological devastation much worse and increasingly dangerous for human survival.

Neoliberalism is encapsulated in the idea that the "magic of the market" is the optimal solution to meeting human needs. Applied to our food and foodways, it animates and justifies not only fossil-fuel use and the private car but also industrial agriculture as the necessary and desirable preconditions of "our way of life." As it turns out, neoliberalism is contemporary US capitalism's medium for defining our way of life as requiring the infinite—or more accurately, the indefinitely expanding—consumption of all forms of material resources and living beings that can be bought and sold. Neoliberalism constitutes industrial capitalism's current charter for its indefinitely increasing fossil-fuel consumption and its transformation of the Earth's ecosystems into "resource" commodities—including industrial food—that humans should buy and consume.

We need to know more about neoliberalism, because if the transformative possibilities that exist within local food activism are ever to be attained at superlocal, regional, national, transnational, or even planetary scales, we first need to come to terms with what the dominance of neoliberalism in the US has meant for local food activism, around both local farming and the provision of food to about one in every seven to eight people in the United States who are "food insecure." This is, in part, what this book is about.

The Postwar History of Neoliberalism and the "Perverse Confluence"

Neoliberalism began as a proactively moralizing discourse and even a doctrine within economics, not so much a description of the world as it is but instead a prescription for how it ought to be (Brown 2015, 2019). Nonetheless, as its conservative advocates have become increasingly dominant politically in the United States, neoliberal discourse has become increasingly hegemonic. That is, its prescriptions have become increasingly taken for granted and implemented by the political elites who govern the local institutions of US public life, a dynamic we demonstrated in a previous book, *Local Democracy under Siege*, focused on local politics in North Carolina (Holland et al. 2007). We found in that work, for example, that around education, the use of space, and the environment, "public-private partnerships" have come into existence since the 1970s in local US politics initially as ways of making government "leaner and more efficient" (cf. Osborne and Gaebler 1992) by combining the logics of government with the virtues of the private sector—either those of corporations or businesses or those of private charities and volunteering individuals (Holland et al. 2007, 1–17, 155–186).

Neoliberalism puts a premium on the language of economics—the view that social action can best be valued in relation to its contribution to successful market performance and accumulation of private property, to whether it is "business-like" or "entrepreneurial" and embodies specific capitalist virtues such as efficiency, cost-effectiveness, profitability, and economic self-interest. Above all, neoliberalism values the self-conscious individual entrepreneur-investor who constantly seeks to maximize profits, at whatever expense to others, in order to benefit society as a whole. In contrast, working-class people are merely expected to be "work ready" to take any employment on offer by such entrepreneurs.

What began as the original impetus of neoliberalism in the 1970s–1980s, to eliminate state social benefits, deregulate "the economy," and "downsize government" ("rollback neoliberalism"), has since the first decade of the twenty-first century shifted to encompass increasingly more areas of daily life in the US as it has become institutionalized ("rollout neoliberalism") (Peck and Tickell 2002). During this shift, something interesting has happened, what the political scientist Evelina Dagnino (2005, 158–159) calls the "perverse confluence" associated with neoliberalism. This is the paradox that even as neoliberalism has attacked collective national public solutions to systemic injustices at the national scale, it has done so by devolving responsibility for governance initiatives to redress injustices "downward" from the scale of the nation-state toward the local scale. Moreover, this change has led to a focus on subsidiarity—a shift in which previous modest individual projects carried out by local private actors, such as charitable initiatives, are elevated and celebrated over "wasteful and inefficient" national-scale programs as appropriate means to redress "imbalances" (i.e., injustices), at least to the extent that these injustices deserve to be redressed (Holland et al. 2007, 107–129; on subsidiarity, see Riches 2018, 33). This in turn has allowed neoliberals to cut back or eliminate national-scale funding to remediate these injustices on the grounds that local "civil society" can do the job better than a remote state bureaucracy because local civil society is "closer to the real problems" that people encounter.

The paradox consists in the fact that modest but new national-scale state funding has been promised for such local projects, thus animating and encouraging local actors, some previously disempowered, to pursue them. Thus, local-level individual or small-scale social projects to mitigate injustices, rather than large-scale social-movement efforts to eliminate them, can enter new political arenas that were previously not open to them.

As we and our coauthors point out in *Local Democracy under Siege* (Holland et al. 2007, 1–17), the rise of neoliberalism or market fundamentalism from the 1980s onward led to a variety of "experiments with democracy" associated with then-new public-private partnerships— ones in retrospect aimed to optimize previously public institutions in the direction of markets, in one direction, or toward charity and the voluntary sector, in the other. In the former case, where public (government-owned) assets were economically valuable and could be allocated in a "business-like" fashion, that is, as profitable for private capitalists, "market-oriented partnerships" have come into existence. These have harnessed public property and funding to invest in—and thus subsidize—private urban "revitalization" projects (e.g., downtown gentrified housing, sports stadiums) and similar ventures that attract global capital (Holland et al. 2007, 157–166).

In contrast, where public institutions could not be optimized to make profits, this has posed a problem for neoliberalism since, if anything, it has proven far more difficult, often impossible, to find privately controlled profitable solutions to social problems such as hunger, homelessness, and child and elder care, previously addressed by the national state. Instead, we have witnessed the rise of "community-oriented partnerships" that have taken the form of nonprofit organizations supported by small amounts of government funding and modestly backed by private corporate and nonprofit foundation capital that have come into existence to solve such problems (Holland et al. 2007, 125–129). Community-oriented partnerships thus seek to meet housing needs through homeless shelters, provide job-skills training through nonprofit organizations supported by foundation funding, offer elder care through publicly funded senior centers operated by volunteers, and staff food banks that collect surplus foodstuffs from retail food corporations and church-based food drives and distribute food to the hungry via soup kitchens (Holland et al. 2007, 127; see 125–128). How adequately these partnerships have solved these public problems is another question, raising many issues. We see the devolutionary impetus toward the local and the subsidiary brought on by neoliberal politics as one "forcing" factor in the efflorescence of local food activism in our four sites, at the same time that activist projects have been chronically underfunded by both state and private sources.

The current period has been one of neoliberal incursions into daily life, as more and more people have been progressively affected by cultur-ally specific neoliberal ideas, discourses, and ways of organizing social life (Holland et al. 2007). It would be surprising if neoliberalism did not affect food activism and the food movement, and indeed we see neoliberalism as having been influential—but not determinative—in the discursive constraints it has imposed on activism. These are constraints that we believe have a limited hold on social action; they have begun to loosen precisely because of the deep skepticism with which a growing number of Americans, particularly younger adults, have come to regard the increasingly empty promises of contemporary capitalism in light of the 2007–2008 global financial crisis, the student debt crisis, and the onset of the generalized crisis of climate change.

Climate Crisis, Ecological Crisis, and the Global Industrial Food Sector

The palpable manifestations of the climate-change crisis and its connec-tions to anthropogenic greenhouse-gas emissions have been increasingly and widely acknowledged by scientists, governments, and the public as a threat to human survival. These have been demonstrated by the scientific evidence that the extraction and consumption of fossil fuels and their products generate huge volumes of greenhouse gases that are warming the atmosphere and oceans of the planet (IPCC 2013, 2018). Even holdout climate-change denialists no longer dare to claim that it is nonexistent but now fall back to claim that it is "uncertain" and that "we don't know enough" to act to slow down greenhouse-gas emissions.

Despite public and news-media attention almost exclusively on emissions by privately owned vehicles and coal- and gas-fired generator plants as major producers of greenhouse gases, an increasing number of experts but relatively few members of the public have become aware of the enormous role that industrial agriculture plays in increasing the planet's greenhouse-gas burden. The current system of globally sourced industrial agriculture, including industrial livestock production, forest deforestation for livestock feed grasses and commodity crops, petroleum-based fertilizers, and long-distance transnational logistics chains that refrigerate, distribute, and retail food accounts for between one-quarter and one-third of all greenhouse gases annually emitted on the planet (IPCC 2019, 10; Poore and Nemecek 2018; Berners-Lee 2019, 22; Lappe 2011, 11).[4]

Even less well known to the public, much less acknowledged by decision-makers, is that scientific evidence increasingly points to a major worldwide ecological crisis. This is the imminent extinction of up to one million plant and animal species within the next few decades due to our current system of agriculture (IPBES 2019, 4). According to an article in *Nature* from May 2019, "About 75% of land and 66% of ocean areas have been 'significantly altered' by people, driven in large part by agriculture. . . . The loss of species and habitats poses as much a danger to life on Earth as climate change does" (Tollefson 2019, 171). An Intergovernmental Science-Policy Platform on Biodiversity and Ecosystems Services (IPBES) report makes it specifically clear that industrial agriculture and fishing—not nonindustrial and subsistence farming—are responsible for increasing species loss, loss in soil fertility and pollinator decline, and decreases in the global monetary value of crop outputs (IPBES 2019, 2–3). The report puts it bluntly: land use change, especially agricultural expansion, "has had the largest relative negative impact on nature since 1970" (IPBES 2019, 4). What makes these species losses and the ecological degradation they represent even more serious is that they are interacting with the ongoing, accelerating processes of climate change (IPBES 2019, 4).

Under these circumstances, there is only one conclusion we can draw, and it has critical relevance to the local food system, those who work in it, and those whom it serves. The conclusion is that the capitalist industrial food economy, although economically and politically dominant, is a major driver of both climate change and the disappearance and extinction of the ecological systems and species on which human life depends. It is a very real question whether over the next century the human species can survive industrial agriculture due to its internal operating logic based on indefinite expansion through fossil fuels on a finite planet and on a recidivist obsession with transforming nature's systems into food commodities, on one side, and into toxins, extinctions, and ecological simplifications, on the other, and then moving on to somewhere else to do the same thing, over and again, with no end in sight.

Under these conditions, it is crucial that careful investigations of the only known alternative to the global industrial food economy—the thousands of largely nonindustrial local food systems that survive (as it were) in its shadow, which are created by those who produce local foods and support those who do so—be undertaken with a fuller awareness of how these systems operate and how those who work within them seek to make them more sustainable and more just. Such is the broader strategic intellectual effort to which this book seeks to contribute.

Tragic Conjuncture: An Ideology (and System) That Exacerbates the Prospects for Survival

Neoliberalism, globalization, the market, and the corporate state are no way to deal with the current compounding and dynamically interrelated climate-change and ecological crises. Short-term corporate performance expectations keyed to the quarterly profit reports of Wall Street corporations externalize the ecological, environmental, climatic, labor, and social costs of private enterprises' "freedoms of the market" that are so deeply cherished by neoliberal advocates. If the paramount consideration of contemporary liberal states' neoliberal politicians is to seek to guarantee corporate profitability, then all other "goods" valued by humans have to be set aside as "externalities" that corporations are not responsible for dealing with.

Furthermore, nation-states are expected to deal with the costs and harms of these externalities to the human population and to the environment—but at a time when neoliberal governance, with its obsession on tax reductions for corporations and the wealthy, shrinks the size of state agencies that are expected to protect the environment and maintain labor standards. Supposed market solutions to environmental problems, such as putative transnational "carbon markets" for "carbon remediation," have been ineffective in reducing greenhouse-gas emissions. Instead, they have been a major source of corporate fraud (Frunza 2015). Nor is there much hope for technical fixes like cheaper solar panels—they may be great for lighting homes, but they have no known translation into energy-rich inputs for maintaining the fertile topsoils of farmland, much less into increasing the amount of water needed to grow the foods that US residents (and others) consume.

The prospect instead looms of a once-in-a-millennium human tragedy as neoliberalism disables the US and other liberal nation-states from dealing with the world's increasingly literal "perfect storms" of climatic change and ecological devastations. The noted journalist and social critic Naomi Klein writes about the "really bad timing" that

> we have not done the things that are necessary to lower emissions [of greenhouse gases] because these things fundamentally conflict with deregulated capitalism, the reigning ideology for the entire period. We have been struggling to find a way out of this crisis. We are stuck because the actions that would give us the best chance of averting catastrophe—and would benefit the vast majority—are extremely threatening to an elite minority that has a stranglehold over economy, our political process, and most of our major media outlets. (2014, 16, 18)

This "reigning ideology" is neoliberalism. Scholars have noted the particular historical coincidence that Klein points to—the era in which globalization and worldwide climate change have coincided (O'Brien and Leichenko 2000). As long ago as 2000, Karen L. O'Brien and Robin M. Leichenko (2000, 227) pointed to what they call "the double exposure": "cases where a particular region, sector, ecosystem or social group is confronted by the impacts of both climate change and economic globalization." They go on to observe what has become increasingly recognized as all too common: the existence of "double losers," "the regions, sectors, social groups, or ecosystems that are adversely affected by both globalization and climate change [and] should be of particular concern" (O'Brien and Leichenko 2000, 230). They give as an example one that is germane to this book: small-scale farmers trying to adopt drought-tolerant seed lines will be prevented by liberalization policies that link farmers' credit to purchasing genetically modified seeds patented by transnational agro-input corporations like Monsanto (now Bayer Crop Science) and Syngenta (O'Brien and Leichenko 2000, 230; see also Gutiérrez Escobar and Fitting 2016). Unfortunately, such events are now being recorded with increased frequency—both in the US and in the global South (Harris 2013).

Neoliberalism as the dominant form of political governance in contemporary nation-states is based on infinite expansion of human consumption on a finite planet, and it is a recipe for collective suicide— even over the short term. People in the United States have now lived with neoliberalism as the dominant way of governing the US and the rest of the world for the past four decades. Although forty years is an infinitesimally short period of time on Earth's geological time scale, what can now be witnessed during its tenure is a shocking acceleration by many orders of magnitude in the processes of change of Earth's climatic and ecological systems.

And what if even repudiating neoliberalism in favor of a "responsible" globally regulated green capitalism will not stop the pace of fossil-fuel-driven destructive change (Mann and Wainwright 2018, 169–171)? Within the global industrial food system alone, complex systems of natural and anthropogenic processes are linked in feedback and feed-forward (deviation-amplification) processes. For instance, intensive industrial livestock rearing (of beef, lamb, and poultry) in the midwestern and southern US and Brazil generates huge methane emissions but also requires large volumes of monocropped corn and soybeans as animal feeds, which in turn require large-scale use of fossil fuels to produce petro-fertilizers, to fuel farm machinery, and to provide energy to transport and refrigerate large volumes of animals, feedstuffs, and processed meats over long distances (Lappe 2011, 3–41). Monocropping in turn is exhausting midwestern topsoils (Philpott 2020). These processes have generated huge volumes of greenhouse gases, leading to climate-change effects like increasing periods of drought that reduce the areas of arable land and lead to disastrous (20–30 percent) crop shortfalls in major farming regions.

Unpleasant side effects "downstream" of these processes include increased zoonotic viral diseases like COVID-19 and H5N1 and related varieties of avian flu spreading from poultry concentrated animal feeding operations (CAFOs) to cities (Wallace et al. 2020); massive "land grabs" of thousands of hectares of fertile African and Latin American farmland by large agroindustrial firms, state-owned companies, and sovereign wealth funds (from China and the Middle East) (Clapp and Isakson 2018, 82–100); and Wall Street speculation in food commodities (Clapp and Isakson 2018, 29–55).

Industrial farming- and food-manufacture-related greenhouse-gas emissions, added to those from outside the food system, are reaching a quantitative threshold with regard to atmospheric and ocean temperature increase, that is, the imminent passing within the next two to three decades of global average mean temperatures beyond the increase of one and a half degrees Celsius since the beginning of modern industry (IPCC 2018). Ongoing increasing emissions along this trajectory have already led to a cascade of weather disasters and bioenvironmental catastrophes that threaten human subsistence and food access—within less than a human generation's time. This is what it means to be living in the era of the Anthropocene, or what more candid scholars call the "Capitalocene" (Angus 2016). Time is running out for one of two things to endure much longer: the survival of human beings or the continuation of neoliberalism as a dominant ideology of human life. Perhaps it is even time to ask whether we are arriving at the defining dilemma of the Anthropocene: Which is to prevail, the survival of industrial capitalism or the survival of the human species?

Neoliberalism, Food Activism, and Food Movements

Coming back to neoliberalism: How is this circle to be squared? Must the defining features of a good society—such as schoolchildren having enough nutritious, tasty, and culturally appropriate food to eat or small farmers receiving adequate incomes for livelihood by providing such foods—always be subjected to the stingy calculus of "cost-benefit analysis" that committed neoliberals demand? Do food activists have to measure their practices as being useful and valuable because they exemplify and promote successful market competition? The answers to such questions are both contentious and significant, because neoliberalism is not a scientifically accurate representation of the way the world is or of who the food activists we studied are but is instead coded into cultural discourses that the food activists we studied have accepted to various degrees.

This book thus poses the question, In what ways and to what extent have the practices of food activists become neoliberal or at least influenced by neoliberal discourse? Some social theorists, especially poststructuralists (e.g., Rose 1999; Dean 1999), have argued that neoliberalism represents a compelling discursive formation that successfully penetrates and transforms the subjectivities of its "subjects" to encompass their sense of reality and determine their behavior. In the name of social order, neoliberalism is said to confine people's horizons of possibility to a social Darwinist reality of possessive individualism and commitment to a mercenary economism that reduces all things to their monetary worth or finds them worthless. Our previous book, *Local Democracy under Siege* (Holland et al. 2007), for instance, had much to say about how neoliberalism deformed local politics but also allowed activists to develop new forms of activism through its "perverse confluence" that we discussed earlier.

Going into our research on food activism, we thus felt great uncertainty as to how activists would respond to neoliberal political and cultural forces. This mattered greatly because, as we propose later, the local food system and the place of food activists *might* represent a major alternative to the fragile but dominant industrial food system, but whether and how food activists were constrained by neoliberalism would determine how far they could go or be willing to go toward transforming our food system if a transition toward a post-/semi-industrial, post-fossil-fuel economy is an imperative for the future. How activists confronted, evaded, accepted, or rejected the neoliberal pressures they faced is therefore an important subject that we deal with in the second part of this book.

With so much at stake, our research findings suggest a reason for real optimism, but how far the promise will go is as yet unclear. Although a few food activists have been seduced by neoliberalism, most have not surrendered to the cynicism by which an extreme neoliberalism has cast its dark shadow on efforts of progressive social change. Most activists have instead reinterpreted market fundamentalism in ways that allow them to continue their commitments to local food-system sustainability and social justice. At the same time, however, most activists also see the market as the central institution by which the local food economy should operate, although they are stymied by its class and racial injustices. Few of the activists we met were able to see beyond this conundrum through to alternatives, although those exceptional crossover activists we met who did see the existence of alternatives provided us with hope about the future of local food systems.

In this book, we describe not only a variety of present trends and social projects among food activists we witnessed but also ones that point to more socially just and environmentally sustainable possibilities.

Food Activism and Revaluing the "Diverse Community Economy"

For many people, local food activism, whether sustainable food and farming activism or charity-based or community food security activisms, may appear to be trivial and even insignificant in comparison to the "real economy" of large-scale farms and food corporations connected by global supply chains. One could easily imagine an interlocutor asking, "Isn't all this small-scale, petty, labor-intensive food production and direct sales all very nice but inconsequential for the US food system? Aren't farmers like Sonja James, who can't sell her pork sausage at a profit at the farmers market, who just wants her farm to be 'economically sustainable,' a bit pathetic? And even more insignificant are those so-called community gardens or promoting discount vouchers for poor people in farmers markets or offering cooking classes or guiding farm tours . . . oh come on!" In such a view, local food activism and activists directly connected to local food production and circulation are cast as the insignificant "others" to the gigantic industrial capitalist food enterprises, the huge number of wage laborers and contract farmers whom these enterprises employ, and the enormous volume of food transactions that occur every day in the United States to bring industrial food to supermarkets and fast-food outlets.

We argue that this unexamined view is an artifact of the neoliberal doctrine that we have just described. In order to explore food activism in this book, we propose to set out a compelling alternative to neoliberalism grounded in the everyday experience of people: the "diverse community economy" (Gibson-Graham 2006). J. K. Gibson-Graham is the name of two feminist geographers who have written under this shared nom de plume to provide a poststructuralist, largely anticapitalist vision of actual, as distinct from ideologized, economic life. In the theory that Gibson-Graham propose, the diverse community economy represents a way of politicizing "the economy" by seeing the economy not as a reified construct of a dominant industrial capitalism but as a site in which diverse local actors can act in awareness of their intrinsic sociality and interdependence with one another as they engage in transactions, labor, and enterprises that are not only capitalist but also noncapitalist and also hybrid ("alternative capitalist" or "alternative market") in character. Within the diverse community economy, kinds of laborers who are not capitalist wage laborers (e.g., unpaid household labor, volunteer), kinds of nonmarket transactions (e.g., reciprocal exchanges of goods between individuals), and kinds of noncapitalist enterprises (e.g., consumer cooperatives, self-employed food producers) play defining, often major, roles within community food production and exchange. Such a grounded vision of everyday economic life, based on empirical observation, is consistent with the ethnographic approach to the local food economy taken by the authors and researchers in this book.

Unlike the case of the unfreedoms associated with capitalism (being exploited, being laid off, losing to market competitors, etc.), the diverse community economy should be seen as a site of conscious decision-making and of ethical discussion (Gibson-Graham 2006, 74). Under these conditions, the quotidian practices of sustainable farming and food activism and in particular *community* food security activism can be studied as sites for decision-making and ethical practice while they add economic and other kinds of value to the local food economy.

Gibson-Graham argue that ethical decision-making implies the cooperation of community residents in order for them to "take back the economy" in four different areas of contingent economic freedom. According to Gibson-Graham (2006, 88),

> An ethical practice of being-in-common could involve cultivating an awareness of
>
> - what is *necessary* to personal and social survival;
> - how social *surplus* is appropriated and distributed;
> - whether and how social surplus is to be produced and *consumed*; and
> - how a *commons* is produced and sustained.

In this book, we therefore consider whether the practices of food activists succeeded within the framework of ethical decision-making when they sought to create viable implementations of the following:

- What was *needed* by farmers, consumers, farmworkers, or community residents
- How social *surplus* within the food system should be appropriated and distributed by local food actors, and which ones, to further sustainability or social justice[5]
- When and whether surplus should be *consumed*—and by whom—or used to meet people's needs or add to the food commons within the local food system
- How a *commons* should be sustained, maintained, or even created (e.g., a land trust for farmland) and by whom?

Food activists' efforts that have successfully confronted these conundrums in ethical debate and decision-making have added enormous economic value to local food economies and those whom they have served, and their social, cultural, aesthetic, and political value are beyond calculation.[6]

Finally, we return to the earlier question: What if the imperative for major reduction of fossil-fuel consumption on Earth requires the end of neoliberal-based economic growth or even the end of capitalism itself? Under these conditions, how might we rethink the value of local food systems within diverse community economies? If the defining features of capitalism are industrial fossil-fuel use and human exploitation, and these are leading to probable human extinction and human misery, surely these components of the current system will be devalued in future diverse community economies? If industrial agrocapitalism has to be significantly downscaled to ensure human survival, would not future local food economies, most of which will be noncapitalist and will not depend on fossil fuels, as well as the work of food activists within them, have to be revalued far more positively than they are now?

The Chapters That Follow

The chapters of the book follow from these concerns. Part 1 comprises chapters 1 and 2. Chapter 1 introduces the alliance of corporations, corporate states, and international financial and trade organizations that form the institutions of the dominant corporate industrial food sector, and it describes the power of this alliance, which has successfully limited the development of the local food sector throughout the United States. Chapter 2 provides a deconstruction of three myths that support the corporate industrial food sector.

Part 2 of the book consists of chapters 3–5. Chapter 3 provides an overview of our findings about sustainable farm and food activists and charity-based food security activists, as well as those activists who undertook both sustainable farming and food activism and food security activism, in four settings in North Carolina. At the end of chapter 3, we also provide an analysis of the gifts that food activists provided local food economies in four sites in North Carolina, with regard to the labor they contributed and successful projects they achieved. Chapters 4 and 5 then move to investigate the influence of neoliberal ideology and discourse on the everyday practices of sustainable farming activists and charity-based food security activists.

In part 3, chapters 6–10 present our findings about four outstanding food activist enterprises and their leaders who "crossed over" by *simultaneously attaining sustainable farming and food objectives in their projects while achieving community food security goals of empowering food-insecure populations and providing them with access to culturally appropriate, sufficient food.* Each of these four outstanding cases comes from one of our four different sites, conveying a sense of the specificity of the "solutions" when activists successfully synthesized the moral logics of both sustainable farming and food and community food security activisms, given the specific political and social conditions of each site. We assess each of these cases with regard to the contributions they make to the diverse community economy that Gibson-Graham (2006) analyzes. Chapter 10 provides a sociopsychological assessment of the transformative potential of the activists we studied for taking the next steps as participants in a scaled up, more inclusive local food movement.

The conclusion first summarizes the analysis of the preceding chapters and then goes on to reflect on the future. It makes the argument that three of the four "cross-over" projects we discovered that synthesized the methods and goals of sustainable farming and food and community food security activism are candidates for what we, following Erik Olin Wright (2010), call "real utopias." This matters for the future food system and perhaps for the future of humanity, given the converging crises of neoliberalism and climatic and ecological destruction, against which large numbers of people are beginning to react in what Karl Polanyi (1957) called opposed "countermovements." In what would be a conflict between countermovements on a global scale, these cross-over projects would join with other real utopias in seeking to affect the outcome, which in turn might determine whether there will be the end of the industrial fossil-fuel era in agriculture. Much could be at stake.

The appendix provides a short analysis of the causal connections between the operation of the dominant corporate industrial food economy and worldwide ecological degradation and climate change—two impending threats to human welfare.

PART I

The Corporate Industrial Food Alliance and the
Myths It Promotes

1

Our Food and the Alliance

What Is COOL for Americans Is Not So Cool with the WTO

We begin this chapter with two stories about the politics of food that determines what foods Americans have access to and under what conditions.

For those who are interested in the future of food in the United States, the first story ends with a curious event that occurred on May 20, 2015, in the Agriculture Committee of the US House of Representatives. On that day, the committee voted 38–6 to repeal the entirety of the US's Country of Origin Labeling (COOL) law for beef, pork, and poultry (Zuraw 2015). This law was enacted in 2002 as part of the omnibus Farm Bill and required supermarket labels on beef and pork to indicate where cattle or hogs were born, raised, and slaughtered as an essential measure to ensure the safety of the meat that Americans eat on a daily or weekly basis (Zuraw 2015). A public survey found that more than 90 percent of the US public surveyed were in favor of the COOL law (Zuraw 2015). So what happened that led to the repeal of such an obviously popular law to ensure the safety of the meat Americans eat?

Two days previously, on May 18, the Appellate Body of the World Trade Organization (WTO) had decided in favor of a WTO arbitration tribunal judgment following the joint decision by three arbitration attorneys that the US COOL law constituted a "technical barrier to trade" and therefore that the US was in violation of its obligations under the WTO Agreement on Technical Barriers to Trade because it provided "less favorable treatment" for imported beef and pork from Canada and Mexico than to "like" domestic livestock products (Public Citizen 2015c). The Appellate Body held that the original COOL law created additional record keeping and verification requirements for imported livestock that did not exist for domestic livestock, and this created an incentive for meat processors to use domestic meat instead of imported sources. In short, the WTO arbitration tribunal, upheld by the WTO Appellate Body, decided that the real significance of the COOL label law was to keep out meat imported into the United States, not to protect the health of US consumers by providing them with more adequate information about where their food came from.

It is not as if the US Department of Agriculture had not already tried to satisfy the WTO by previously modifying the COOL law in 2013 so that it included less stringent requirements about listing the locations of different stages of livestock processing (Public Citizen 2015c). Instead, the Appellate Body upheld the 2012 original WTO tribunal ruling against the COOL law. This ruling originated from a complaint filed with the WTO tribunal by Mexican and Canadian meat corporations with the support of their respective governments against the US COOL law.

The May 2015 ruling initiated an internal WTO process to determine the level of trade sanctions—basically a very large fine amounting to tens of millions of US dollars or even more, in the form of trade tariffs that Canada and Mexico would be authorized to impose on US imports to their countries in retaliation for this "technical barrier to trade."

So it was the fear of retaliation from the WTO that led the chair of the House Agricultural Committee, Mike Conaway (R-Texas), on May 15, 2015, to announce to the press that he was willing to repeal COOL as soon as Congress reconvened, and that is precisely what happened, thirteen years after it had been passed by Congress.

One interesting additional fact: the US meat-processing industry was also opposed to the COOL law, despite what the WTO tribunal judges had concluded were terms in it that favored their very own industry. To the contrary, this industry had been lobbying vigorously against the law for the previous decade (Zuraw 2015). This was probably because meat processors in the US often use imported cattle and hogs in their products, even though consumers assume that their meat animals are born and bred, fed, and slaughtered in the USA. In a globalized world, where animals, like people, can move and be moved, how do consumers know where their meat comes from?

Moreover, the WTO arbitration tribunal process that led to the COOL law's repeal was not very democratic. Basically, WTO arbitration tribunals work when three lawyers trained in international arbitration law—one appointed by the complainants (here the governments of Mexico and Canada), one by the defendant (the United States), and the third by common choice—reach a decision based on their judgments in interpreting the WTO trade agreements, here the WTO Agreement on Technical Barriers to Trade (Public Citizen 2015a). Other than the trade considerations in this agreement, these lawyers are not expected to bring in "extraneous" considerations such as the health of three hundred million US residents or their right to be informed about the food they eat. Moreover, these lawyers usually have previous corporate experience in the relevant industries and can even sit in judgment on complaints against corporations that they had previously been employed by. Although it is clear that the WTO is looking out for the meat-processing industry, it is not clear who, if anyone, is looking out for the US consumer/eater.

Transatlantic Food Fights

Our second story about who determines the food that people eat has to do with a trade agreement in which US-based food corporations sought to determine what kinds of food the people of Europe should eat. On September 5, 2014, years before Brexit, the British newspaper *The Guardian* reported that "Britain and other European Union member states are under increasing pressure from North American business groups to open their borders to imports of genetically modified food as part of negotiations for a new Transatlantic trade deal" (Harvey 2014, 1). The article went on to note that "documents from various US and Canadian government agencies and business trade bodies suggest strong pressure is being brought to bear from US industries to allow GM products and other foods into EU markets that would violate the EU's current standards, in the name of free trade" (Harvey 2014, 1). The business trade organizations involved were some of the heavy hitters in the US food industry and included the North American Export Grain Association, the National Grain and Feed Association, and the American Soybean Association—business federations of large-scale US grain growers and processors (Harvey 2014, 1). They were also supported by Monsanto and other corporations in the agrobiotech industry.

As a result of such pressure, European environmental organizations warned that the EU would be all too willing to give in, as the new Transatlantic Trade and Investment Partnership (TTIP) was then being negotiated between European, US, and Canadian governments "with the active participation of dozens of large businesses" (Harvey 2014, 1). Notably absent were the representatives of the US and European publics. What was at stake in this transatlantic food fight?

Since 1998, the European Union has not approved the import or planting of genetically modified (GM) foods or seeds for use on European farms because there have been strong pressures from its member states and their citizens not to approve such imports. The EU responded by passing the 2003 laws requiring that no GM foods/crops be imported until labeling and "traceability" rules were put in place (Smythe 2009, 100). The EU "traceability" mandate required documenting the production, handling, and transport of a food product so the specifically located crop it came from and the specific traits it had could be publicly identified. This mandate applied to GM cultivar seeds in particular (Smythe 2009, 120). There has been strong concern among consumers and farmers in the EU that European non-GM crops not be contaminated by GM crops introduced from overseas, and it is long standing, with the first EU prohibitions on GM foods dating to 1990 (Public Citizen 2003).

This is what excited the ire of the US and Canadian governments and their food-industry organizations and led them to pressure the European Union. Not long after the European Parliament passed the 2003 laws requiring GM labeling, which effectively prohibited the import of GM foods, the US, Canadian, and Argentinian governments on behalf of their respective GM-crop industries filed a complaint at the WTO against this legislation. Because the EU applied its 2003 laws evenhandedly by prohibiting GM crops among both European and foreign producers, the US and its allies could not argue that the EU laws were a "technical trade barrier." Instead their WTO complaint was filed on the grounds that the EU prohibitions were a violation of the WTO Agreement on Sanitary and Phytosanitary Standards, which strictly limited member countries' food-safety policies and standards by requiring them to demonstrate that there were clear "scientific risks" that justified food-safety policies such as the EU prohibition on GM foods (Public Citizen

2003, 1–2). Failing that, according to this agreement, such laws constituted a "trade barrier."

In 2006, a WTO arbitration tribunal found in favor of the three complaining governments and their food industries against the EU law, on the grounds that it required unreasonable "delays" in approving GM imports and that the food-safety safeguards instituted by six EU member countries were found not to be based on appropriate "risk assessment" of safety concerns (European Commission 2009, 2). While the EU did not (and has not) changed its 2003 laws, the EU agreed to meet with the complainants to engage in "technical" discussions of the issues involved. This is how things stood up to the 2015 negotiations over the TTIP, in which the US and its agribusiness corporations renewed their attack on the EU's GM foods policy. It is worth noting that the use of Bayer Crop Science / Monsanto's Roundup-resistant GM corn and soybeans is always coupled with Monsanto's Roundup pesticide, whose active ingredient, glyphosate, has been classified by the International Agency for Research on Cancer as "probably carcinogenic to humans" (IARC 2016).

According to surveys, a majority of both US and European consumers are in favor of laws that segregate GM from non-GM foodstuffs and require them to be labeled so the consumers have a choice of which to buy (Public Citizen 2003). Nonetheless, the WTO tribunals have found in favor of the US, Canadian, and Argentinian governments and their agribusinesses to mandate that consumers do not have a right to such precautionary food-safety laws or even to be informed about what their food contains. It thus appears that rather than the people who actually have to eat the food that the system provides, the US government, agribusiness corporations in biotech and food processing, and the WTO are in charge of the food system. These circumstances raise questions about why the US government and large-scale food corporations are so militantly set against popular opinion and in favor of preventing people from being informed about the food they consume and why so many people allow this condition to exist.

Most Americans have not followed these two relatively obscure stories or understood the limits they impose on the ability people have to decide what foods to eat. We even hesitated to start this chapter with these stories because telling them, given their actors' use of bureaucratic jargon, can create an off-putting effect for many people (including the authors), an effect we call the "MEGO effect": "my eyes glaze over." Yet it is important that the events these stories relate be known because of their implications. These stories strongly suggest that most Americans do not have control over choosing the foods they eat but that powerful institutional entities like food corporations, governments, and trade-agreement enforcement agencies do.

This chapter examines the various forms of power that the global corporate food industry, nation-states, and international organizations have and exert, and to what ends, in shaping the US food system. Despite widespread public approval of local food and farming and at least the appearance of tolerance by food corporations of the local food and farming sector, we will see just how limited and constrained the local food systems of the United States are and, in historical context, why they are constrained by corporate power. Under neoliberalism, the reach of the industry is simultaneously global in scope but also intimate in its effects on the US food system. These effects not only constrain how the food needs of Americans, including large number of food-insecure people, are met but also limit the present and future growth of sustainable local farming and foods.

We first examine the different kinds of power that food corporations exert over the core agencies of the US state—those departments that dispose of enormous public wealth through funding, including the Department of Agriculture. Then we turn to the power and influence of food corporations over what we call the "residual US state," where a co-alition of public and private entities (food corporations and nonprofit organizations and groups) exert governance and set policy to provide "emergency food" to food-insecure populations in the US.

Welcome to the Alliance

An alliance of transnational food corporations, the US government, and international trade and financial institutions (e.g., investment banks, the WTO, the US Mexico Canada Agreement (USMCA, successor to NAFTA), the International Monetary Fund) governs the globalized industrial food economy and actively sets the agendas around food choice that are presented to US consumers or eaters. Through this alliance's governance of the globalized industrial food economy, it has come to play a dominant role in the world's food-provisioning systems, which has allowed it to limit the development of the millions of local food economies that represent its only viable alternative. Although this alliance has internal tensions and there are many marginal spaces in which local food economies function and their activists operate, it is extremely powerful, and it operates at multiple scales and in multiple arenas of the US political system—and beyond it. It shapes the food choices that people make in ways that are both consequential and increasingly invisible to them.

For instance, people may possibly be able to choose their food using their own judgments of its quality and healthiness at the supermarket, but the conditions under which it was grown (e.g., how the laborers who grew it were treated) are almost always invisible to them. They can go to the supermarket and choose organic Fuji or Gala or Golden Delicious apples to buy but have no idea whether any of these were grown under fair labor conditions. As indicated by the preceding two episodes that restrict the rights of eaters to know what is in their food and where it comes from, the population is only nominally, if that, in charge of the food on offer to it. Therefore, it is imperative to examine how the US population's supposed food choices are made and how they are constrained. People's choices are particularly shaped by the practices of this alliance of corporations, government agencies, and international trade organizations. In what follows, for brevity, we refer to it simply as "the alliance."

That people believe otherwise and see themselves as free actors with a great deal of responsibility and agency in their food choices is what we call "the myth of food consumer sovereignty." To those who believe this, this chapter aims to demonstrate that food choices are not just individual decisions but are also shaped by agendas set by large-scale institutions that do not consult us about the food choices they present to us. This chapter seeks to illuminate the power structures of this alliance that subvert consumer sovereignty—impelling and influencing people to make some of their choices about food more probable and even certain (where some foods have market monopolies or are even addictive), and they prevent people from implementing their choices or even make them impossible. They do this by preventing other choices from coming into existence: through industrial agriculture and industrial technologies of food production, transport, processing, and distribution, combined with successful political lobbying, the alliance has kept the alternative—thousands of diverse local food economies in the US—from growing and flourishing.

Corporate Capture of "Core" and "Residual" Components of the US Corporate State

In order to understand how this alliance exercises power over the food that Americans eat and how it is produced and marketed and how such power leads to deleterious effects on human health, economic welfare, and survival, we need to expand our field of vision to the national scale to situate the alliance within the context of political and economic changes over the past thirty to forty years in the United States. Most frequently, scholars have referred to these years as the period of "neoliberalism," "market fundamentalism," or "free-market ideology," and we (Holland et al. 2007) and many other authors have written about this extensively. However, it is necessary to further specify the conditions under which the food alliance exercises power by examining what some scholars are calling the US "corporate state" and the role of the alliance within it (Kapferer 2005a, 2005b, 2010).

The US corporate state is organized around the activities of corporate and state elites that occur in the spaces of contested renegotiation of the boundary between government, which is "public," and capitalist corporations, which are "private"—a renegotiation that has been particularly pronounced with the ascendancy of neoliberalism. The increasing ambiguity of this boundary allows a borderland to be occupied jointly by corporate elites and state officials who play by rules of their own making that completely lack accountability to the US population. In such institutional borderlands, corporate elites connected to one another (by ties of ownership and management, by kinship, marriage, shared educational background, and/or ideological preference) share decision-making with government officials in areas previously reserved to the latter. Through these arrangements, these elites take control of state resources and institutions—financial, agricultural, regulatory, and the like

(Wedel 2009, 73–110)—operating at various scales. Corporations have thus assumed an overweening role in the processes of governance. This is what "state capture" by corporations means.

Under current conditions, corporate capture of the state in the US takes two different forms. On the one hand, corporate elites gain direct access to the financial and other resources held by government, often, in the process, preying on the population while they engage in what amounts to the authorized looting of public resources. Corporate capture through official mechanisms (e.g., massive subsidies for planting certain crops) occurs through oligarchic practices within what we call "core" state agencies (see later in this chapter). Relative to the liberal US state before the 1970s, these core state institutions now show an intensified technical-functional integration between the owners and managers of private corporations and the elected and unelected political elites, officials and staff of US government bureaucracies and of such international organizations as the WTO, World Bank, and International Monetary Fund (IMF).

At a broad level, the privatization of core state resources and agencies is at the heart of much of this integration: state funds for administration, food provisioning, policing, imprisonment, immigration regulation, military, and other functions are profitably extracted by corporations through unregulated contracting-out arrangements (Wedel 2009). More specifically, *core state agencies consist of government bureaucracies in the US government* (in the Commerce, Treasury, Interior, Defense, and Agriculture Departments) that have substantial control over state resources, financial, material, and discursive. The political elites that lead these bureaucracies are combined in informal and official coalitions with the owners, managers, and representatives of some of the largest corporations in the US. The latter include top managers and owners of transnational investment banks and corporations, including food corporations (Johnson and Kwak 2010; Sklair 2001); oligarchic networks of wealthy families (Domhoff 2022); the national security complex that knits together the Pentagon and private military, police, and security corporations (Barry 2010; Singer 2003; Scahill 2008); and a large number of private foundations and conservative think tanks (Hardisty 1999; National Council for Research on Women and Schultz 1993).

Thus, for example, US international food aid to developing countries facing food shortages has been contracted out to US food corporations (grain traders, grain millers) and their allies (food transporters), which in turn contract with large-scale growers to provide this "in-kind" aid to developing-country governments and to food-related nongovernmental organizations (e.g., CARE, World Vision) that distribute such food on the ground. The provision of such aid is enormously profitable to the food traders managing these peculiar food-market supply chains in the guise of a foreign aid program. Alternatives—such as providing these countries direct cash payments that allow their governments the flexibility to import the food they need on their own terms or even (heaven forbid!) assisting their own farmers to grow more food for the population—have been ardently fought against, and successfully so, by these corporations, for five decades (Clapp and Fuchs 2009).

Transnational food corporations' capture of *core* state agencies takes place at multiple scales: *directly at the scales of the nation-state and of global/transnational financial institutions and indirectly at local and state scales as well.* On the other hand, corporations, including food corporations, have a different relationship to what we call the "residual" (or "left-over") part of the US corporate state that is not core agencies. Residual state programs of social provision are popular among citizens, ensure social order, and provide legitimation to class rule: hence the existence of "entitlement" programs (e.g., Social Security), welfare (e.g., Temporary Aid to Needy Families, or TANF), and the USDA's Supplementary Nutrition Assistance Program (SNAP), formerly known as (and still called by many people) "food stamps." Where privatized provisioning systems can be profitably administered by corporations through contracting out the administration of services to "clients," as in the case of TANF, they fall within the core state, but other, less profitable services are relegated to the residual state.

In this chapter, we first investigate the mysteries of the globalized industrial food economy and how its corporate actors wield power over US consumers/eaters by examining its relationship to the core agencies of the US state. We then turn to how the increasingly most powerful of these corporations—large-scale food retailers—play an overriding determinant role in the emergency or charitable food sector of the residual state, what is increasingly being called "food bank nation" (Riches 2018).

The global food system is complex, governed by huge corporations producing, processing, moving, and exchanging vast quantities of foodstuffs, is transnational in scope, and operates at multiple scales. We start with the organization of the food system, and then we discuss the forms of power that its corporations exert over food and people's lives with food.

Food corporations, operating both on the national scale and transnationally, belong to four sectors of the global food industry. Moving through the food supply chains from "farm to [eater's] fork," these are (1) large-scale agroindustrial producers, (2) food traders and processors, (3) providers of agroindustrial inputs (seeds, fertilizers, and pesticides), and (4) food distributors and retailers. The first sector has shown large increases in the size and scale of operation of farms since the 1960s; the other three sectors are significantly concentrated, with the largest corporations in each sector holding a disproportionately high market share for their respective sectors (Clapp 2016). However, above all, global mega-retailing chains like Walmart, Tesco, Carrefour, and a few others have such concentrated buying power over their upstream suppliers that they increasingly determine the structuring and setting of standards for the global industrial food economy (Clapp 2016, 117–126).

Forms of Corporate Power

In what follows, we look at the ways in which the market, instrumental, structural, discursive, and scalar powers of transnational food corporations operate to structure the food system that people in the US now have.

According to Jennifer Clapp (2016), there are several sources of influence that these corporations have brought to bear on the US federal government. To begin with, the concentrated *market power* of these corporations gives them influence within the US national state because, since the 1950s period of the Cold War when then–Secretary of Agriculture Orville Freeman referred to "food as a weapon" in US foreign policy, the geostrategic significance of food, who has it, and who controls its production and trade have been of paramount concern to the US foreign-policy establishment. If food is a strategic weapon used by the US state against its perceived foreign enemies, then those who manage it strategically—transnational food corporate elites—have great influence over policy makers.

Domestically, US official concerns about potential social unrest among the working population and the appearance of sufficient and cheap food provision for a growing postwar US population, which has even up to now associated a full dinner table with material prosperity and fulfillment of the American dream, have clearly provided food corporations with significant access to state officials. The ability of the US working population to buy what appear to be reliable and inexpensive supplies food, moreover, is not only consistent with but a necessary bulwark for the neoliberal policies that have globalized US industries, disenfranchised labor unions, and driven down wages and benefits for millions of people in a global "race to the bottom." No governments have legitimacy if they cannot ensure a reliable and affordable food supply to their populations. US political elites realize this and attend carefully to the expressed interests of food corporations as a consequence.

The market power of food corporations extends beyond monopoly setting of prices for food consumers. The oligopoly buying power of retail corporations like Walmart allows them to dictate price ceilings to those sectors that are "backward" (earlier) in the supply chain—agricultural producers, processors, traders—and increasingly to set "private standards" for farmers not only for food safety but also for characteristics of food favored by retailers, for example, their appearance and shelf life (Clapp and Fuchs 2009). Retailers respond to the issue of whether these "standards" are in line with what US eaters want and need only in the most legalistic sense: if it is legal, they can do it. Thus does market power percolate "down" to US farmers.

Transnational food corporations deploy their *instrumental power* to influence and even shape decision-making by US government officials in the realm of food in the desired directions. Individual food corporations as well as their business federations (e.g., the North American Millers Association and the Biotechnology Industry Organization) employ large numbers of lobbyists and consultants to shape (and even literally write) bills before US Congress, particularly the Farm Bill and other bills of interest to the corporate food sector, such as those dealing with agrobiotech intellectual property rights (Food and Water Watch 2010; Clapp 2016, 113–115). According to a study by Food and Water Watch

(2010, 1–2), between 1999 and 2009, the fifty largest agricultural and food-patent-holding corporations and two of the largest biotechnology and agrochemical trade associations spent more than $500 million lobbying the US Congress about laws covering genetically modified foods, intellectual property rights over seeds and animals, food corporation subsidies, and related issues. The same study indicated that over the same decade, food and agrobiotech political action committees contributed more than $22 million to the electoral campaigns of congressional candidates, most of them incumbents (Food and Water Watch 2010, 2). Food corporations' direct influence over US lawmakers, as when a corporation contributes money to a politician's reelection, constitutes such instrumental power.

The *discursive power* of the global food industry is evident in its ubiquitous advertising, the appearance of its executives at congressional hearings, and its issuing of reports and press conferences on issues of importance to it, such as its claims for the safety of GM foods and for the nutritional value of "fortified" manufactured foods it sells (Clapp 2016, 129–131). Lobbying combines the industry's instrumental power to fund the reelection bids of incumbents with its discursive power, as when food corporate lobbyists set out the actual language of the Farm Bill before Congress. The promises of food abundance, low food prices, and new jobs in food manufacturing or cutting-edge biotechnology in a politician's electoral district, when appended to campaign funding for reelection, helps smooth the swallowing of the pill of having a lobbyist impose such precise language on a politician's tongue.

Beyond these lobbying and electoral campaign finance activities, we see the *structural power* of the food industry manifested through its owners' and managers' interpersonal and institutional connections. Upper-level managers of transnational food corporations have moved through the "revolving doors" between corporate board rooms and high offices within the US Department of Agriculture, the core state agency that legally regulates the food industry (Mattera 2004; Clapp 2016, 126–129).[1] In 2004, among the forty-five top officials of USDA, the biographies of about one half show previous employment with agribusiness companies and their trade associations, or with agribusiness lobbying firms or research groups including university research centers funded by agribusiness corporations; "there are approximately as many industry people among appointees as there are career civil servants" (Mattera 2004, 10). As Philip Mattera (2004, 10) puts it, "Thanks to its political influence, Big Agribusiness has been able to pack USDA with appointees who have a background working in the industry, lobbying for it, or performing research or other functions on its behalf. These appointees have helped to implement policies that undermine the regulatory mission of USDA in favor of the bottom-line interests of agribusiness."

Large-scale food manufacturers and their allies (e.g., food processors such as grain millers) have a vested interest in US domestic food policy both broadly and more specifically with regard to the agricultural subsidies that allow their commodity crop growers (i.e., large-scale farmers) in the US to continue to survive, if not prosper, given the vagaries of the world prices for farm commodities. Here the existence of the US "farm lobby" of large-scale producers, grain traders, food manufacturers, meat processors, agro-input corporations (producers of farm machinery, GM seeds and pesticides, and petroleum-based fertilizers), and its influence over congressional and executive decision-makers is well known—except to most consumers.

As we have seen, the farm lobby's key objective has been to maintain the flow of billions of dollars in agricultural subsidies to farmers, which take the form of crop price supports and crop insurance for the major commodity crops—corn, soybeans, and wheat but also for dairy products and for the meat-processing industry—stipulated in the omnibus Farm Bill that comes up for renewal every five years. The subsidies translate into the enhanced bottom lines of corporate food processors, particularly those with products based on corn and soybeans, due to their cheaper prices, which local food advocates often claim undercut smaller-scale competitors selling nonrefined foods.

Moreover, the farm lobby has hobbled attempts by local food systems advocates in the Community Food Security Coalition to include expansion beyond a few million dollars of USDA's programs to support local community-based agriculture in every omnibus Farm Bill that came up for renewal from 2002 to 2012—a direct attack on US local food systems (Salmon 2012).

Last, the alliance shows its capacities for exerting *scalar* power—the ability to "jump scale" to exercise influence at the transnational scale, thus shaping national food policies. Neoliberal globalization since the 1980s has meant the surge to transnational prominence of US and Western food corporations within institutions of global governance in finance (World Bank, IMF, US Export-Import Bank) and trade agreements (USMCA, World Trade Organization). The financial influence from the instrumental power that food corporations exert not only gains them entry to decision-making in the US Congress but also provides them access to the new mechanisms of global food governance within these financial organizations and trade agreements.

At the global scale, food corporate representatives lobby international meetings of bodies associated with the WTO, USMCA, and related trade agreements. Food-corporation executives have played key roles in US trade delegations within the Department of Commerce.[2] According to Clapp (2016, 127), industry groups attend international meetings on environmental or food standard agreements, and governments sometime include corporate representatives on their negotiating delegations. As ActionAid International (2006, 3) points out, business lobbyists from firms like Walmart, Coca-Cola, and McDonalds dominate the US Trade Policy Advisory Committees, while 93 percent of a total 742 external advisers to the US Commerce Department who attended meetings with US trade negotiators and had access to confidential WTO briefing documents represented corporations like Burger King and Monsanto, rather than representatives of labor, citizen, or environmentalist organizations.

Once food corporations have gained access, especially those based in the United States, they have been able to exert influence and even make decisions about US policy around globalization, "free trade," and the implementation of neoliberal packages of privatization, "efficiency," and "deregulation" as these apply to treaties around food, like the 1994 Uruguay Agreement on Agriculture, which the US government strongly promoted and used its economic and political power to leverage other nation-states to adopt. Thus, the international agreements of the WTO have allowed transnational food traders and processors that export grains and processed foodstuffs from the US to gain open access to foreign markets, for agricultural input producers like Monsanto to force their penetration of new markets in genetically engineered seed stock among small-scale farmers in the United States and overseas while requiring that their "intellectual property rights" be protected, and for major transnational retailers to compete legally with domestic retail firms in the global South (Clapp 2016; Desmarais 2007; Weis 2007).

Corporations producing genetically modified seeds and pesticides used with them such as Bayer Crop Science (previously Monsanto) are also able to affect the deliberations of international organizations such as the Food and Agriculture Organization / World Health Organization's Codex Alimentarius, which sets international food-safety and labeling standards, in order to prevent the compulsory labeling of GM foodstuffs. So far, Monsanto and other GM-crop input producers have been successful in this effort, employing a combination of instrumental and discursive power first to gain access to these global-scale deliberations and then to help structure debates through visions of what is appropriate and necessary, for example, the determination of scientific "appropriate risk assessments" that do not demonstrate harm from GM food consumption—surprise! As is the case of the US COOL law repeal discussed earlier, this has placed the onus on the advocates of mandatory labeling, who instead invoke the "precautionary principle" (that food not shown to be safe should be assumed unsafe) and the right that consumers should have to know what is present in their food (Smythe 2009, 107–115).

Corporate input into treaty provisions affecting food and agriculture corporations has continued over the past several years in the formation of trade agreements still being negotiated. Dozens of representatives of transnational food and agroindustrial corporations and their business federations (e.g., ADM, Cargill, Walmart, National Cattlemen's Beef Association, US Grains Council, Dupont Crop Protection, Dow Agro-Science) were among the 605 corporate representatives consulted secretly by US trade representatives during the more than eight years of negotiation that led up to the Trans-Pacific Partnership trade agreement put before the US Congress in late 2015 (Public Citizen 2015b) before it was rejected by incoming President Trump.

These arrangements account for why the WTO could successfully challenge the US federal law that requires labels on meat and seafood products to specify their countries of origin and could seek to overturn the EU's prohibition on the import of GM foodstuffs—the two episodes discussed earlier. Transnational food corporations have even (since the middle of the first decade of this century) exerted a new form of scalar power by "jumping down" in scale, as when Monsanto successfully used its influence in 2005–2006 in California to gain passage of a state law that would *preempt municipal and county laws* passed against GM crop cultivation, which led fifteen other states to pass similar laws (Roff 2008; Carolan 2011, 180–183). This is why, for example, the companies processing and marketing your beef raised in Brazil are not required to label them as such and those manufacturing GM-engineered corn chips can have their corn grown in your county.

These arrangements also affect the lives of hundreds of millions of farmers across the world, including US local farmers—and therefore the foods that Americans can eat. A "corporate food regime"—global for the first time in history—is characterized by the dominance of the alliance of international food corporations, the US corporate state, and international trade (WTO, USMCA) and financial institutions (World Bank, IMF) during the neoliberal period (McMichael 2013, 41–61). The WTO's Agreement on Agriculture (AoA) "liberalized" food prices internationally by stripping all tariff and other protections from domestic food prices in the global South of Latin America, Africa, and South and Southeast Asia, at the same time that it gave foreign corporations the "right to export" food to these countries.[3] The outcome was "the decoupling of subsidies from prices, removing the price floor, and establishing an artificial 'world price' (substantially below production costs) for northern grain surpluses dumped on the world market at the expense of non-corporate farmers everywhere" (McMichael 2013, 54).

Another outcome as small farmers in the global South were driven off their own farms by this dumping (e.g., two million corn farmers in Mexico by NAFTA after 1993) was that a large number of them were pressed as cheap stoop labor into plantation-based vegetable, fruit, and nut production for export to the global North. The result was that US and EU small- and medium-sized farmers could not compete with the low prices for which these food imports from the global South were sold in Northern retail markets (McMichael 2013, 54–55).

Consumers in the US and EU quickly became habituated to these low prices. Is it surprising that local farmers in the US with their high costs of living and of farm inputs have set prices for their foods too high for a very large number of US residents to afford? Tomatoes from Mexico, anyone?

The Residual State: Food Charity, "Food Bank Nation," and the Management of Food Scarcity

Within the residual state, in lieu of direct corporate control, community-oriented public-private partnerships provide "emergency food" to large numbers of food-insecure people in the US (on such partnerships, see Holland et al. 2007).[4] Neoliberalism mobilizes the "voluntary," "charitable" sector of civic society to compensate for austerity cuts to social benefits, cuts associated with the US corporate state, which is committed to reducing corporate taxes. In these circumstances, civic- or community-oriented public-private partnerships are formed around regional coalitions of underfunded state agencies, nonprofit and faith-based organizations, private foundations, food retailers and distributors, and farmers, to provision and operate hierarchies of food collection, storage, and allocation sites that administer and provide "emergency food" to food-insecure populations *at the local scale*. The apex nodes of these hierarchies are regional food banks affiliated with the national nonprofit network of food banks Feeding America (previously America's Second Harvest), where food collection, storage, and wholesale allocation of food occurs, with secondary nodes being numerous local food pantries and soup or community kitchens that receive the emergency food from food banks and distribute it to food-insecure people.

All of this "voluntary" activity takes place despite the fact that the federal SNAP (food stamp) program, the Women, Infants, and Children (WIC) voucher program, and the school meals program administered by the US Department of Agriculture provide the vast majority of foods by dollar and nutritional value to poor food-insecure people—by at least an order of magnitude greater than the food charity sector (Joel Berg, executive director of the New York City Coalition Against Hunger, personal communication). Although the private charitable sector that distributes food to poor people in routinized, repeated acts of charity quantitatively represents no more than a small supplement (with regard to relative quantities of food provided) to these government programs, this does not mean that it is not significant for feeding large numbers of working-class people who are food insecure. In many locales in the US, the palpable presence and influence of the private charitable food sector, with its highly visible food banks, food pantries, and community kitchens, are far more prominent socially and politically than these government programs. In the United States as in other Organization for Economic Cooperation and Development (OECD) countries, this private charitable food sector has come to be known by the name of "food bank nation" (Riches 2018).

In the United States, for a brief decade of the 1970s into the 1980s, poor Americans received food benefits as a right ensured by the state, but much has changed since then: food provided to them has been transformed either into a "work support" paid in cash by the state to employed workers or into an unrequited gift handed out as food by the charitable sector of food banks, food pantries, and community kitchens (Poppendieck 1998; Dickenson 2020, 24–39). Since the 1980s, the need has remained great. The percentage of the US population belonging to households with "low food security" or "very low food security" was 14.5 percent in 2012, the year our North Carolina fieldwork ended (Coleman-Jensen et al. 2013, 6, fig. 1). People whose incomes were at least 130 percent below the "federal poverty line" ($23,283 in 2012; Coleman-Jensen et al. 2013, 10, n9) qualified for SNAP, while the WIC food program had similar eligibility requirements for women and their children (USDA 2012b, 1).

The distribution of emergency or surplus food in the United States to food-insecure working-class people has a structure to it created by neoliberal political forces from the 1970s onward. The public-private partnerships that make up the food charity sector of the residual state have come to complement SNAP and WIC and other US state programs (e.g., for school meals) funded by the USDA as a core agency of the corporate state. A new division of labor has emerged between US federal programs that provide monetary food aid (SNAP, WIC) directly to eligible (i.e., "deserving") recipients and the charitable food sector, which relies on surplus food provided by the USDA as well as that donated by corporations, individuals, and farmers and then distributes it as charity to poor people—"deserving" or otherwise. Over these decades, and accelerating particularly since Bill Clinton's 1996 "welfare reform" (Personal Responsibility and Work Opportunity Reform Act, or PRWORA), SNAP benefits have increased in real dollars provided to recipients, not decreased—unlike what many welfare advocates had feared (Ben-Shalom et al. 2011; Moffitt 2015). That is the good news—for some food-insecure people.

The bad news for many others is that over the same time, SNAP benefits have shifted from being a right to which any food-insecure resident (including noncitizens) and their dependents are entitled toward being a reward for those legal working-class citizens (and their dependents) who are willing to be employed at whatever "jobs" employers wish to offer them—whether these jobs are full- or part-time, permanent or temporary, pay wages high enough or not to live off, provide benefits or not, are physically hazardous or mentally stressful or not—as long as these jobs exist as wage labor within the formal capitalist labor market. While some adults do not have to be formally employed to receive SNAP benefits if they are unable to work due to a "physical or mental limitation," take care of a child under the age of six or of someone "incapacitated," or are "studying in school or in a training program at least half time," all other adults between sixteen and fifty-nine are considered "able to work" and must meet the "General Work Requirements" for SNAP. These stipulate that they must be formally employed in a job at least thirty hours per week or participate in workfare or in a SNAP Employment and Training program; and they must take "a suitable job if offered" and "not voluntarily quit a job or reduce their working hours below 30 a week without a good reason" (USDA 2019b). Ethnographic evidence (e.g., Dickenson 2020) suggests that the latter provisions have been interpreted to mean that a "suitable" job for recipients is any job offered and that there is no "good reason" allowed a SNAP recipient to walk out of a job. In contrast, no work not involving a legal employer—say, working as a gig worker for Uber, being self-employed as a repairman or caterer, or caring for a neighbor's children on a regular basis—counts as a "job" on SNAP eligibility forms (Dickenson 2020, 9–10, 12).

In contrast, the charitable food sector takes in, stores, and transports the food it receives from the USDA, corporate donors, restaurants, individuals, and a relatively few farmers and allocates it to everyone else in the working class who is poor and receives food not as a right but as a gift, but with implicit strings attached (see later in this chapter). Recipients include those who are unemployed or not formally employed, those employed in the formal labor market whose combined wages and SNAP benefits are insufficient for them and their family members to get by on or to have enough food to eat during the month, and, since 1996, immigrants who may be working but are prohibited by PRWORA from receiving SNAP benefits (USDA 2019a).

Increasingly those people in the working class designated by the SNAP program as "able-bodied adults without dependents" (ABAWD) but who are unwilling to be employed—or whom SNAP officials perceive as unwilling—have become the special objects of opprobrium in the post-PRWORA US. If ABAWD are discovered to be unemployed, are unwilling to accept a job offered by an employer, or quit a job, they can lose their (and their family members') SNAP benefits after three months and not be eligible to regain them for another three years (USDA 2019a). People who are cast in the role of ABAWD irrespective of their abilities, skills, and social and economic contributions at, say, self-employment, self-provisioning, or unpaid domestic labor or the labors of community care (e.g., operating a neighborhood kindergarten cooperative) become the devil incarnate of conservatives and neoliberals among the political elite: they represent the figure of the "undeserving," "idle" poor.

This division of labor between US state benefits (in money for food) and the food provided in the private charitable sector has everything to do with the neoliberal obsession with forcing working-class people to offer their bodies to capitalist employers, no matter how low the wages are or how degraded, unsafe, or coercive the conditions they labor under: the working-class body must be "work ready" for exploitation (Dickenson 2020, 139). For those who are willing to accept such jobs and "tough it out" under these conditions on behalf of their children and other family members, the US corporate state provides the SNAP benefit as their reward. When working people are unwilling to be subject to the exploitation and abuse associated with a job, when their wages when employed are still so low that they do not earn enough income to eat adequately, when they work informally for income "off book" (e.g., as gig or temporary workers), or when they are not legal citizens irrespective of their work, they still have access to the private charitable sector.

When working-class people enter, undergo the rigors of, and exit the precarious, often temporary and part-time, poorly paid service jobs in the capitalist labor market that neoliberal globalization has brought to the US economy, they frequently move between these two food-provisioning alternatives, as recent research indicates (Dickenson 2020). In Maggie Dickenson's (2020) superb ethnography on the place of

"America's food safety net" under current conditions among food recipients of SNAP and food charity recipients living in a largely white community in North Brooklyn, New York City, she illustrates these shifts. Some of her informants were formally employed at jobs with low wages and still received SNAP benefits, while others were fully employed and yet unable to apply for or receive benefits because they infringed program regulations or failed to have documents required by welfare workers to process their SNAP applications. Others received SNAP benefits, but their wages were so low that they still needed to patronize food pantries in the charitable sector. Others were laid off from or quit an unbearable job and as a result lost their SNAP benefits and then had to join the lines outside food pantries. Others found informal waged work (e.g., janitor, child care) but received their wages "off book" and were not eligible for SNAP and then had recourse to food pantries. Others spent time in USDA-certified job-training programs doing "community service"— anything from collecting trash on the streets to assisting nonprofit organizations; still others ended up having to enroll in tedious labor in workfare programs that would still allow them to receive SNAP benefits.

In the course of working people's adult lives, they are moving back and forth between these two institutional systems for receiving food, depending on their own personal circumstances, but within an overall trend since the 2007–2008 financial crisis toward increasingly economically precarious formal employment in service jobs that are temporary or part-time. These jobs increasingly are structured such that vulnerable SNAP recipients are subjected to hyperexploitation when they seek to retain their SNAP benefits, which also cross-subsidize the corporations that offer these jobs in the emergent services sector. As a complement, the charitable food sector is an unsatisfactory fallback when all else fails.

SNAP benefits often do fail. The Institute of Medicine of the National Academy of Sciences found that SNAP-benefit dollars provided to recipients were often inadequate due to SNAP regulations or their interpretation (National Research Council 2013). The maximum SNAP benefit that the USDA allows may be too low because it does not allow for recipients to buy higher-cost food items that allow time-stressed recipients to prepare meals more quickly. The USDA does inaccurate, lowball calculations of proportions of incomes needed to pay for housing, medical, and other expenses, especially in the case of people living in urban areas with high housing costs. It calculates its SNAP benefit reductions for those who are formally employed such as to overestimate their food costs compared to nonfood costs. When determining SNAP benefits, it consistently ignores food inflation costs and interregional cost of living differences (National Research Council 2013, 177–179). Janet Poppendieck observes that even USDA economists' research indicates that SNAP benefits are often insufficient to meet the monthly food needs of recipients and their family members and that many food-insecure people who qualify according to their incomes are ineligible to receive SNAP vouchers due to bureaucratic obstacles. She writes, "These findings lend great credibility to advocates' assertions that benefits are too low to enable participants to achieve food security, that eligibility thresholds are too low to make the programs available to all in need" (Poppendieck 2014, 181).

Throughout the history of the SNAP program since Clinton's 1996 "welfare reform" law (PRWORA), there seem to be consistent attempts by congressional conservatives to enact SNAP restrictions and by USDA officials to interpret them based on willful ignorance of working-class lives and on the assumption that low-income people do not deserve either sufficient incomes to get by on or sufficient benefits to purchase culturally appropriate food adequate to their nutritional needs.

Over time, an increasing number of people have joined the lines of people or vehicle queues to seek out food from the privately operated but publicly authorized food charity system to provide themselves and their family members with enough food to get through to the end of the month. It was only then that they encountered people like the charity-based food security activists we visited, interviewed, and worked with.

Our argument is that if these government supplemental food programs are structurally designed to be insufficient in the amount of food they provide to food-insecure people, then there are implications not only for their economic and nutritional status but also for the working conditions of food security activists in the private charitable sector. What does it mean, for example, that food security activists work at intake desks or distribution tables across from food-insecure individuals who have been subjected to state-induced traumas around food provision for most of their lives? Does it perhaps in part explain the distrust and emotional distancing that some food security activists manifest toward food recipients—for example, the suspicion voiced by some activists that people of a certain race tend to engage in "shopping" across multiple food pantries to "double dip" in the food benefits they receive—given that they have often been portrayed by conservatives as belonging to the "undeserving" or "lazy" poor?

What goes on when food activists, sometimes themselves in economically precarious conditions, have to confront a large number of people who, with dependent family members in tow, have periodically or chronically found themselves without sufficient food due to capricious administrative regulations and ritualized verbal abuse by political conservatives and neoliberals—and then suffered at the hands of the already insufficient private charitable sector?

Strategies of Capture of Residual State Agencies: The Local Scale

In addition to USDA-collected surplus foods such as those allocated through the Emergency Food Access Program (TEFAP), large food retailers (e.g., Walmart, Food Lion), food distributors (e.g., Sysco), and food processors (e.g., ConAgra) are the main sources of the "emergency food" that are collected by regional food banks in North Carolina and elsewhere in the US. Without such large-scale supply sourcing, the current emergency food system, based solely on government food programs, would simply be unviable. To get a sense of the relative contributions of the USDA TEFAP and a few other state food sources to food banks compared to the corporate and private sector, we can refer to one quantitative datum favored by food banks and the USDA: the number of meals distributed in a year. In 2021, Feeding America, the nonprofit national network of food banks, distributed more than 6.6 billion meals to recipients. The USDA's food programs provided 2.5 billion of these, while food retail corporations and local businesses donated food for 2.4 billion meals; the remaining 1.1 billion meals came from two sources: food drives conducted by individuals and families, churches, schools, and businesses, and local farmers who donated excess crops or allowed them to be gleaned from their farms (Morello 2021). While we have not been able to look back to our 2011–2012 year of fieldwork in North Carolina, there is no reason to believe that the proportions of state-provided versus corporate/business-donated foods supplied to food banks were not approximately the same then: both were the largest and roughly equal providers of food to the charitable food sector.[5]

Food distributors and retailers make food donations not primarily out of altruism but because of the enormous and very generous tax deductions allowed them for such donations, savings to companies by thus avoiding food waste disposal expenses, and the public-relations goodwill that donations create for them as generous "community partners." The power of corporate food retailers and distributors operates at the local level through the presence of food-corporation managers on the boards of directors of regional food banks, where they can make sure that the accounting procedures necessary to track their tax-deductible donations are followed and that tonnages of donated food passing through food banks to food pantries, community kitchens, and backpack programs are accurately monitored—yes, tonnages of specific foods, not their nutritional value.

While food banks remain dependent on such large-scale food distributors and retailers, most of the foodstuffs these businesses donate are processed foods, thus reinforcing the long-term consumption of these inferior foods by an impoverished population already ironically beset by a combination of food insecurity and obesity, driven by overconsumption of refined carbohydrates and sugars, which are ubiquitous in the industrial foods they consume. As retailers' logistics systems improve in efficiency, relatively fewer of these manufactured foods are getting to recipients, although those that do represent valuable tax deductions for the donors. Meanwhile, food banks, food charities, churches, and community groups seek desperately for other sources of food to make up the shortfall, such as raw foods captured by gleaning operations and community food drives. This is especially the case when the quantities of TEFAP aid supplied to food banks fall due to high farm commodity prices, since TEFAP is not as such a guaranteed source of surplus food but instead is the USDA's instrument for purchasing surplus (overproduced) food to keep off the market in order to increase demand and maintain higher food prices.[6]

The largely volunteer (and frequently themselves low-income) workers at food pantries and community kitchens find themselves preoccupied with monitoring their clients to parse out scarce foodstuffs fairly and to prevent "church shopping" or "pantry shopping," as recipients try to make ends meet while engaged in desperate triaging of their needs for food versus needs for medicines, rent for shelter, child care, and so on.

Throughout, corporate donors come across as "good community partners," take pride in and boast of their preventing systemic food waste, and publicize their food donations as generosity to blunt popular criticisms that they are antiunion and provide only low wages and few benefits to their workers.[7] Again, the discursive power of food corporations to seize the moral ground, and to define appropriate limits to charity, is evident.

A political-economic analytical perspective on corporate sponsorship and influence over the charitable food sector provides a more critical corrective. This comes from viewing the enormous food overproduction and waste generated by the globalized corporate industrial food sector as a major problem confronting food retailers in several ways. First, the physical disposal of huge amounts of food waste (absent the charitable food sector to donate to) would impose an enormous financial cost on the industry. Second, the conspicuous presence of massive amounts of food wastes in local landfills would create supply-side moral, political, and thus public-relations questions that would challenge the corporate management of food scarcity: Why should *anyone* in the US population not be able to pay for food when so much waste comes from the food industry? Why should people have to pay even as much for food as they do? In a period when an increasingly large number of people are becoming aware of climate change, it would also be bad for retailers' "green" reputations if food they failed to sell ended up in landfills, generating the potent greenhouse gas methane, but donating it to food banks and local food pantries solves this problem.[8]

The overproduction of industrial foods that generates such waste also creates a potentially catastrophic problem for national retail food markets: Without somehow withdrawing a significant percentage of industrial food from these markets, how would the prices of foods be maintained even as high as they are? The donations that food retail corporations provide Feeding America and food pantries—while receiving substantial tax credits in the process—do precisely this task of maintaining food prices at profitable levels for retailers.

Joshua D. Lohnes (2021, 360), drawing on his extensive ethnography of food banks in West Virginia, concludes that his data reveal "how food waste transfers into the charitable food economy are tightly controlled . . . to ensure that they do not disrupt the scarcity logics and profit imperatives" of the industrial food economy. He contends that food banks are now thoroughly integrated into this economy, which "accumulates wealth by revaluing waste as hunger relief," and that "it is critical to attend to the politics of production structuring charitable food circuits, not merely the politics generating the demand for emergency food" (Lohnes 2021, 360–361).

Graham Riches (2018, 109) summarizes who most gains from the institutionalization of what he calls "food bank nation": "Within a culture of philanthrocapitalism, Big Food conglomerates, agri-business, producers, supermarkets and restaurants acting from a sense of solidarity and corporate social responsibility profit literally and figuratively from demonstrating goodwill, a sound business sense and competitive branding. It is a form of corporate social investment. . . . Over time Big Food, Big Philanthropy and Food Banks have become the gatekeepers of today's charitable food safety nets in the OECD. The corporately captured food bank nation has developed a life of its own."

* * * *

In this chapter, we first described the ways in which the alliance of corporations and corporate states that compose the US globally sourced industrial food economy exercises its power to advance the industrialization of the US food economy and to limit the development of local and regional food systems in the US. We showed how the powerful corporations within this alliance have captured key agencies of the core and residual parts of the corporate state, as the state sets food policies around food production, distribution, and consumption at international, national, state, and local scales in the US. Second, we examined the processes by which the food-insecure population in the US has had its own food needs subordinated to the imperatives of capitalist employment ("the job") and profitability (for corporate donors) by means of the complementary operation of the SNAP program of the USDA and the private charitable food sector.

2

Three Myths That Prop Up Industrial Foods

The Myth of Consumer Sovereignty

Why have the desires for locally grown and fresh foods over the past two to three decades had so much uptake among the US population, leading some to become food activists? In part, this is because of three prevailing myths about food that an increasing number of people, including the food activists we studied, are calling seriously into question—even while others still accept them in whole or in part. We start with the myth of consumer sovereignty.

In February 2011, Patrick Linder, our Durham research associate, undertook what proved to be a wide-ranging conversation with Pete Schmidt, executive chef and executive director of FoodBase, a small nonprofit organization that sold natural (nonindustrial) composts and pesticides. A middle-aged man well known around the Durham local food scene, a graduate of the Culinary Institute of America, and an enthusiast about local food cooking, urban agriculture, and Cuban agroecology, Pete at one point remarked,

> Americans are going to continue to love things that taste good. . . . I'm a chef. Chefs are alchemists, so . . . I need to do this and this and this to make this taste great so these people will eat, love it, and order it the next [time], or . . . I need to do this this and this so this client, this party, . . . and other people will hear about what great food we do. . . . That's alchemy, you know? It's . . . really about making adjustments to a food group and pairing it out with others so that it's like the delicious thing that all works together. . . . So when you're doing that with food—that's what they do with fast food, that . . . they are geared towards those things that are really hard for us to resist: fat, sugar, salt—those three things. I mean those are the primary things.

Pete's comment captures a view shared by many local food activists about our industrial food economy—that as a form of commerce, it is based on taking advantage of Americans' attractions to "those things that are really hard for us to resist: fat, sugar, salt." However, many people would find his perspective, even as a chef, off-putting, even patronizing. Many people, including some who support local foods and farming, see issues around food primarily as a matter of personal choice and responsibility—about what each person desires to eat, what people have been open to learning about diet and health, and their willpower to make responsible choices to eat as well and as healthily as they can.

After all, many people assume that they have a wide range of choices when they go to the supermarket. Particularly in the case of the category of the yummy, sweet, high-calorie, high-refined-carbohydrate processed foods that populate the center aisles of the supermarket, they face a myriad of apparent choices, with hundreds of new brand names appearing every year. Even beyond that, in the outer aisles in the vegetable, seafood, and meat sections of the supermarket, their choices appear to be abundant and, for the most part, affordable as well. Thus, most Americans see themselves as having a very large range of choice about the food they consume and, if anything, feel confused because they seek good food from so many choices but on the basis of so little information.

Even so, an increasing number of people now look beyond the supermarket to other sources of better, healthier, and fresher food choices, for example, to local farmers markets. Within supermarkets, over the past several years, there has been a distinguishable move by consumers away from the central aisles of manufactured foods (called by industry insiders "the morgue") toward the peripheral aisles of "fresh" produce and baked goods (Taparia and Koch 2015).

Among the local food and farming activists whom we studied, this vision of choice and a right and responsibility to choose good food underlie their actions. They feel that they are increasing their own and others' choices of foods they eat as consumers, choosing to support small farmers as good producers of food, and some activists as gardeners or farmers produce good food on their own. They see themselves as working to make locally grown, fresher, and healthier produce and fruit and meat animals more available in their community. We cannot emphasize this point too strongly because it is key: local food and farming activists have accepted the idea that they have the right to food they consider to fit with their values, and they are willing to expend a great deal of energy to build and support its production. They see themselves as choosing something better than that offered them by the industrial food corporations. This vision of choice defines their activist positions.

Still the question is, What is a person making a choice *about* when they choose to purchase and eat one food rather than another? Many see this as an almost unrestricted range of choices. Whether inside the supermarket or beyond it, many Americans accept what we call "the myth of food consumer sovereignty." This myth provides food corporations with the perfect excuse for continuing to do what they do most efficiently: make money by selling inferior manufactured foods that are low in nutrients and high in calories and synthetic food additives but foods that, in Pete Schmidt's words, are made from "things that are really hard for us to resist." The rationale offered by industry is, "We are just selling people what they want to buy, nothing more. We are only acting in accordance with the tyranny of the market, where people tell us what they want, and we are as much victims of that market as anybody else. It's not our fault and not our problem if people make bad food choices."

To those who accept the myth of food consumer sovereignty, our earlier analysis of the power of the alliance of global industrial food corporations to influence what Americans eat, and under what conditions, should be a wake-up call. These episodes point to the not well-appreciated fact that personal choices are much more shaped, frequently constrained, and sometimes prevented than we could ever imagine by remote institutions that few people know much about, like the alliance of global industrial food corporations, the WTO, or the US Department of Commerce. These institutions are more powerful than most people realize and operate at multiple scales (i.e., in multiple arenas) of the US political system (Clapp 2016). They shape our food choices in ways that are both visible and increasingly invisible to us.

For instance, people may seek to choose healthy and nutritious food by examining the labels on supermarket packaging, but the conditions under which the food was grown (e.g., how the laborers who grew it were treated) are almost always invisible to them. One of us can go to the supermarket and choose organic Fuji or Gala or Golden Delicious apples to buy but have no idea about their environmental effects or energetic costs (e.g., due to transport). Therefore, it is imperative to examine how Americans' supposed food choices take place and how they are constrained.

That people believe otherwise and see themselves as free actors with a great deal of responsibility and agency in their food choices is what we call "the myth of food consumer sovereignty." In truth, food choices are not just individual decisions but are also shaped by agendas set by large-scale institutions that do not consult people about the food choices presented to them. Indeed, people do not have the choices that they think they have or that they deserve. This book in part seeks to investigate these two contentions by illuminating the power structures of this alliance that subverts consumer sovereignty—these structures impel and influence people to make certain of their choices about food more probable or even predictable; they prevent people from making other choices at all or make them very difficult to implement.

However, some people take the myth of food consumer sovereignty even further and as part of a political philosophy toward human beings. This is the idea of individual responsibility for one's food choices, as in all other things. A person's supposed freedom to choose among the plethora of food offered by the industry is translated to individual responsibility for the consequences of their "choices." "Be an informed individual consumer." "Be responsible for yourself and your family for the food you eat."

This, the myth of individual responsibility, a core position of the neoliberal program of political transformation in the US and elsewhere, has been promulgated by thirty to forty years of corrosive attack by political leaders on collective values and ideas such as the common good and on government for upholding and supporting these values. When Margaret Thatcher in 1987 said to the British Conservative Party conference, "You know, there is no such thing as society. There are individual men and women, and there are families," and implied that each person had better start looking out after themselves with only minimal regard to a few others ("families") within the charmed circle of individual self-interest, her pronouncement clearly articulated this myth of individual responsibility: "Look out for number one."

This is the credo of neoliberalism in its purest, crudest, and most candid version. The extent to which this corrosive version of the myth of consumer choice extends to food and many other necessities of life and affects large numbers of people and how they see other human beings is something that everyone should be concerned about.

The Myth of Profit-Driven Techno-Abundance

So for me, we need to measure our food system in Charlotte and how we provide access to healthy food for all humans in a way that's mutually exclusive and collectively exhaustive. . . . So the people that we saw this morning are gonna get a decent lunch at Urban Ministry Center. But the ones that are living in the woods, what are they gonna eat for dinner? The kids who are out of school this summer, whose teachers are bus drivers and teachers' aides at Charlotte Mecklenburg Schools, so the parent's not making an income, and the kid's not getting a free or reduced lunch or breakfast, there is food insecurity. So maybe the success of the system could be assessed by what is the level of food security by the population.
—Joan McGivens, Charlotte Benefit Bank

Our practice is now—as far as large-scale agriculture—is not sustainable. You know, Earth wasn't meant to hack modified corn on it year after year after year after year. Animals weren't meant to be in concentrated feed lots, you know, in their own—in the conditions they're in. And I think that, like, small farms—I don't know. As, like, cheesy and naïve as it might be, I think small farms can save the world.
—Dave Carpenter, chef, F.A.R.M. Café, Boone, North Carolina

Much of the US population, if not Joan McGivens and other food security activists, accept the myth that profit-driven, large-scale industrial technologies applied to nature will bring about food abundance for all—abolishing the Malthusian threat of natural food scarcity. We call this "the myth of profit-driven techno-abundance," because it is *profit-driven, large-scale industrial technologies* that the alliance invokes in this myth. This myth is the dominant mainstream assumption that the food industry presents as the triumph of "our technology" over an indifferent nature.

This myth is most often invoked to refer to the genius of certain private corporations to overcome the early nineteenth-century dilemma, stated by Thomas Malthus, that increases in the human population will always outpace the capacity of farmers to increase the food supply, leaving too many people with little or no food. Corporate agribusiness, casting itself in a heroic mold, claims that it overcomes this dilemma by its unique innovations in science and technology, which allow the production of high volumes of foodstuffs, thus bringing about universal abundance. Profits are necessary to support and incentivize innovation, so the story goes. By implication, humanity should be grateful to corporations by allowing them the innovator's profits that they so richly deserve for undertaking the effort.

This myth is a dangerous one constructed on a hidden disaster. The myth of profit-driven techno-abundance hides the Faustian bargain that the corporate industrial food economy has made, one an increasing number of people, shedding their prior willingness to ignore it, increasingly feel they urgently have to come to terms with. This is that the Promethean powers of science and technology that have brought about such enormous productivity in volumes of industrial foods that global food corporations brag about in their public-relations brochures have come at the too-high cost of harnessing fossil fuels—coal, oil, and natural gas—to the industrial model of food production that they are so proud of.

At every step in the production of industrial food—from its rationalized production on vast monocropped plantations on lands whose deforestation has produced enormous volumes of carbon dioxide as a by-product (e.g., Amazonian beef and Indonesian palm oil production); its application of huge amounts of synthetic fertilizers manufactured by the Haber-Bosch process, which consumes massive volumes of coal, natural gas, and oil (Huber 2022, 90–91, 102–103); its synthetic fertilizers, which, when applied to the soil, generate nitrous oxide, a potent greenhouse gas; its CAFOs for beef and other livestock, which emit massive volumes of methane; its transnational transport and refrigeration logistics technologies, which emit huge volumes of carbon dioxide; its promotion of food overproduction and wastage; and the partial consumption of its foods and their resulting deposit as waste in landfills, where anaerobic decomposition generates even more methane; and one could go on—the global industrial food sector forces climate change. The current industrial food system has become the largest source of greenhouse-gas emissions on Earth (see the appendix for sources). The global industrial food economy creates between one-fourth and one-third of all greenhouse gases (Berners-Lee and Clark 2013, 153; Poore and Nemecek 2018; Oosterveer and Sonnenfeld 2012, 89). It is thus one of the largest drivers, and probably the single largest driver, of anthropogenic climate change, as we detail later in the book.

While the industrial food system might be charitably regarded as the *past* source of abundance, given the modern capitalist organization of industrial agriculture, it represents a present and a future not of abundance but of destruction and scarcity arising from climate chaos and its catastrophic environmental and ecological effects—one in which capitalism's organization of the food system cannot be tolerated much longer. The industrial food system's god is not Prometheus the bringer of fire but Shiva, the all-destroying god.

Moreover, industrial agriculture's systemic use of fossil fuels to boost productivity and corporate profits appears to be a necessary feature of the internal logic of the capitalist accumulation process that drives industrial food production (Malm 2016; Huber 2022). For example, the only other known sources of industrially used fertilizers than Haber-Bosch synthetics are highly finite, diminished natural sources of nitrates, like guano or subterranean nitrate deposits in South America (Huber 2022, 90–91). Neither the United States nor the human population of the rest of Earth can afford this kind of "abundance" much longer. Changes have to be made and an alternative found.

The myth of profit-driven techno-abundance is doubly dangerous because it serves as rhetoric to forestall its most feasible alternative, when proponents of corporate industrial agriculture attempt to trivialize small-scale farmers as being unable to "produce in quantity" for the multitudes around the world who need food. This claim is so taken for granted that the possibility that millions of small farmers might grow sufficient food for the Earth's human population, when combined with other reforms, is never seriously considered. However, scientific research about the potential for scaled-up food production by millions of farmers engaged in small-scale agroecological farming (Badgeley et al. 2007; Aznar-Sanchez et al. 2019; Pesticide Action Network North America 2022), combined with measures to reduce food waste, shift in diets (e.g., less meat eating), and increasing yields in low-yield farming regions (Foley et al. 2011), suggests it should be. Placed on the defensive, advocates of local food and small-scale farming, including ones we encountered in our research, feel vulnerable to this argument, but the argument must be recast overwhelmingly in their favor due to the connections between industrial food capitalism and climate change. The "alternative" to agroecology—the capitalist industrial farming that "we" now have—is *the* problem, not the agroecology still practiced by hundreds of millions of farmers, which generates many orders of magnitude less greenhouse-gas emissions than the industrial food economy.

There is more: the question of equitable distribution. Irrespective of what quantities of food are produced, consider how many people will benefit from the food. Will those who have the power to allocate resources (not food but the income to buy food in the US capitalist economy) do so justly so that food sustenance is widely and fairly shared among the human population? To put it this way is to completely reformulate the problem, which is simply to state that *a large proportion of this high volume of industrially produced food is not shared with those who most need it but only sold to those who can afford to pay for it*. It is this inequitable sharing of food that must be understood. The politics of food and of food sufficiency/insufficiency is all about class, about who has access to capital and the income required to buy (certain) food on the market and who does not.

USDA statistics found that in 2020 more than 38 million people, or 11.8 percent of the US population of 325 million people, lived in "food insecure households" (Economic Research Service 2021, 7, table 3.1). Judged by their incomes, the vast majority were poor people from the working class, and large proportions of them lived in households with children and were racial and ethnic minorities (Economic Research Service 2021, 17). However, among this population of 38 million people, only 41.6 percent of their households were covered by the SNAP program, the largest government program that provides food to the food-insecure population (USDA 2021b, 39, table 9). As discussed earlier, many of these households contain adults who are prevented from receiving or from continuing to receive SNAP benefits because the USDA considers them "able-bodied adults without dependents," who must first be required to be legally employed in a "job" or in "training" for one but are not doing so, for a variety of valid reasons.

For all those millions who "fall through the cracks" and do not qualify for SNAP benefits, as well as those who do qualify and receive them but have incomes still too low to ensure food security for their families, neoliberalism's private-public partnerships of "food bank nation" (Riches 2018)—the US complex of food banks, food pantries, soup kitchens, and kids' food "backpack" programs—has sought to step into the breach to provide this population with "emergency foods" on a recurring *permanent* basis. There are many questions that need to be asked not only about how well food bank nation succeeds in adequately providing sufficient food to food-insecure people but also about the social and cultural costs to them of doing so and to the US population as a whole.

Dave Carpenter brings up other relevant questions related to the technological basis of the globally sourced food system that the majority of people in the US rely on. What does he mean that "the Earth wasn't meant to hack modified corn on it year after year" or that "animals weren't meant to be in concentrated feed lots . . . in the conditions they're in"? Put another way, what are limits on the plasticity of the genotypes that define the lives of domesticated plants and animals that evolved under complex ecologies, and what are the costs of manipulating them so that they can be "processed" at huge scales? His questions raise other issues about farming, nature, scale, and commerce—in particular, the sustainability of their current combination—that we address in this book as we discuss what we learned from food activists about the food system Americans have and the systems they might have in the future.

The Myth of Cheap Food

I always tell people when they ask me that question, "Jessie and I started farming because we were tired of food we are afraid to eat." We met a farmer years ago who challenged us to learn more about where our food came from. . . . We decided, well, we're gonna farm. If we can't change the whole system, we can at least change—we can define what we do. And we can produce without all the petrochemicals and without all of the injustice to animals and injustice to people and to supply a wholesome and healthy product that is raised sustainably.

—Sonya James, Felicity Farms, Lincoln County, North Carolina

If you want to compete in the top tier [of fine restaurants], you have to do it. And not just because the other guy's doing it but because, like I said, like, that's where the quality product is. Food that's fresh out of the ground tastes very different than food that sat on a truck.
—Dave Carpenter, chef at F.A.R.M. Café

As Sonya James, a small-scale farmer and restaurant owner residing in the Charlotte metropolitan region, pointed out, there is the issue of the healthfulness and nutritional value of the foods that corporations market to the population—an issue that is subordinated (in proper neo-liberal fashion) to the third prevailing myth about food in the United States: that in the US, "food is cheap." It is true that the food budgets of Americans on average only amount to 10 percent of their disposable incomes (2013), and that historically average relative food costs for US families have gone down steadily since 1960 (when they were almost 18 percent) (Barclay 2015), lower than elsewhere in the developed world and certainly than the developing world. An economist employed by the USDA interviewed by NPR put it this way: "Food is a still a good bargain for the American consumer" (Barclay 2015).

But is this in fact so? Costs still exist whether or not they are included in the itemized budgets of food that families purchase: these costs take a variety of forms—economic, social, political, medical, psychological. In economic terms, costs not included in a specific budget itemization are called "external costs" and can be given a dollar amount. That said, there are other costs that cannot even be quantified but can be enormous, as will become clear shortly, when we write about the costs of obesity and the diseases it brings. The only thing that makes US food "cheap" is that most of the economic, social, and political costs of producing, market-ing and consuming it are externalized—that is, pushed into the budgets not only of the local and regional food sector but also of other sectors of society than those called "the food economy"—costs to the livelihoods of small and midsized farmers, yes, but also costs to Americans' health, their labor rights and subsistence standards, and the environment, in-cluding the Earth's climate. These take the form of tangible damages that are often so enormous that they cannot be adequately accounted for.

For example, the epidemic of obesity in the United States since the 1970s is directly related to increases in the use of high-fructose corn syrup (HFCS) as a major ingredient in the soft drinks, processed foods, and fast food sold to most Americans (Bray et al. 2004).[1] The food stock for HFCS production is industrially grown corn—one of the most subsidized crops in US agriculture. There is a clear and direct relationship between obesity, high blood pressure, and a significantly increased incidence of circulatory diseases such as heart attack, stroke, and kidney disease among people who are obese. These diseases have to be treated, or those who are afflicted will die, often in great pain. Yet the US health-care system is enormously expensive for patients when they come in for treatment of these diseases via surgery, diagnostic tests, and drugs. The amount spent on obesity and obesity-related pathologies in the US health-care system is astounding.[2] "Poor diet is a risk factor for four of the six leading causes of death in the US: heart disease, cancer, stroke and diabetes. When combined with obesity, these diseases have been estimated to cost US$556 billion per year" (Carolan 2011, 75), with the costs as of this writing now far higher.

Looking at the full circle of causality, then, obesity in the US is directly connected to the use of high-fructose corn syrup by the processed food industry, which through political means has managed to offload the social and economic costs of this foodstuff's effects on obesity onto other sectors of society—those who suffer from these diseases, their families, the health-care system, and government programs (and indirectly US citizens/taxpayers) that pay for much health care. What would happen if the full economic and social costs of obesity, which processed, high-calorie, low-nutrient food has contributed so much to bring about, were actually incorporated or internalized fully or in part into the retail price of such foods? This is but one example of how the costs related to the kinds of food produced for Americans and consumed by them have been externalized and left out of the full cost of food reflected in its market price.

The costs are actually much greater than calculated by economists and Wall Street pundits. Despite attempts by the insurance industry to try, how can one quantify the cost in pain and suffering that is imposed on a person who has suffered a massive heart attack and lives an enfeebled existence or on someone who is paralyzed and made unable to speak by a stroke? How can one calculate the pain and suffering caused to the human population on Earth if an antibiotic-resistant bacterium that is lethal or toxic to humans develops an immunity to antibiotics due to meat producers' habitual use of antibiotics to dose their livestock to lessen their costs of production (Khachatourians 1998) or if another zoonotic pathogen like the COVID-19 or H5N1 viruses emerges from concentrated industrial animal production to start a future global pandemic (Wallace et al. 2020)?

The experience of Sonya James and her partner, Jessie, arising from the fact that they "were tired of food [they were] afraid to eat," which led them to seek "a wholesome and healthy product that is raised sustainably," has within it the intuitive recognition that industrial agriculture's proposition that "food is cheap" is deceptive. This is the recognition that the food that people need often costs more than they can afford and is more than what the dominant food system provides them with.

The comment by Dave Carpenter, chef from F.A.R.M. Café, that "food that's fresh out of the ground tastes very different than food that sat on a truck" reminds us that the trend of eaters who strongly prefer freshly grown local produce is already well under way in the US but also that not all who would like to eat it can afford to.

* * * *

The argument of this chapter is that the three myths of consumer food sovereignty, profit-driven techno-abundance, and the cheapness of food provide the cultural and ideological foundations of the dominant globally sourced industrial food system but are deeply flawed. Many sustainable food and farming activists across the US and elsewhere in the world are now questioning and some actively challenging these myths, which still have an implicit hold on other activists, including some we encountered in North Carolina. How activists' decision-making and ethical values have explicitly and implicitly challenged these myths while they have had to come to terms with the corrosive and divisive logics of neoliberalism in the era of climate change is set out in the chapters that follow.

PART II

Food Activists Confront Neoliberalism's Ecological
Devastation and Economic Inequalities

3

Activism for a More Sustainable and Just Food System

In March 2011, our Durham research associate Patrick Linder was interviewing James Sokolov, activist, owner-worker, and founder of Cornucopia Landscaping, a permaculture landscaping cooperative, who told us how difficult it was for local organic farmers to make a living income in North Carolina. At one point, Patrick came back to an earlier comment by James: What did he mean when he said that when James was in his early twenties, "so many people [were] seeing the writing on the wall about food and they decided to become farmers"? James replied,

> For a lotta people in our generation and, say, from the late '90s on, antiglobalization and 9/11, . . . events have been accelerating at a pace that is causing people to look beneath the surface a lot more. . . . What you end up having is a lotta people seeing—hearing a story, saying something like, "Well, when Cuba collapsed, farmers became the best-paid besides doctors." . . . [For] people from a certain class who are able to analyze and look at that writing on the wall . . . of failing imperial oligarchy, it's a rational choice to make sure you can get a livelihood and a living, and . . . [what] we have a lot of, and more and more each day, is people looking to step into that sort of essentialized farmer role. They see that their office job is not gonna last forever. There's layoffs constantly.

Shifting to North Carolina, James spoke of those who tried to take up local organic farming and what they learned from it:

> My brother worked briefly for Bill Dow, who is the first organic farmer in North Carolina, and when he quit, he was like, "My heart's not in it. I just . . . ," and Bill Dow was like, "Well, you're not the only one." He's like, "I don't know where my retirement or any of that stuff is gonna come from. All the food that I sell goes to restaurants, and still working x number of hours a week with no security." And this is the first organic farmer

in North Carolina, so it's like if you suddenly get all these people trying to compete for that dollar, there's definitely gonna be a scramble to get in line for something that, in actuality, is kind of an [expletive]. You end up you're selling your soul. When you're farming on that big of a scale, it's not necessarily sustainable without a lot of inputs, unless you really, really, really almost got a lifetime of knowledge.

Moral Logics, Cultural Production, and the Local Food Movement

James's perspective as a millennial, intellectual, cooperator, perma-culturist, and sustainable farming activist led us to ask, What drives small-scale, organic farmers and the activists who support them (e.g., as volunteers on their farms) to undertake and persevere in the extraor-dinarily difficult and complex labor they do, and to what ends? Our response that James's comments provoked was to focus in this chapter on the "moral logics of food," that is, the discourses shaping people's views of what they ought to practice with respect to food and farming, as they are constrained by the social and material world that encom-passes them.

However, the moral logic of food that James and other activists hold to is only one of many different moral logics of food. In this chapter, we first distinguish between the moral logics of the corporate industrial food economy, of sustainable farming and food activism, of charity-based food security activism, and of community food security activism. Second, drawing on the ethnographic observations and interviews of our team's research associates, we present three exemplars of each of the latter two moral logics and one exemplar of a "cross-over" synthesis of these two approaches. Third, we describe specific characteristics of the projects of the food activists we studied, and fourth, we summarize sev-eral personal traits that unified or divided food activists. Finally, we re-turn to the model of the "diverse community economy" to demonstrate that food activists have made invaluable and distinctive contributions to our local food economies.

By a "moral logic," we refer to a culturally specific set of discursive claims and embodied practices stating how relations between individu-als, community, and the people in power *ought* to be with respect to some perceived value—for example, food, water, employment, or health care—as a life necessity, a medium of identity, and a cultural good.

We see "moral logics" as a central cultural dimension of any movement that seeks to effect social change, like the local food movement. In contrast to earlier sociological approaches to social movements predicated on "rational action" (McAdam 1982), our approach in this book is allied with recent sociological approaches to social movement theory focused on the places of cultural production, context-based decision-making, biography, emotions, and motivation within and operating through social movements (e.g., Payne 1995; Jasper 1997; Kurzman 2004, 2008; Polletta 2002; Amenta and Polletta 2019).

We also follow the work of other anthropologists who have adopted cultural approaches to social movements, with specific attention to individual and collective identity (re)formation (e.g., Escobar and Alvarez 1992; Alvarez et al. 1998; Satterfield 2002; Graeber 2004; Nash 2005; Escobar 2008), as well as in previous work by the authors (Holland et al. 1998; Holland et al. 2007; Price et al. 2008). The theorization of moral logics within social movements thus seeks to capture the processes of meaning-making that not only frame but also help shape movement strategies, goals, discourses, and practices on an everyday basis.

However, as James Sokolov's comments suggest, the local food movement is distinct from what most sociologists, anthropologists, and geographers have studied as "social movements," whether the labor, civil rights, and women's movements or other identity-based movements, for example, anti-HIV movements. Unlike these social movements, the local food movement is not strategically focused on attaining national state- or industry-wide political changes, either reformist (e.g., legal) or revolutionary. To the contrary, it is remarkably decentralized, and its localism is conditioned by neoliberalism's policies of devolution and subsidiarity and by neoliberal branding of a region and its specific (food) commodities (Holland et al. 2007; Heller 2013). Local food activism remains self-consciously decentralized ("local"), reform-oriented, and remarkably differentiated from place to place with tenuous translocal connections to other locales (Nonini 2013)—although these are not absent and have been made in the past (e.g., the Community Food Security Coalition movement) or are being made in the present, in the case of the One World Everybody Eats and the slow food movements.[1]

Nor, in part due to the movement's local adaptations, has a widely shared theoretical critique of the global industrial capitalist food economy emerged among local food activists, although almost all food activists we interviewed offered criticisms of the qualities of the industrial food it produced and its negative effects on human health and the environment. Nor has the local food movement engaged in social protest to obtain its objectives. Its activists have largely remained within the legal and political limits on their activism, consistent with their actions to expand the presence of sustainably grown foods in established markets and to make them more available within legal channels to food-insecure people.

We also seek to contribute to a new approach that decenters the study of social movements away from leaders and their strategic accounts of movements and instead pays more attention to the diversity of place-based or situated actors who opportunistically improvise to incorporate movement concepts, relations, and practices, often appropriated from afar, into their everyday lives (see, e.g., Wolford 2009). Local food activists in North Carolina appear to have drawn on movement elements from the sustainable food and farming movement and community food security movement, which originated in California and the Midwest (Allen 2004) and from other diverse social sources (e.g., civil rights and antipoverty movements), and reworked them in accordance with specific circumstances. The concept "moral logic" captures the ensemble of these mobile meaning-bearing elements.

All social collectivities, including movements, operate in accordance with moral logics of how life should be led, for example, what humans' relationship to food and through food to one another ought to be, as these arise and are articulated dialectically relative to opposed moral logics. The moral logics of the local food movement have arisen in opposition to the moral logic of the globally sourced industrial food sector.[2]

On the one hand, the moral logic of the global industrial food sector focuses on and positively values private profits, industrial methods of food production and processing for faraway markets, and "tough-minded CEOs" and celebrates the provision of "cheap" food for mass and niche markets of consumers (Roberts 2008) and, as a means of avoiding waste, for distribution as charity to food-insecure people. Within this moral logic, consumers and corporations are seen as interdependent through the exchange of money for food commodities, where consumers owe corporations (i.e., their managers, shareholders) money in return for the "value added" from the global supply chain of food commodities whose adequacy is defined by sufficient quantity and consumers' "taste" (Roberts 2008). The "community" is a purely instrumental relationship between individuals and corporations within the market setting, and what is "local" is irrelevant, except perhaps when it is invoked to elevate the brand of the food retailer as a "good citizen" when it donates surplus food to needy poor people in a locale.

On the other hand, the moral logics of the local/alternative food movement make paramount the ideas of "local," "livelihood," "sustainability," "healthy foods," and "small farmers" as defined by ecological and social interdependencies (among farmers, eaters, crops grown, livestock reared, water and farmland resources, etc.) and of "social justice" and food as a right for all food consumers and producers. In addition, for community food security activists, "community" is also implicated as a self-defined local population committed to a degree of self-sufficiency in producing, accessing, and eating culturally appropriate and sufficient food. Within these moral logics alternative to the corporate industrial food economy, food not only must suffice in quantity but also must "sustain" by its quality (as "healthy" or "nourishing"), and such foods' production must be "sustainable" through cooperative interactions between eaters, producers, and the ecological systems (food chains, etc.) and natural conditions of production (i.e., "the environment") that grow food (Allen 2004).

In contrast to alternative food activists, charity-based food security activists show a moral logic that situates them between the moral logics of the corporate industrial food sector and the alternative food movement. While adhering to a notion of "community," they define it hierarchically in terms of charitable organizations that constitute it and thus have the right to determine how food insecurity within a local population should be addressed; these organizations provide food as a gift that they define in quantitative terms to a local population irrespective of how this population's members qualitatively define their own desired relation to food.

Transactions around food involve both market-based and non-market-based exchanges. Within the moral logics of alternative food activists, what is "local" and what is a "community" are each defined by these interactions and by reciprocities and debts between its members. For some activists, these exist not only between human members but also between humans and nonhuman members (including plants, animals, and biophysical entities, such as "the soil" or "the watershed"). "Local" is not defined by proximity as such but by the supposed capacity of the eater/consumer to "know where their food is coming from," because they can meet the farmer who grew it at, say, a farmers market. Personal relationships, not confidence in a large-scale commercial system, form the basis for trust.

The moral logics of the global industrial food sector and of alternative food activism are opposed because in each case the understanding of interdependencies and the implementation of the (material and social) technologies that sustain them define what the members of a local community owe one another, ultimately a moral question—and these moral logics fundamentally differ in how they answer this question.

In contrast to the neoliberal discourses and practices that constitute the moral logic of the global industrial food sector, local/alternative food activists draw their moral logics of food from diverse dissenting discursive sources and have reinterpreted these sources both within scholarship (among academic activists) and more broadly (Beus and Dunlap 1990; Dahlberg 1991; Vallianatos et al. 2004; Wekerle 2004; Allen 2004, 2008; Anderson 2008; Community Food Security Coalition, 2001, 2008; Winne 2008).

Jay Thomas, one of the founding members of the Crop Mob in the Durham area, when interviewed by our research associate Patrick Linder, stated what he called a fundamental "predicament"—one we discovered throughout our fieldwork:

> And I think it's, in a lot of ways, a really—I mean, it's a foundational problem because from a farmer's perspective, working in sustainable agriculture, they need to be selling their vegetables for more than they are [able] in order to make ends meet. And from a justice perspective, un-

derprivileged folks need to be able to buy the food from those farmers for less than they're selling it for now. So I see that at the root of the problem. And it's a real—it's a predicament rather than a problem . . . that . . . has a solution, and a predicament doesn't necessarily have a solution.

The moral logics of local food activism that Thomas's insight points to thus took two distinct versions divided over the question of what constitutes a just market price for food in a class-divided society. On the one hand, the moral logic of sustainable food and farming activism centers on the question of what the relationships between humans and the environment and local ecologies with respect to food ought to be and has become identified with "sustainable agriculture" and the needs of farmers. On the other, the moral logic of community food security activism focuses on what the relationships among humans with respect to the food they need to live ought to be—a question of social justice—and identifies with the need and rights of people to eat (and when possible, to grow their own food), regardless of their class or race identities, as defining a "sustainable" food system. In contrast to the community food security version, the sustainable agriculture moral perspective defines a different kind of "community" where environmental and economic sustainability overlap, so that the community is primarily composed of people who make a living from cultivating land and protecting its ecologies, with the livelihoods of farmers mediated by a market that ensures their continued livelihoods (Rodale 1983; Berry 1977/1996). Such, for example, was the predicament described by James Sokolov when even the most experienced organic farmer could not escape economic precarity through the market. The community food security moral logic, in contrast, defines a community by the just and equal distribution of food (and when possible, the capacity to grow it) among its members in ways that are culturally appropriate and maintain the dignity of all its members, articulated in terms of "food justice" (Wekerle 2004), "food democracy" (Hassanein 2003), or "food citizenship" and "food sovereignty" (Anderson 2008).[3] The moral logic of charity-based food security activism is one that mediates between a community defined hierarchically by food charities linked to the corporate industrial food sector and the local community of related people defined as such by community food security activists.

In what follows, we first present exemplars of food activists in our four sites who were animated in their practices by these two moral logics, as well as by a "mix" of both moral logics that some activists manifested.

Exemplars: Sustainable Farming and Food Activism

The Crop Mob of Durham and Nearby

Patrick Linder, one of our research associates for our Durham site, interviewed Jay Thomas, a founding member of the Crop Mob, about its ongoing project by young farmers of exchanging labor on different farms in the Research Triangle region of North Carolina. Jay described its origin at a meeting of young farmers in October 2008, three years earlier. The farmers had come together to discuss the major problems facing them: gaining access to farmland, receiving fair wages, getting health care. One impatient participant said that she "hated coming to meetings" and wasting her time talking "about the same things over and over again and not really ever get anything done." Her intervention led Jay and the others present to find a way of "building community" among farmers by working together instead of "just sit[ting] around a table talking."

Out of this, Crop Mob came into being. Jay characterized the work of Crop Mob as building "the kind of community that would be needed to make sustainable agriculture for the people who practice it because . . . it's super, super labor-intensive work. . . . So the Crop Mob is a social event where the farmers can get together and socialize while getting stuff done and being productive . . . because it can be real easy to feel like you're out there all by yourself toiling away and . . . like you're not part of a movement when you've gotta do all of your work by yourself." Before a Crop Mob Sunday, one of its members invited other "Crop Mobbers" to show up at their farm to join them in laboring together to prepare their field for cultivation, for example, removing rocks, tilling, adding compost, while enjoying each other's company. At the end of the day, after the work was done, the host offered a dinner prepared for all those who had worked together, and the day would end in celebration of the collective achievement. The genius of Crop Mob lay in members rotating as hosts, and thus each member who wished would be able to draw on the collective labor of the entire group to open their field for planting by a single day's work.

Over the next year, young farmers' enthusiasm for Crop Mob Sundays grew, and Jay as a founding member and a committed organizer took on the task of preparing the host farmer before the convergence of the Mob, when it was not even clear how many young farmers, as well as those who might simply be interested even as "tourists," might show up. Jay recalled,

> At that time, it was twenty-five or thirty people [who] might show up, and that seemed pretty daunting. But we went from twenty-five to thirty to forty to fifty to seventy to eighty to, at times, one hundred people showing up for Crop Mobs, which . . . does get to be pretty daunting [laughs]. And so, yeah, definitely around the time when the *New York Times* and there was a lot of media around the Crop Mob, . . . [we] were having up to one hundred, to one hundred and some people show up to Mobs, which is pretty crazy. You need to be really—the farm needs to be really prepared to facilitate that many people working.

The nationwide media attention led an increasing number of young farmers, as well as the simply curious, not only to attend Mobs within the Research Triangle region but also to set them up more widely, "to the point now where there's like sixty-nine groups around the country." Eventually, however, "the buzz died down," and the group was "getting back down to more sustainable and reasonable numbers of thirty, forty, fifty people showing up at Mobs." Jay summarized Crop Mob's success: "It's really great that communities have taken this idea and ran with it and made it into something that fits their community and that works for them."

Decision-making by Crop Mob members was avowedly nonhierarchical, although Rob observed, those who participated developed "certain practices . . . in order to organize [them]selves, or in a lot of ways prevent . . . [them]selves from becoming an organization": "There was no money involved, . . . we weren't looking to raise money, and we wouldn't accept money if somebody wanted to give us money." Jay noted that "a lot of folks involved in the Crop Mob . . . would be self-described anarchists," while many, including Jay, were suspicious of the "nonprofit industrial complex," as they witnessed "nonprofits falling apart because their funding was being pulled" during the Great Recession of 2007–2008. In Rob's view, the success of Crop Mob over the previous three years largely validated Crop Mob members' perspective on the possibilities of its self-organizing nature.

Elma C. Lomax Incubator Farm, Cabarrus County, Charlotte Region

Sarah Johnson, our research associate in the Charlotte metropolitan region, brought to our attention that there were relatively few food-related organizations committed to a sustainable agriculture paradigm within this highly urbanized region. She noted that the Incubator Farm was the only project she was able to identify "that is focused on getting more people working as farmers; most are focused either on increasing consumers' awareness or providing more opportunities for producers and consumers to connect with each other." In Cabarrus County, north of the concentrated population of Charlotte, a "community food assessment" (Pothukuchi et al. 2002, 6–15) undertaken in the county convinced many people, as Sarah said, that "without more producers—and specifically vegetable growers—local food isn't feasible, because they can't meet demand within their own county, let alone the demand in Charlotte."

In response, the Incubator Farm was set up in 2007. According to Sarah, it "was a thirty-acre farm that was bequeathed to Cabarrus County with the stipulation that it could not be developed. The decision to turn it into an incubator for beginning farmers came from the county Agricultural Extension agents, who worried about aging farmers: the median age of farmers there is fifty-eight, and while interest in locally produced food has increased considerably in the last two years, many are worried that there just aren't enough farmers to support a local food system."

Approximately a dozen beginning farmers were offered nine weeks of classes in organic farming, and then each was leased half an acre of land for three to five years at $100 per year, far lower than land rents prevalent in the area. Operating expenses for equipment and expenses (e.g., to power the heated greenhouse) were funded by the county government. A knowledgeable "mentor farmer" was recruited to join Extension agents in guiding the beginning farmers. Several of the beginning farmers applied to the program because of the failing economy of the

Great Recession—they had either lost jobs or were unable to keep their small businesses operating.

Sarah characterized the combination of agricultural and commercial methods taught to beginning farmers: "The model is a mix of collaboration and individual businesses: they all agreed to maintain organic practices; all the farmers share commonly held equipment; the first group all had the buy-in of helping to build the greenhouses and sheds; and some of the farmers have joined in working the high tunnel together. But each farmer has their own business plans: most do a mix of selling at markets and to local businesses as well as a CSA. They each decide what to plant and how to maintain their plots."

As of the time of Sarah's fieldwork, the future organization of the Incubator Farm was uncertain. The mentor farmer had formed a partnership with one of the beginning farmers, and they had both left the farm to lease land elsewhere and start up their own business. Meanwhile, the influx of the next cohort of beginning farmers was about to occur. As a result, Sarah said, "the questions being worked out are how long can people stay on the land and whether the farms will stay organic."

Watauga County Farmers Market, Appalachian North Carolina

The Watauga County Farmers Market has been open in Boone, North Carolina, since 1974. In 2011, the market described itself as follows: "Up to seventy-five vendors offer quality locally farmed produce and plants, as well as locally made crafts and edibles. . . . We are a 100% local, producer-only market with the High Country's finest selection of quality produce from local growers. Seasonal fresh fruits and berries. Local jams, jellies, and honey. Fresh baked breads, cakes and pastries from our country kitchens. Fresh farm eggs. Fresh and aged goat cheeses. Locally farmed meats and livestock" (Watauga Farmers Market 2011). Jen Walker, our research associate in the Watauga and Ashe Counties field site, recounted a Watauga farmers market board meeting she attended in May 2011 that revealed the ways in which commercial and community food security moral logics in the neoliberal US were often at odds, even in markets where organic and sustainable foods were central to the mission of this farmers market.

The meeting began at 7 p.m. by setting out an agenda of nuts-and-bolts issues that required decisions similar to those facing many farmers market boards in rapidly growing North Carolina cities. Since stall space was at a premium in this thriving market located in downtown Boone, not far from Appalachian State University, a major regional university, the question of whether new members of the market should be accepted and allotted vending-stall space came up. One applicant for new membership was rejected as "too commercial"; the other was accepted because she grew some of the ingredients in the herbal vinegar she sold.

This was followed by other items: the overuse of parking space by vendors in the market parking lot; the market's proposed T-shirt competition and whether the market's current logo should be retained or a new one solicited for the shirts; the appearance of a new banner for the farmers market (manager's) booth and whether it should have an electronic benefit transfer (EBT) notice on it (stating the market' s acceptance of SNAP and WIC electronic vouchers); the market's purchase of a digital billboard in order to advertise its attractions to the public; and several other items.

As the meeting went on beyond 8:30 p.m., it became clear that the most controversial item of board business had been deferred until last. This was an appeal by Teen Adventures (TA), a local nonprofit organization devoted to working with local high school youth, which a few Saturdays before had been allotted a stall to sell meals to market patrons as part of its fundraising venture but had subsequently been evicted from the market. According to the market manager, "there were loud, angry complaints" from other vendors objecting to the presence of TA as a nonprofit organization selling at the market. The market manager had immediately prohibited TA from being a market vendor in the future, and angry words were exchanged.

TA's executive director and a staff member decided to appeal this decision by submitting a full application to the board that evening so that they could sell their meals in the market in the future to raise funds and as part of their community outreach. They were kept waiting outside the meeting room for an hour and a half and then decided to walk into the meeting room on their own, but members of the board refused to acknowledge their presence until their item of business came up. When called on to speak, the Teen Adventures executive director pled his case that its catering license, approved by the county health department, allowed it to operate anywhere in the county including the market, and he offered the board TA's membership application with fees paid for the rest of the season, one Saturday per month. After two minutes, he was abruptly cut off by the board president. In the ensuing discussion, two people in the audience spoke up in defense of Teen Adventures and its application, one a market patron stating that as a former board member of Teen Adventures, he and his wife were angry: "We're not going there anymore if this is the way this organization is treating [Teen Adventures]." Soon thereafter, the board president asked the TA representatives to leave the room.

Board members proceeded to argue among themselves about the merits of TA's application. The board president saw it as part of the earlier question raised about allotting scarce stall space to new members but also indicated that he considered Teen Adventures to be "really controversial"—probably a reference to its work with local teens. Two other members suggested that the board allow TA to sell its meals at the market one Saturday a month without offering it membership. The market manager conceded under questioning that the "loud, angry complaints" she had invoked came probably from a single vendor, who happened to be the market's immediate-past board president, just voted out of office, who objected to the presence of any nonprofit organization because it "unfairly" competed with other vendors, given that its tax exemption conferred an unfair advantage to it compared to vendors operating commercial businesses. One board member wondered "why there's so much anger towards nonprofits" at the market, which provoked the board president to speculate about the competitive advantages that nonprofits had and their "unfair" threats to real farmers. He noted that one of the other board members was paying off a mortgage to operate his farming business, while if TA's application were approved, it might mean that even the nearby research station could be entitled to sell its excess produce at the market—since "in his mind, they have no expenses because they're funded by the state, so they could undercut everyone." This led another board member, an Agricultural Extension agent in his day job, to laugh at this claim as a manifest impossibility. Another board member quipped that "competition is the American way." Soon after this, having been in session until past 9 p.m., the board concluded its meeting, without reaching a decision.

Exemplars: Food Security Activism—Food as Self-Determination, Food as Charity

Interfaith Food Shuttle, Durham

The Interfaith Food Shuttle (IFFS), headquartered in Raleigh, North Carolina, is a food bank established in 1989. By 2010, it supplied food to approximately 220 food pantries, soup kitchens, mobile farmers markets, and cooking classes within a seven-county area including Durham and distributed six million pounds of food in 2009 to thirteen thousand people (Wilson 2010, 1). It operated with twenty-five paid staff people and hundreds of volunteers, with thirty-eight thousand hours of volunteer labor supporting it annually (Wilson 2010, 2). IFFS was distinguished from most food banks for its strong focus on collecting perishable food from food retailers and local producers and distributing it rapidly to its donor organizations (Wilson 2010, 2). This contributed to its national reputation as an innovator (Poppendieck 1998, 107–110, 116, 266).

The observation by our research associate Kevin McDonough of a meeting in Durham illustrates one IFFS project, a mobile fresh-foods market in a food-stressed neighborhood. Participants included two IFFS staff, Gerald Payton (food recovery and distribution services) and Amelia Martin (coordinator of the West End Community Garden); Kevin (volunteering with this garden project); and Cindy Jones, Gina Pierce, and Tracie Elder (representatives of three nonprofit organizations and churches engaged in distributing emergency food in the neighborhood). This observation captures the labor-intensive, ad hoc, logistically complex work of providing emergency food to Durham's poor neighborhoods, here, the West End neighborhood in Southwest Central Durham, where a large proportion of residents were low-income African Americans.

Kevin writes,

Amelia and I arrived at the Durham Service Center of the Inter-Faith Food Shuttle (IFFS) at Northgate Mall a little before everyone else.... The impetus for the meeting was to find some way for IFFS and Immaculate Conception Church (ICC) to contribute food to some groups at Lyon Park [West End Community Center, or WECC]. Gerald Payton had been in touch with Amelia and had agreed to meet with Cindy and Gina of the ICC. Part of the inspiration for making these connections was that the IFFS mobile market that used to serve the neighborhood near ICC and Lyon Park (West End neighborhood) had closed over the summer.

Kevin explained why this was important to residents' food security: "Mobile markets are essentially drop-off points for produce in different neighborhoods that IFFS works with. A truck comes full of food, . . . and volunteers help neighbors pack up the food they want. [Amelia] had been in touch with some of the staff at IFFS . . . in the hopes that they could open a new mobile market in the neighborhood and continue the 'flow of food' to the neighbors that went to that market."

IFFS's Durham Service Center consisted of four narrow rooms of converted retail space in the old (and now repurposed) Northgate Mall. These rooms were organized either for table space for volunteers to put together bags of assorted foods donated to IFFS or for storing and organizing long rows of bags of food packed on shelves for future deliveries for IFFS's programs in Durham, such as Grocery Bags, Backpack Buddies, and Kid's Café. Not long after the others arrived, Gerald arrived in a truck filled with grocery bags of food that needed unloading. After those who had already arrived helped Gerald unload these bags and stored them on shelves in the front room, the meeting began.

Kevin described the meeting as one in which the volunteers from the church and neighborhood organizations in the West End neighborhood emotionally expressed their concern about the cessation of IFFS's food deliveries to the neighborhood's poor population.

[During the round of introductions starting the meeting], Cindy told us that she was born and raised on Kent Street. Cindy got somewhat emotional when she began talking about food service stopping at WECC. She was upset that the people who came to that mobile market weren't getting anything now. Amelia interjected that IFFS delivered for more than twenty years to that spot at the WECC. Gerald said, "Twenty-five or so years." Tracie [Lyon Park Community Center] asked about how best to recapture the individuals that had been receiving food from the mobile market. She said that First Calvary [Ministries] has a food pantry service that is open on Tuesdays and Thursdays. This led to a discussion about how to recapture the service that was provided to the West End. Gerald and Amelia and Tracie began estimating how many people were served by the mobile market when it was active. Gerald said, "At most fifty." Cindy said, "I came once. There were a *lot* of people—at least fifty."

What appears to be a small difference in numbers between Gerald and Cindy indexed the contrast in power and status between givers of food and receivers of it: Gerald was an employed full-time IFFS staff member, while Cindy drew on her experience as a neighborhood resident and member of ICC and possibly as a charitable food recipient. Kevin went on to record how Gerald positioned his authority in the context of IFFS's growing mission to feed the hungry in Durham: "He said the Food Shuttle came to Durham in 2009 and extended its food distribution programs. He explained about Backpack Buddies. He told about the IFFS farms in Raleigh, the Culinary Job Training Program, the Young Farmer Training Program, Kid's Café, Cooking Matters (cooking classes for youth, teens, adults, and families). Cindy cut Gerald off after he had been delving into these explanations for several minutes by saying, 'It's just so much! We need to back up!' Everyone laughed at this."

A main topic that came up in the meeting that followed was the difference between "closed" and "open" markets, both of which IFFS operated in different neighborhoods of the city where food-insecure people lived. Gerald explained that "the mobile market at WECC was a closed market because it served the same people every week and every person/family received a specific set of bags of food that was for them and had their name on it." Amelia, the other IFFS staff member, asked, "What is the need in the West End?" She proposed that instead of a mobile market at the WECC, IFFS could distribute food to Maplewood Square, an apartment complex in the West End neighborhood. Cindy of ICC offered that there is a "willing seniors group" at Lyon Park that could bag and receive food. Amelia noted that the open market meant that it was open to anybody within limits to take what they wanted. Kevin described, "Gerald confirmed it was. Open markets operate with a truck that brings produce, and people are helped selecting what they get. . . . Gerald added that one benefit of a closed market . . . is that 'you don't have people coming back a second or third time and trying to get more portions than are listed on the board.'"

This meeting in the IFFS Durham Service Center illustrates the delicate negotiations that occur in the case of IFFS around who receives charitable food, when, and under what conditions—negotiations that invoke much that is left unspoken. For IFFS staff like Gerald, there is the matter of IFFS's mission and his own: rationing out donated food across neighborhoods and families so that everyone in need receives some food. For Cindy and Amelia, the question is about who is in more need and how much and which kinds of donated food they should receive. Thus, the question of whether people should be allowed to come "back a second or third time and [try] . . . to get more portions than are listed on the board" was not a trivial one but instead pointed to a fundamental difference in the moral logics of charity-based activists and community food security activisms. There is an underlying argument here between two positions about who gets to determine who gets to eat and under what conditions.

This argument took subtle forms. Cindy, a West End resident, referred not only to the unmet needs of West End seniors who were eligible to receive charitable food from IFFS but also to their being "willing" to bag food. Cindy is alluding to a well-known exchange of labor for food. When the labor of, for example, a "willing seniors group" or of other poor people who receive donated food is occasionally called on to break out the food into bags or boxes for distribution, store it, and even pass it out under the eyes of staff, such volunteer laborers are often allowed (or exercise as their moral right) to receive an informal but nonetheless material "wage" by taking more food than they would otherwise be given or to take some foods that they prefer over others, much as in the case of the "open market" that Gerald criticized (cf. Dickenson 2020, 102–104). How might this situation be defined in such negotiations—by charitable staff as undesirable "leakage" or theft through greedy takings by a few individuals or as food taken as a very small but fair nonmonetary wage by unpaid volunteers? This exemplar shows the tensions in moral logics between charity-based activism and community food security activism—a theme taken up later in the book.

Black Women's Health Network, Charlotte

Black Women's Health Network (BWHN) originated in St. Louis but had a branch in Charlotte started by Dorothy James. According to our research associate in Charlotte, Sarah Johnson, "The goal of BWHN is to address the issues of chronic disease related to diet and lifestyle especially among Black women, who are disproportionately likely to be affected by them."

Sarah observed that Dorothy sought to make it possible for lower-income African Americans to gain access to local foods: "In the past year, she has developed a plan for a series of produce stands at several Charlotte African American churches—she believes this will make them more accessible to the community, as well as making it clear that they are the desired customers." The program's name was "A Time to Harvest." The church's involvement will be to set up a market in its parking lot or inside in bad weather. "The churches will also do marketing and advertising, and BWHN will be hiring seniors and youth in the church to work at market [and] set up stands," Sarah explained. Dorothy hopes that these markets will be held seasonally on Saturdays. Sarah observed that the model of developing networks for distributing local food through BWHN produce stands set up in African American churches has been used by BWHN for years and that Dorothy "sees this as key to reaching community."

Making local, fresh food available in this way was not without its challenges, as Sarah described: "One of the issues is that of finding a producer willing to sell at the markets and at prices that would be affordable for the lower-income members of the community; she eventually found a couple of local farmers who were willing to sell to her on consignment at wholesale prices, but it took several months. Dorothy also had a small battle to get Twenty-First Century Farmers Market to accept her, since her market model is not the classic farmers market."

Sarah summarized what she thought most important from this project: "Dorothy . . . is explicitly talking about how race interacts with local food efforts." In African American neighborhoods lacking supermarkets ("food deserts"), residents' nutritional needs, the absence of fresh and nourishing food, and increased health problems due to poor diet were part of a broader syndrome associated with class and racial discrimination in Charlotte. BWHN sought to address this syndrome.

Happy Hills Community Garden, Rocky Mount

Willie Jamaal Wright, our research associate for our Halifax, Edgecombe, and Nash County site described to us the distinctive Happy Hills

Community Garden (HHCG) and its remarkable manager, the late Law-rence Farmer, in Rocky Mount, eastern North Carolina. Willie Jamaal introduced the garden and its neighborhood: "The Happy Hill Commu-nity Garden . . . is located in the Happy Hill community of Rocky Mount. Those who live in the community affectionately refer to the neighbor-hood as 'The Hill.' Unfortunately, the community is known notably for its crime and poverty rates." In 1999, Hurricane Floyd flooded the Happy Hills neighborhood, as a result of which an apartment building there was abandoned by the landowner. The city demolished the building, and the land lay vacant for many years. In 2007, the Parks and Recre-ation Department decided to convert the plot into a community garden. Until 2011, it was the city's only community garden, but few Happy Hills residents tried to gain access to it until Lawrence Farmer became its informal garden manager.

Farmer was a retired Rocky Mount fireman, who felt a duty to give back to the city that had supported him for so long and was a Nash County master gardener with extensive gardening experience. Not only during his first year at HHCG did he spend thousands of dollars of his own money for supplies, seeds, and other expenses on the garden, but he had also since then paid the $750 annual fee to the city for access to the garden plot and water yet refused to collect the $15 user's fee from residents seeking access to garden space. When Willie Jamaal asked him whether he charged residents a fee, Farmer said, "in his thick southern accent, 'I don't rent it [garden plots]. I let 'em have it for free . . . 'cause I rent [the garden] from the city and I give it to the community.'"

The fee waiver made the garden very popular among Happy Hill residents, to whom Farmer gave first priority before opening it up to residents from other neighborhoods. Residents were also drawn in by his welcoming manner, such as holding a spring opening dedication day, and by seeing Happy Hills neighbors cultivating the garden. Wil-lie Jamaal described the friendly social relations, mutual support, and exchange of knowledge that Farmer helped create among gardeners: "[They] are a diverse bunch of agrarians and include Black and white men and women, couples, families and singles, the elderly and the young. I had a plot in the garden and was surprised by the willingness of everyone to engage one another and share their knowledge, supplies, and services to gardeners who needed them. Maybe this show of support was an aspect of the spirit of gardeners, that even in a racially divided town like Rocky Mount, we could come together in this space to create our own community."

Farmer viewed this collective effort of the gardeners as a significant turning point. As Willie Jamaal observed, "Folks wouldn't ever be speaking to one another if it wasn't for the garden. On any given weekday or weekend, morning or evening, it was not uncommon to see people talking with one another, taking tours of each other's plots, sharing knowledge, stories, seeds, and harvests." Farmer's fee waiver and his friendly manner were crucial not only to gardeners' participation but also to the food *and* financial security of poorer residents in Happy Hills. Willie Jamaal said, "The importance of this garden to individuals on fixed incomes is profound. From my observations, I learned that many of the gardeners used the food they grew to supplement their food needs and free up funds for other household costs." Willie Jamaal described Mr. Hawkins, a resident of Happy Hills, who was in his sixties and under pressure from his employer, Sears, to take early retirement yet opposed retiring because he would lose part of his pension and was in a precarious economic situation. Willie said,

> I often spoke with Mr. Hawkins while in the garden about the progress of plots. I would share seeds and growing methods with him, and he would reciprocate by allowing me to harvest as much of the bountiful basil I desired. . . . Because of his economic situation, Mr. Hawkins relied heavily on the food he produced in his HHCG garden plot and the small garden plot in his backyard. He also shared with me his financial struggles. One Saturday he told me how he blanched and froze much of the squash and greens he harvested during the summer season to support him during the winter. Also, many of the meals he ate during the spring/summer seasons were comprised of what he grew in his garden.

Willie Jamaal asked Farmer why he felt that the garden was important to residents. Farmer replied, "For one thing, the economy. With the economy we got now, we need all we can git. I think it [the garden] does help 'em as far as the economy goes." This was in the depths of 2007–2008 financial crisis, when Edgecombe County's unemployment rate edged toward 14 percent.

Exemplars: Mixed

In a few exceptional "crossover" situations, the concerns of sustainable farming activists and community food security activists came together to address a key predicament in this book, one to which Jay Thomas only began to hint at earlier: How would a population achieve lasting capacity to produce and consume local, fresh food that is culturally appropriate and in sufficient quantity irrespective of their incomes or wealth? Most food security activists assumed that the only "crossover" solution to this predicament could take only one form: a market solution—that is, that "a fair price" had to be found—a "solution" consistent with neoliberal views. However, a few activists undertook projects that sought and discovered alternative solutions that did not depend solely on market mechanisms but instead arose from new cultural productions and creative social arrangements with respect to markets. The exceptional community food activists who undertook small-scale food projects to confront this predicament are featured prominently later in this book.

Even more rarely, crossover might be attempted at a higher scale. Food policy councils were an innovation in the early 2000s in large cities, typically ones that had benefited from globalization. Food policy councils were initiatives by political and economic leaders in these cities, in a commonly used phrase, to "bring all stakeholders in the food system to the table" as part of a long-term regional development strategy to further the economic synergies between a city's population's demand for food and its hinterland farms' capacity to supply that demand. Both farmers and low-income urban eaters/consumers had to benefit (Winne 2008).

One such exemplar was the Charlotte Mecklenburg Food Policy Council (CMFPC), founded in 2009 and led by a charismatic activist Marilyn Marks and studied by our research associate Sarah Johnson.[4] CMFPC brought together farmers' organizations from throughout Charlotte's metropolitan region, food retailers including farmers markets, neighborhood residents' organizations, city/county health department nutritionists, the county school district, philanthropists, outstanding local food activists, and others to attempt to set out a set of coordinated policies that would both benefit the surrounding rural farming sector by stimulating demand for local/regional fresh food, on one side, and, on the other, increase access for Charlotte's lower-income city populations who faced food insecurity.

Sarah's research in this case was particularly detailed and insightful, for she spent many hours with CMFPC's founding director, Marilyn Marks. Marilyn had a prior career in banking, and that made it important to her to be "looking at the economics." Drawing on this perspective, Marilyn diagnosed one of Charlotte's key preoccupations: "Particularly with a city that's money-driven and bank-driven like Charlotte is, we need to show how much money is leaving this area because we are not using local foods. We need to find out what it would mean to us to be able to put some people to work at farms that have lost their jobs. We need to see how much stronger a city and greener a city we would be if we were inviting and encouraging local farms."

Marilyn told Sarah that her previous activism as a gleaner in the Society of Saint Andrews had led her to lead CMFPC with a "continuing emphasis on access issues: 'At this point our food policy council is probably more focused on accessibility of food to everybody more than sustainability.'" Putting the two together, Marilyn went on, "I do value . . . the economic part of me, so it's really interesting for me to want to do this study on food deserts and then the next thing I want to do, . . . I want to show that we can make money selling local food here and that we can have a huge market and growers—they can grow, they can lease more lands so that we can make money." Marilyn's interest in addressing food deserts had led CMFPC to undertake a "community food assessment" (Pothukuchi et al. 2002) to determine where supermarkets and other retailers had failed to provide the residents of specific neighborhoods with adequate and affordable fresh produce.

As Sarah's fieldwork continued to unfold in 2010–2011, she was able to witness the completion of the community food assessment and its use by the CMFPC under Marilyn's leadership. Sarah wrote, "The Charlotte-Mecklenburg Food Policy Council formed . . . to address policy-level issues in the local food system. Thus far, its primary focus has been on the community food assessment phases 1 and 2. Other efforts have included partnering with the West End Market, advocating for changes in zoning for mobile markets to make them more feasible, and acting as an advisory committee for county commissioners on food-related issues. They have also recently partnered with a newly formed youth policy council on healthy eating."

Characteristics of Food Activism: Social Projects, Work, Joys, and Discontents

What were the characteristics of local food activism that we observed across our four sites?

First, there was the importance of sponsorship. The projects and labor undertaken by food activists were often dependent on their receiving grant funding and other required project resources from the nonprofit organizations created by the community-oriented public-private partnerships that have arisen due to neoliberal reforms over the past three to four decades (Holland et al. 2007, 125–129). This means that local governments often had to join forces with private corporations and foundations, community organizations, and volunteers in order to address major social problems that hitherto were dealt with via large-scale state programs and benefits. As a result, nonprofit organizations for local food provision came into existence from these partnerships. They were governed by coalition partners, funded modestly and year by year by city or county governments, private foundations, or for-profit corporations, and they had barebones budgets that depended on successful grant applications. They employed a modestly paid full-time executive director, one or two full-time or part-time paid staff members, and a few or many unpaid volunteers who together did the work. The functions of such nonprofit organizations included providing emergency food, creating and implementing modest plans for expanding local food production, and much more. Unless these organizations had their own facilities, for example, church kitchens, they outsourced these tasks to individuals and small groups deemed to have expertise or experience in these areas, whom the organization put to work on short-term shoestring budgets, few if any nonmonetary resources, and rarely at a living wage.

Second, these were small-scale projects. Whether backstopped by nonprofit organizations or self-funded, the vast majority of food activists consisted of one or a few concerned individuals who undertook their labors in a relatively small-scale "project" in order to achieve some improvement in growing, distributing, or providing wider access to locally grown foods. Activists might be starting a small urban farm, serving on a citywide planning committee for a new farmers market, offering workshops to grow shiitake mushrooms, or offering cooking classes to minority youth.

Third, these were do-it-yourself projects. Even if activists received organization funding from grants, these projects were "do it yourself" or, when a few individuals came together, "do it yourselves" in nature. The labor and time spent, supplies consumed, and spaces to organize activities—whether farm plot or meeting room—depended largely on what an individual or small group might already have or find or put together to work with each other. Or they depended on some small amount of money that a person might have to rent a plot of land, on what one or two people could pool together in tools and materials from their garden sheds, or on what working together they might be able to scrounge, for example, mulch from a neighbor's yard or compost from a friend's kitchen waste. Projects therefore had a short lifetime, could be put on hold for weeks or months due to the sickness or personal problems of the principals, or simply would not achieve even a limited goal or would be abandoned.

Fourth, these projects could be physically and intellectually demanding of activists. These do-it-yourself projects called on individuals' available physical labors, intellectual efforts, financial resources, technical skills, and interpersonal/networking abilities and, in this sense, could at times approximate a full-time job with little or no pay. Volunteers or WWOOF ("woofer") interns working on a local organic farm were one example.[5] However, most activists had the latitude to change their pace of work and the deadlines of the goals to be achieved or in some cases might simply abandon the project.

Fifth, most labor put into a project went unpaid. Rarely did a grant cover the amount of labor that an activist put into a project at a living wage level, and much such labor simply went unpaid. An exceptional minority of activists worked as modestly paid staff members for non-profit organizations, churches, or local governments, with Agricultural Extension agents and county health department nutritionists on state and local government payrolls receiving full-time salaries and benefits at the high end. Most activists who worked with organizations did so as unpaid volunteers. As a generalization, the vast majority of food activists working on these projects were unable to depend on what they generated from this work as adequate income for them to live from. This had implications we explore in the next section.

Sixth, activists preferred to "walk the walk, not talk the talk." "Getting your hands dirty" was often a literal characteristic of many such social projects (e.g., starting a garden), but even when it was not, the key idea was that people were engaged in meaningful effort in order to secure a physical outcome, a real change in the material world, and they were not there "to just talk" in the sense of socializing instead of working together to achieve the shared objectives. This meant that when individuals undertook physical labor on such projects, it was acceptable to converse, but meetings and discussions outside the contexts of doing the labor were downplayed. Extended discussion of hypothetical outcomes, elaborate planning schemes, or theoretically intriguing possibilities, especially when it took time away from the focused physical tasks at hand, were discouraged. Activists' patience with extended meetings, when these were held, wore thin, given the sense of the urgency of a project.

At times, expectations by project leaders to move directly into physical labor when, say, starting a garden or urban farm would take an almost anti-intellectual edge ("let's stop wasting time talking") but would be seen as communicating urgency. However, a lack of discussion between activists could limit and even sabotage the outcomes of certain projects when interpersonal dynamics between participants were potentially conflictual and when the project was stressful but remained undiscussed—as when people from different racial and class backgrounds came together to work on a common project. On the other hand, in the absence of such conflicts, the sense that "we're here to get things done" allowed small groups to focus productively on what they needed to achieve in the time available.

Seventh, projects depended on improvisation and experimentation and were occasionally even playful. Local food projects, because they were poorly funded, depended on activists' capacity for improvisation, ingenuity, and experimentation. Most projects undertaken by local food sustainability and food security activists were grounded projects made feasible only through what they opportunistically acquired from nearby microenvironments—this plot of land, that market stall, this meeting place in a church or community center. Activists improvised not only about the physical landscape but also socially. What did it mean, for example, to "connect farmers to consumers in order to increase a farmer's profile" at a farmers market? A farm visit by a group of consumers? Text messages? An internet presentation? At times, a project could allow for or generate playful experimentation or just play—as was the case for the Crop Mob in Durham, which we described earlier. It is perhaps for this reason that many projects we visited often had some kind of potluck meal or food accompanying a work session.

Eighth, projects were metonyms that stood in for a desirable "food system." Activists spoke of the physical projects they participated in, for example, gardening, as metonyms whose materiality stood in for "good healthy food" or as "providing the food people need." It is almost as if they were saying, "See? We're building our local system right here, right now!"

Ninth, many but not all projects were market-oriented. Projects, particularly those undertaken by sustainable farming and food activists, tended to be market oriented; that is, they were designed to "bring produce, honey, livestock . . . to market," in order to demonstrate the potential for generating income from the project. Other projects, for example, opening up a school garden, were done out of other motives, to achieve other goals, such as educating youth about "where their food came from."

Tenth, projects always served as grounds, occasion, context, or pretext for other culturally specific objectives to be achieved, in addition to the specific tasks required by the project: sociality, care of children or elder care, cultural expression (e.g., playing music), playing games, telling stories, therapy, heritage, and much more. The social and cultural values that a specific project expressed to activists and other participants were reinforced by its perceived economic value to the community, for example, a garden providing fresh produce, but flourished independently of it.

Eleventh, longevity of projects was rare. A very few exceptional projects led by charismatic activists were able to persist over the long term, allowing them to scale up their projects, build financial capacity, employ staff, and start their own nonprofit or for-profit organization. For most food activists, however, such recognition was not something that they experienced. An example would be the Interfaith Food Shuttle in Durham, described earlier as an exemplar of community food security activism.

Finally, local food activists' projects were rarely connected to geographically higher-scale social movements as such but were aligned with the local movement around food we described earlier. While the activists we studied shared a sense of their common affinity in seeking to improve their "local food system" and many manifested the *transformative potential* for participating in a scaled-up national movement around food, there were only an exceptional few who sought out translocal social movements beyond their locales for aid, ideological and technical support, and solidarity.

Who, Then, Were Food Activists?

What can we say about the food activists whom we have presented in the exemplars in this chapter and whose work will be examined in the chapters that follow?

First, food activists tended to be in liminal or interstitial positions in the economy. On one end, there were relatively large numbers of self-subsidized retirees, former university students, and "second careerists"— many of them in transition from one status to another as food activists who served as underpaid or unpaid organizers, volunteers, and interns. A few sought to become farmers and were looking for farmland or just setting out to farm but were not as yet fully occupied with farming.

At the other end, an exceptional few were relatively economically secure, full-time salaried government employees or managers for nonprofit organizations. Clergy, the "threadbare calling," received modest full-time salaries and benefits often paid "in kind," and they fell in between, as did chefs and self-employed small-scale food vendors, with seasonal or part-time incomes allowing them free time to work as activists. We know that a few activists were successful small-scale business owners (e.g., in high-tech) or lived on inherited income. Many others relied on part-time jobs, and we surmise that some activists depended on the income of a domestic partner to get by. In contrast, very few food activists were full-time farmers or food processes or vendors: Who among them could find the time?

Second, activists had been moved to undertake their work due to religious or political convictions, past work experiences, or university course work. Some saw themselves as active Christians who sought to alleviate hunger, some as deeply concerned by social justice or environmental issues, others by their specific experiences as volunteers in food-related nonprofit organizations or on farms. Prior to being activists, some had worked as chefs or as full-time food workers, and this had animated their activism. Some previously were university students whose course work led them to develop a practical interest in protecting the environment or alleviating poverty through the nonprofit sector.

Third, most sustainable food and farming activists were racially white, while community food security activists were more multiracial as a group—they were African Americans, whites, and Latinx. Since sustainable agriculture activists tended to identify with the plight of small farmers, while community food security activists were concerned about dilemmas faced by food-insecure populations, both were constrained in different ways by the racial associations of rural versus urban spaces that neoliberal and conservative ideologies have successfully imposed on many North Carolinians. In our two urban sites, Charlotte and Durham, this difference tended to coincide with distinctions between racialized and classed spaces—far more whites than African Americans were vendors and customers at farmers markets, while relatively more African Americans were active in churches and food-distribution sites to alleviate the food insecurities of poor urban residents. In contrast, as our research associate Willie Jamaal Wright pointed out, in our most rural site, in coastal eastern North Carolina, sustainable farming activism was barely evident; this was probably due to the preponderance of "commodity" or industrial farming in the region.

The implication of this was that there was an implicit social division between the two kinds of activists around the key predicament posed earlier: What is a fair price for food, for the farmer and for the eater?

Food Activism and Revaluing the Diverse Community Economy

For most people in the US, it seems obvious that capitalism constitutes "the real economy"—one measured by gross domestic product, quarterly corporate revenues, millions of people employed or laid off, millions of stock shares transacted daily, and huge volumes of commodities produced per month. The feminist geographers Gibson-Graham have diagnosed this conventional wisdom that brackets out all noncapitalist economic activity in their brilliant 2006 book, *A Postcapitalist Politics*, as "capitalocentrism" or, as we prefer, capital-centrism. Gibson-Graham (2006, 55) observe that "in its current hegemonic articulation as neoliberal global capitalism," capital-centrism "has now colonized the entire economic landscape, and its universalizing claims seem to have been realized."

However, it is one thing to say that almost all people in the US come into contact with capitalist institutions as employees, consumers, managers, or (a few) owner-investors. It is something else altogether to claim that their lives are subsumed or exhaustively defined by their relationship to these capitalist institutions. Despite their a priori ideas about "the real economy," the vast majority of people manifest an understanding through their actual practices that everyday economic life is far more than what capitalism seeks to reduce them to within it. This is the gist of Gibson-Graham's (2006, 79–99) profound insight of how people actually do get by—and of how at some level most realize they do get by—through their participation in the "diverse community economy."

Gibson-Graham (2006, 79) insist that the diverse community economy must be viewed not as a homogeneous social entity but as one whose members are interdependent through their connected economic and social relationships with one another. Gibson-Graham (2006, 84) observe that some economic practices conspicuously show human interdependence (e.g., cooking a meal for one's children), but whether conspicuous or not, interdependence between people is a defining characteristic of society. Even when one is under the illusion that one is an autonomous, self-interested individual (e.g., an "entrepreneur"), "community is not eradicated by the mediation of capitalist commodity relations. What is at stake is whether or not interdependence is acknowledged, recognized and acted on—or obscured or even denied" (Gibson-Graham 2006, 84). Instead, people must share in their discovery of the "coordinates for negotiating and exploring interdependence" (Gibson-Graham 2006, 86). Once this occurs, the community economy can be treated as "a site of decision, of ethical praxis" (Gibson-Graham 2006, 87).

To illustrate this idea, Gibson-Graham provide an inventory of the kinds of labor, transactions, and enterprises that constitute the diverse community economy and illustrate individuals' actual interdependence. In people's everyday lives, they expend different kinds of labor, engage in different sorts of transactions with one another, and work together in different forms of enterprise to produce needed goods and services, as well as to produce surplus (social wealth) within the economy.

Gibson-Graham (2006, 63) examine the different kinds of labor and what they produce in the actual community economy: unpaid labor (housework, family care, neighborhood work, volunteering, self-provisioning, even the work of the enslaved/incarcerated), capitalist wage labor, and "alternative paid labor," for example, self-employment, work in a cooperative, reciprocal labor, in-kind labor, and workfare. Responding to the work of feminist economists, Gibson-Graham make an observation that most capitalists seem incapable of acknowledging, namely, that once unpaid housework, volunteer work, and other non-market-oriented labors that sustain households and locales are taken into account, "empirical work on the subject has established that, in both rich and poor countries, 30 to 50% of economic activity [in outputs] is accounted for by unpaid household labor" (Gibson-Graham 2006, 57, which also includes references to studies cited). Moreover, Gibson-Graham break down productive activity within the 1990 US economy to show that only about 50 percent of all productive activity consisted of the production of commodities, but because some of these commodities were produced by the self-employed, workers in cooperatives, and the incarcerated, *capitalist* commodity production accounted for no more than 40–45 percent of total hours worked, while another 50 percent of production hours came from combined household and government sources (Gibson-Graham 2006, 68–69).

As to transactions within the community economy, Gibson-Graham note that the most prevalent forms of transaction do not occur in markets but are the nonmarket transactions "that sustain us all." These are transactions of goods and services produced and shared within households, transactions of people interacting with nature (e.g., self provisioning through fishing or gardening), people and organizations giving away goods and services, people stealing things, taxes appropriated by the state, and so on. While money provides the means for the commensurability for market transactions, the commensurability of nonmarket transactions depends on cultural values and norms (Gibson-Graham 2006, 60–61).

Similarly, the community economy has diverse forms of enterprise. Gibson-Graham define an enterprise as an economic unit for producing some good or service while simultaneously producing, appropriating, and distributing the surplus labor of those who work within the enterprise (Gibson-Graham 2006, 65–66). Within the diverse community economy, enterprises are noncapitalist, capitalist, or "alternative-capitalist." Noncapitalist enterprises include cooperatives, self-employed producers, household-based enterprises where men appropriate the labor of women, or even slave-based/incarceration enterprises, for example, prison farms. Capitalist enterprises take several forms—family firms, individually or partner-owned capitalist firms, publicly listed corporations, and so on—where the surplus labor of workers (or women) creates the basis for the enterprise's "profits" (surplus). The production of surplus and its disposition is a central question within the diverse community economy: How is surplus to be used, and toward what end, and on whose behalf?

Table 3.1 follows Gibson-Graham (2006) in setting out a provisional analysis of the local food economy in terms of the diverse kinds of labor, transactions, and enterprises that characterize it based on our findings—which the exemplars presented in this chapter illustrate. *The key conclusion to take away from table 3.1 is that the transactions, labor, and enterprise forms that compose most of the local food economy are concentrated not in capitalist market transactions, wage labor, or for-profit privately owned enterprises but in nonmarket and alternative transactions, unpaid and alternative paid labor, and noncapitalist and hybrid enterprises.*

If we consider only the forms of labor deployed in the local food economy, several conclusions follow from table 3.1. First, the local food economy functions as well as it does because of a huge volume of unpaid labor expended by food activists. This is the labor of people cultivating gardens; self-provisioning for themselves, family members,

TABLE 3.1. How the Local Food System and Food Activism Map onto the Diverse Community Economy

Transactions [with rules of in/commensurability]	Labor [how compensated]	Enterprise [who appropriates surplus]
Market transactions [economic commensurability] Sales at farmers markets, CSAs, farm stands, truck stands ["laws" of supply and demand] Farm-to-institution sales (schools, hospitals) [monopolized] Donations by corporate food retailers to food banks [calculable state tax credits for donated food, calculable savings by avoiding waste disposal fees]	*Waged and salaried labor [protected by union negotiation, unprotected, personally set]* Farmworkers, processing workers [unprotected wages]: workers at restaurants using local foods (e.g., chefs, managers, waiters) [unprotected wages] Salaried, part-time: farmers market managers [unprotected salaries] Salaried, full-time: government employees (Agriculture Extension agents, nutritionists) [compensated by negotiated salary + benefits] Salaried, full- or part-time nonprofit organization employees: clergy; garden managers; food bank/food pantry/kitchen managers and staff [unprotected salaries, wages]	*Capitalist enterprises [family, business owner(s), board of directors, shareholders]* Local family farm [business owner/family "head"] Local farm with farmworkers [farm owner] Food processors with workers: cheesemakers, charcutiers [business owner] Restaurants and catering services using local foods [business owner] Retail and food distribution corporations (as charitable food donors) [boards of directors to shareholders]
Alternative/market hybrid transactions [social commensurability] Use of SNAP, WIC vouchers by recipients "Double Bucks" used by recipients Barter of volunteer labor for food	*Alternative paid labor [various paid compensations]* Self-employed: farmer [takes a wage from produce sold] Self-employed: local food processor, restaurateur [takes a wage from goods sold] Cooperative [co-op wage + share] Reciprocal labor exchange: farmers (aspiring) [reciprocated labor] In-kind: some volunteers Grant paid (temporary, part-time): for local food-related labor	*Alternative/capitalist hybrid enterprises* Nonprofit enterprises dependent on outside funders [board of directors]: food banks, farmers markets, church-based food pantries and kitchens, homeless shelters, meals on wheels programs, nonprofit urban farms Consumer/worker food cooperatives [consumers, workers] Nonprofit food retail businesses dependent on sales + outside funders

TABLE 3.1. (*cont.*)

Nonmarket transactions [incommensurable]	Unpaid labor [nonmonetary, subsistence goods compensation]	Noncapitalist enterprises [wages to owners]
State allocation: land (e.g., gardens), water [citizen entitlement]	Neighborhood work to start garden, compost-ing, etc. [nonmonetary compensation]	Farmers (self-employed) [sur-plus taken by self as wage]
Gift-giving of food (informal) [cultural reciprocity norms]	Self-provisioning gardeners [compensated by produce + nonmonetary]	Food processors (self-employed) [surplus taken by self as wage]
State transfers (SNAP, WIC, TEFAP) to food-insecure recipients [requires hold-ing formal job, citizen entitlement]	Volunteers on farms, staffing farmers markets, etc. [non-monetary compensation]	Cooperatives [surplus taken by co-op owner-workers]
Charitable food distribution to food-insecure people [cultural norms of asym-metric exchange]	Volunteers in food pantries and kitchens for food-insecure people [nonmon-etary compensation]	
Gleaning [traditional right]	Interns (Woofers) on farms [room and board, nonmon-etary compensation]	
Incarcerated giving garden starts to gardeners	Incarcerated [food and board, nonmonetary compensation]	
Donations by local businesses to gardeners	Organizing/advocacy of local foods & farming [nonmon-etary compensation]	
TEFAP Food from USDA		

and neighbors; volunteering on local farms and volunteering for neighborhood food pantries and kitchens; and organizing and advo-cating for local food, farming, and community food security.

Second, alternative paid labor is also well represented in the diverse local food economy. Self-employed farmers and food entrepreneurs, consumer and worker cooperatives, reciprocal labor exchanges (e.g., Crop Mob), in-kind or volunteer labor, and food activists working from small-scale grants provided by nonprofit organizations were all present, and several are identified in the exemplars given in this chapter. By "al-ternative paid" labor, we mean that the person laboring is inadequately compensated for their labor and so has to find a second job, rely on their partner's or their own saved/inherited income, or face the prospect of trying to get by on an income below a living wage.

Third, capitalist labor relations within the local food economy are nonetheless present and cannot be ignored. We have anecdotal evidence that on local farms in our four sites, farmworkers were employed (and exploited), as were food workers and artisans in food-related enterprises. However, we found very few informants referring spontaneously to the crucial labor of farm laborers and food workers within food production or to the adverse conditions under which they worked—only four informants out of the hundreds we interviewed. Although without question local farmers worked extremely long and physically demanding hours irrespective of whether they hired farmworkers (or put to work unpaid family members), the bucolic myth of the self-reliant, individual small farmer who "does it all himself" appeared to be alive and well in North Carolina.

A key question this book asks is, How are the kinds of nonmarket and hybrid transactions, of unpaid and alternative labor, and of noncapitalist and alternative enterprises that are essential to the local food economy to be valued?

* * * *

In this chapter, we discussed a key concept in this book, the "moral logics of food." We first set out the distinct moral logics of the corporate industrial food system, of sustainable farming and food activism, and of two kinds of food security activism: community food security and charity-based food security activisms. Second, drawing on the ethnographic observations and interviews of our team's research associates, we presented three exemplars each of sustainable farming and food activism and of food security activism and one exemplar of community food security activism that was a "cross-over" synthesis of these two approaches. Third, we described characteristics of the projects undertaken by the food activists we studied. Fourth, we summarized several personal traits of food activists in our four sites. Finally, we returned to the model of the "diverse community economy" to demonstrate that food activists have made invaluable and distinctive contributions to our local food economies and beyond them to the US food system. We next turn to how sustainable food and farming activists and food security activists each in very different ways encounter the pressures of advanced US neoliberalism on the lives they lead and the work they do.

4

Sustainable Food Activists Engage "the Economy"

If they do want us to stand out front of the place and dress in overalls, well, sure, okay. But if they just want [that] it was raised grass-fed for this amount of time and this and this, aged for this long, I mean, we can give them that, too, but it's like we are in this in-between line, where it's like we wanna be as efficient as we can be and low-cost efficient, but we also wanna be able to give them something more than just US Foods or somebody else is gonna give them.
—Noah Smythe, wholesale agent for Appalachian Growers, Watauga County, North Carolina, when asked by Jen Walker what restaurants and other buyers sought from him

Noah's reply to our ethnographer Jen not only reflected the ways in which stories about local farming are forged from mythic elements of an imagined American bucolic life but also points to the tensions that the farmers he bought food and meat from face from "the market," precisely what Noah as local food wholesale buyer embodied: the pressure on farmers to be "efficient" and "low cost" but also sell "something more" to customers than what a corporate industrial foods distributor would provide.

Noah's reflection led us to ask the question, How do food activists confront the conditions of contemporary neoliberal rule and its exaltation of "the market," and how do these conditions in turn constrain the thinking and actions of food activists? As we have seen, neoliberalism celebrates the place of markets, private property, entrepreneurship, and individual material self-interest in bringing about the good society, while it denigrates the state as the bête noire that corrodes these outstanding institutions and dispositions—that hobbles them and makes them less vital in performing their inherent telos of improving life. Neoliberalism also holds that the state unfortunately rewards those who show themselves to be undeserving, incompetent, or lazy with respect to dealing with market pressures, finding business opportunities, accumulating capital, or displaying an appropriate egoism in daily life, and it demands that these abuses be curbed. Thus, the less state regulation of and interference with markets—especially labor markets—and with the use of private property, the better. "Privatization," "deregulation," and the elimination of government programs that support the unworthy are key terms in the lexicon. "Efficiency," "profitability," and "entrepreneurship" are also key concepts that refer to phenomena that advance the goals of enhancing the operation of markets and the accumulation of private property.

It is unclear how, and in what forms, the practices and discourses among local food activists working to promote small local farms, farmers markets, CSAs, incubator farms, or community gardens could be characterized as neoliberal or viewed through the optic of neoliberalism. How does one, for example, reconcile the neoliberal impetus to "reduce costs" for school meal programs by having them "pay their own way" (Poppendieck 2010) with "farm to school" projects of sustainable food activists who seek to bring fresh and nutritious food to schoolchildren? How can such circles be squared? Must the defining features of a good society—such as schoolchildren having sufficient good-quality and culturally appropriate food to eat or small farmers receiving adequate incomes for their livelihoods by providing such foods—always be subjected to the niggardly calculus of "cost-benefit analysis" that neoliberals demand? Do food activists measure their practices as being useful and valuable because they exemplify and promote successful market competition? We found that the answers to such questions are far from obvious.

To the extent that neoliberalism and economistic thinking have come to be key constraints in how food activists figure their worlds and engage in activism, we would expect them to emphasize market competition, efficiency, productivity, and profitability, applied both narrowly to the specific "economic" practices they engaged in (e.g., promoting the sales of farmers' produce in farmers markets) and metaphorically to other practices within the areas of life that touch on food activism.

As noted earlier, we expected there to be differences around neoliberalism and neoliberal subjectivities that distinguished sustainable food and farming activists from food security activists. In this chapter, we concentrate on the former. Among sustainable food activists, we found no universal uptake or acceptance of economistic and other neoliberal logics, imagery, or practices. What we did find among these activists and what we report on in what follows was a focus by activists on "the market" for local food combined with their disruptions of neoliberalism's narrative around the virtues of markets. Although on cultural impulse most activists assumed market solutions to the problems of livelihood faced by small farmers growing local food, the conditions around local food and farming that activists spoke of constituted a significant blockage to their uptake of a vision of neoliberalism's triumphant market solutions. Time and again, when it came to the challenges of small-scale local farming "scaling up" to meet the needs of the population, this vision of a Panglossian market was found to be deeply unsatisfying. For instance, in one of our sites, one activist said, "It's more important that small farmers receive higher prices for their food than that the population pay lower prices for it." But neoliberalism's promise is that the population will receive lower prices—to compensate for the gutting of social programs and falling wages and incomes. However, this fails to meet the prescription of what the good society is about. Thus, sustainable food and farming activists greeted neoliberalism's narrative that the market is ubiquitous and universally valuable with frustration, denial, and a search for alternatives.

What we did find were a few individual activists who used the economistic discourse of neoliberalism, while it was relatively absent among other activists. In general, there were very few sites where economistic discourse tended to be voiced or where economistic practices tended to be encouraged. In those few sites, we found that economistic language around efficiency, profitability, productivity, and entrepreneurship focused on how local small-scale farmers might gain access to, adapt to, and be appropriately integrated within local food markets such as farmers markets, but it was accompanied in one form or another by disruptions to neoliberalism's narrative of virtuous markets. We have discovered that activists employed terms like "efficiency" and "productivity" not because they subscribed to an overriding economistic logic but because these terms indexed or pointed to more profound quandaries among activists, such as the pressures that local food activists felt were imposed on farmers and local institutions to "scale up" to compete with large-scale enterprises.

Neoliberalism's Allure: "You Know, We're Not There to Save the World. . . . We Are There to Make a Place for Farmers and Artisans to Come and Sell a Product"

As noted earlier, it was rare to encounter among sustainable food and farming activists a space of practice where economistic discourses were voiced, but it did occur in one of our rural sites. George Everett and Judy Lodge were a married couple who were recent arrivals in the area, both having come from another southern state, and George was still active as a part-time attorney with his out-of-state practice. George and Judy had established themselves as farmers; George had become president and vendor at one of the region's farmers markets, while Judy was president of Appalachian Growers, an organic farmers cooperative. Judy had been a leader in a prominent local nonprofit organization that supported sustainable agriculture. As possessors of scarce skills such as legal expertise and as past leaders in the area's food organizations, George and Judy were seen by other local food activists as a powerful couple, and several were skeptical of their motives with respect to promoting the local food and farming economy as a whole.

When our ethnographer Jen asked George what the main goal was for the farmers market that he oversaw, he replied,

> The goal of the county farmers market is pretty simple. It's to . . . make a place of business for farmers and artisans to come and sell their products. There's a lot of 'em . . . goes with that simple statement. But with thirty-six years, we've made a place that's really unique. . . . It's a bazaar and has a lot of synergy to it. There's kinda at this point a goodly amount of produce and bakers and meat producers and artisans and crafters. And there's the necessary volume of customers that make it worthwhile for those people to come and do that, . . . practice their trade. . . . Fewer venders, then the customers can't come and get what they want. . . . People come to the farmer's market not to buy tomatoes, not buy . . . peaches, . . . not to buy this. . . . They come to buy groceries, which is what makes it successful.

When asked further by Jen about additional goals of the farmers market, George mentioned that another goal was to educate customers about "the economics of it." He went on,

And there's just this—you know, it's a huge social—one giant social interaction going on up there. I really love to see it. . . . I just—my undergraduate work is in economics, and I love to see that market, . . . true little market working. And when you watch people, Judy and I come up with something new . . . [laughs], next two or three other people have it, . . . and then we've got to come up with something else new, or somebody else does something that we like and—just to watch, you know, the market. . . . The participants try to differentiate their product and get their customers. I just love watching it. It's great.

George exalted the farmers market as a virtuous institution in its own right, bringing together diverse vendors, who only happened to be farmers, with many customers, who only happened to be buying locally grown produce, livestock, and handicrafts. It was coincidental that local products were what the market was about; indeed George reduced them all to "groceries." It was the market that mattered, not what was bought or sold in it. George saw the market working its wonders of allocation, putting competitive pressure on vendors to innovate ("then we have to come up with something else new, or somebody else does something that we like"). The market in George's account was an admirable self-regulating mechanism, beautiful in its own right ("I just love watching it. It's great"). When asked what the most important goals of the farmers market were, George emphasized that it was to "try to avoid mission creep and stuff": "You know, we're not there to save the world. We're not there to save the children. We're not there to save the whales. We're there to make a place for farmers and artisans to come and sell a product and make it attractive enough so enough of the vendors come to attract enough of the customers. So it's really to keep a market going and sustainable to me that's the most important."

In all of this, George argued, the market creatively functioned to make use of the individual economic self-interest of participants like George and Judy themselves. Explaining to Jen what his own interest was in participating in the farmers market, he mentioned what he saw as his own important role for local food activists seeking his assistance as an experienced attorney who had set up and led numerous nonprofit organizations prior to coming to the state: "We tried to narrow our focus and work with those organizations [e.g., the farmers market] that help our enterprise here on the . . . farm and the rental cabins be sustainable. . . . So it's a very self-motivated and motivating by our self-interest to help those organizations be successful where we sell and do things."

George's vision of the good society brought into existence by the market—its provision of valued goods to those who could afford them, the benign competition among market actors leading to innovation, and the harnessing of individual self-interest to these moral ends: this is the allure and promise offered by neoliberal discourse.

"As If" Entrepreneurs or Banal Figures?

There is no figure more exalted in the ruling conservative circles of neoliberalism than the "entrepreneur." What economic conservative, after all, has not read the glorious chronicle of John Galt in *Atlas Shrugged*? And what figure is more celebrated in the business press than the entrepreneur, the businessman who takes risks and recombines the factors of production, engages in transformative change through disruptive innovation, and thus shifts paradigms for entire industries, in what Joseph Schumpeter (1942) called the "creative destruction" of capitalism? Who among us has not heard Jeff Bezos, Elon Musk, or Steve Jobs glorified in such exalted terms? We therefore expected the food activists we interviewed to have much to say about farmer entrepreneurs, entrepreneurs in marketing food, processing it, preparing it as restauranteurs and chefs, and so on.

We were surprised to find that although yes, entrepreneurs and entrepreneurship were brought up by food activists, it was not all that frequent: only a total of forty-five references to either were found within our more than two hundred interviews of food activists. Moreover, unlike other terms of the economy, we did not find the discussion of entrepreneurs concentrated in gatherings among food activists but rather scattered in the interviews of several individuals. Nonetheless, what these food activists said straightforwardly broke down into either of two ways of talking about the "entrepreneur." On the one hand, activists spoke of entrepreneurs with approval—as those who could otherwise make money but always lacked certain critical resources (capital, space, training, etc.) to accumulate it. They were "if only" or manqué entrepreneurs: they could fulfill their entrepreneurial destinies "if only" they had access to market spaces, training, credit, and the like. Then, once these deficiencies were remedied, they could emerge as "successful" entrepreneurs. Or, on the other hand, activists spoke of entrepreneurs as those people whom others in authority had *already* declared to be so merely because they engaged in profit-making. That is, "entrepreneur" was an ascription applied to any businessperson seeking money profits by participating in the local food-supply chain. Thus, there were two senses of entrepreneurs among food activists: the exalted figure that society failed to provide opportunities for realizing their potential to accumulate capital or the pedestrian figure engaged in "business" who was an entrepreneur merely by virtue of trying to make a profit by successfully selling a product.

Of the total of forty-five references to entrepreneurs made by food activists we interviewed, sixteen references exalted entrepreneurs or entrepreneurship, gave prescriptions as to how to cultivate them, or listed resources they lacked to explain why there were not more of them. Take, for example, Noel Page, regional director of a national nonprofit organization devoted to economic development in the poorest regions of the United States, including Appalachia. He made the largest number among local food activists of references to entrepreneurs, five out of forty-five. At one event, he spoke before other food activists about the loss of the "culture of entrepreneurship," and this was what his organization wished to correct. He also mentioned that the focus of his organization was on entrepreneurship, not on jobs, and that historically the mountains have been composed of people who were "jack of all trades" people and that this was an asset. He went on to say that entrepreneurship should be taught starting at a young age and that the need was to "make a job instead of take a job." At every step in his organization's collaboration with local groups, he said that entrepreneurs needed to be supported.

We found a few other examples of "entrepreneur" exaltation. In Charlotte, at a planning meeting for a new public market, a consultant remarked that opening up spaces for markets would help entrepreneurs not only because the small businesses at markets were smaller than those operating out of storefronts but also because "that level of entrepreneur is the most interesting one," because they have creativity and enthusiasm. At the same meeting, the head organizer of the market project, arguing for the virtues of the market's planned location near public transit, remarked that it is "not just about this space": "We can incubate small businesses and reduce the barriers for entry" for entrepreneurs. Or, in Durham, the opening of an agricultural processing center was justified by its assistant manager as serving as a food-business incubator for local food entrepreneurs, and to assist entrepreneurs, the center made office space available to them in the front of the building.

As to the second use of the term "entrepreneur," this was more frequent than the first use. People were said to be entrepreneurs if they had been so designated by others, for example, one man who explained that he was attending a meeting to plan a new location for a farmers market because, in part, he was "the entrepreneur in residence" at a nearby university: "We worked hard to help fledging businesses, whether they be a business or social enterprise, and we put a good bit of effort into sustainable business or how could sustainable enterprise flourish."

Far more common was simply to refer to *any* businessperson engaged in the selling of food commodities as an entrepreneur, and in fact, the majority of our references to entrepreneurs were of this sort. For example, a community garden project at one of our urban sites had a "youth entrepreneurship program," and its leader referred to what the youth were being taught as "occupational vocational skills" and said that in addition to agricultural skills, their training in entrepreneurship would consist of a field trip "visiting a century-old farm . . . or visiting a[n] . . . insect and reptile specialist to learn about beneficial insects or visiting a blueberry farm." He referred as an example to starting a beehive and said, "If we want to do it as a youth entrepreneurship component, then the youth can sell their honey. . . . It's the easiest thing in the world to sell."

To summarize, discourse around entrepreneurs and entrepreneurship among food activists tended to sort into these two meanings. Either our consultants saw a deficit—the need to have more entrepreneurs whose scarcity was due to the obstacles they faced (lack of access to capital, space, or training, or other "barriers to entry" posed by the food system), which activists sought to remediate—or they applied the term "entrepreneur" to any person who bought and sold food products. There were exceptions, as in the case of the founder of one Charlotte organization that supported local small farmers by buying their produce for collective consumption by organization members. He told our ethnographer of his "epiphany": "The in-box creates an atmosphere where you're reacting to the outside world coming at you, and it takes you off of the position of taking action of your own choice. So we are as a culture being moved to one of reaction from action. And the whole entrepreneurial piece is act, don't react. You hire people to take care of this stuff so you're not distracted." Yet it is difficult to see how this insight, however valid for white-collar professionals like himself, might apply to small farmers and other small-scale operators in the local food economy.

A Note on "Self-Employed" Business and Neoliberal Disdain

As we have seen, in our interviews there was little elaboration of the meaning of what it was to be an entrepreneur or how to engage in entre-preneurship in the local food economy in the customary neoliberal way. There was little talk that emulated CNBC talking heads about "taking risks" or using new or "disruptive" technologies that might change an industry or bring about business "success." Growing local, fresh, and nourishing food uses labor-intensive methods, requires long-term cul-tivation and nourishing the soil in relatively small plots of land, and demands that one personally sell one's crops, poultry, and so on directly to consumers. It is not easily subject to the strategic transformations that the successful capitalist entrepreneur is supposed to undertake, such as scaling up, becoming more efficient, hiring lots of labor, and above all becoming wealthy.

Nonetheless, there is an important issue that needs to be discussed. Most of our consultants who were small farmers or worked with them to sell farmers' locally farmed produce, meat, or processed food items— through farmers markets, CSAs, roadside stands, and the like—were engaged in a kind of labor, or undertook a kind of enterprise, that neo-liberal discourse displays a deep disdain for. Local farmers who sold at farmers markets either began as self-provisioning producers who shifted from provisioning solely for self and family members into being "self-employed" independent business persons, or they began as self-employed businesspeople who sought to produce a surplus to transform into money income through the marketplace. The produce grower, poul-try or hog farmer, or food processor who grew or processed their own foodstuffs and used organic and other nonindustrial methods to do so almost always did most of the labor, with perhaps the help of a partner, a part-time worker, a young woofer intern, or volunteers, particularly during periods of harvest. They were not a person who was interested in hiring laborers full-time, managing a payroll, expanding acreage, and becoming more "productive" in order to become rich. Who would rationally try to engage in such small-scale, intensive local farming in order to become wealthy?

Sonya and her partner owned a farm in Lincoln County that raised organic hogs and poultry, operated their own "food cart" (mobile food service), and had just opened a small restaurant near Charlotte. For Sonya, being a farmer meant that her business had to be "economically sustainable." "That we can economically sustain the business, because it is a business; it's not a hobby. I'm the farm's only full-time employee. I do have a part-time guy that helps me five mornings a week and a lot of folks that volunteer to help me sell at farmers markets, and they're a godsend because I wouldn't be able to make ends meet if I didn't have them." She spoke of what "making ends meet" involved in addition to having good "production techniques"—this was "not just farmers working together, but it's farmers working with city people; it's farmers working with county planning people [e.g., farmers market inspectors]; it's farmers working with their neighbors; it's being a good businessperson and a good neighbor."

Like Sonya, many of our consultants spoke of the need to be "economically sustainable," thus shifting the meaning of sustainability away from referring only to sustaining the environment to one of sustaining a livelihood in which the farmer comes to have a fruitful long-term relationship to a local eco-economic environment in which a market mediates between the labor of livelihood and the capacity for the farmer to flourish within a cash economy. Put more bluntly, not all markets are capitalist—something that becomes clear when neoliberalism proves unable to suppress the well-based historical finding that rural markets in agricultural goods long predated (by several centuries) modern industrial capitalism, as in the case of rural China (e.g., Brook 1998).

Neoliberalism actually disdains farmers like Sonya and the thousands of local farmers like her because, frankly, they are its "losers"—they have not managed to become wealthy by the accumulation of capital, which is the *only* accomplishment that ultimately matters to most neoliberals. A fundamental neoliberal premise is that only the person who accumulates capital (or perhaps inherits it, but that is an uneasy issue) can define market "success" and therefore must be listened to, esteemed, admired, even worshiped as the "successful entrepreneur" (Holland et al. 2007).

Neoliberal thinking has retained its grip on small local farmers and our food activists precisely because of its hold on the US population through its exaltation of possessive individualism, combined with a capacity for exploiting one's fellow human beings. To speak seriously of maintaining "a business, . . . not a hobby" as Sonya did performed a ritual obeisance in the direction of a dominant neoliberalism that exalted "entrepreneurship" while at the same time leaving her, like almost all other local farmers, cast as manqué entrepreneurs. Neoliberal intellectuals have merely shown diplomatic taste by refraining to point out that manqué entrepreneurs do not qualify as successful entrepreneurs: the only measure of success for neoliberalism is wealth, and wealth only comes to capitalists. What our food activists may in fact be telling those who are listening is that it is time instead to shed neoliberal thinking and its capitalist defaults, which have led us to economic depression, fossil-fuel-driven climate change, and industrialized food that is neither nutritious nor equitably distributed.

Our consultants repeatedly demonstrated to us one central insight: there are noncapitalist ways of managing to make a livelihood—of harnessing our labor even if unpaid (Gibson-Graham 2006, 62–65) or in "independent" noncapitalist enterprises that serve profoundly important social goals (Gibson-Graham 2006, 65–68)—even if the labor is hard and the livelihood challenging. Engaging in food-based self-provisioning and self-employment to make money income are worthwhile in their own right and show highly esteemed kinds of caring and productive labor (e.g., making sure that one's family members are well fed, bringing nutritious food to market for others' consumption) and highly developed and knowledgeable skills around cultivating crops, rearing livestock, butchering, minimizing weeds, avoiding pests, dealing with weather vagaries, and so on. These forms of productive and caring labor and biotechnical skills have, after all, been the basis of the successful adaptation of the human species to living on planet Earth up until the advent of industrial capitalism approximately four hundred years ago—and they may well be reemerging in human postcapitalist futures as well.

Nonetheless, neoliberal advocates can take solace from one of our findings. This is that even though activists could tell us little about what made a local food entrepreneur an entrepreneur, not one of the forty-five references to entrepreneurs or entrepreneurism voiced any criticism of these ideas. This cultural cliché and icon is safe, if a bit shopworn due to overfamiliarity and repeated incantations of its efficacy. For neoliberalism, the disrupted narrative of the virtues of the market held few charms for activists.

Preoccupations with "Efficiency" Are Preoccupations with "Scale": Limits of Capitalist Accommodation to Local Food and Farming

When we interviewed local food activists in our four different sites, like many other terms in economistic discourse, "efficiency" as applied to local food and farming or to their connections to later steps in the supply chain leading to the consumption of local food in the community received relatively few mentions. We found a total of only fifty-one references to efficiency among our more than two hundred interviews, and it is informative that fully two thirds of these (thirty-three) references to efficiency were made by twelve informants who were wholesalers, farmers market managers, local goods market managers, vendors, farm managers, and interestingly, Agricultural Cooperative extension agents.[1] It is perhaps not surprising that those activists (and a few affiliated non-activists) whose professional roles placed them squarely within the sphere of commerce in local produce represented spaces of practice that articulated concerns about efficiency. Moreover, eight of these twelve informants came from our two urban sites, and only four came from our two rural sites. Thus, our discussions of efficiency with our informants were highly localized and relatively infrequent when they occurred. But that is not to diminish their importance, for, we shall argue, those informants who did mention "efficiency" tied it integrally to a problem faced by local farmers and food activists: the predicament of trying to "scale up." Viewed in this context, this problem shows the limits of capitalist accommodation to local, small-scale, labor-intensive farming and leads us to ask whether these limits are tolerable for local farmers and local food activists.

We would like to contrast what the two informants who had most to say about "efficiency" were preoccupied about because of what it reveals of these limits. Noah Smythe, quoted earlier, was our consultant who most frequently mentioned concerns about "efficiency" (nine of the fifty-one mentions). He worked as a wholesale agent for Appalachian Growers (AG), the farmers' cooperative for which George served as president in one of our rural sites. When our ethnographer Jen interviewed Noah, she began by asking him whether with his recent hiring the goals and purposes of AG had changed at all. Noah replied that it had, that now the cooperative as a wholesaler was committed to expanding beyond its previous four-county area:

> Now there's all these farms that're increasing their production in eastern or western North Carolina, and so we're gonna try to draw the gap, kinda connect them all and then still hopefully make sales through that. So . . . that's probably a new thing there. So a more expansive kind of market is probably added on to the goal of just encouraging small, medium, large farms to grow at this wholesale level and make these sales to people entrusted. So we're kinda people who're more interested in keeping the price down and stuff like that. Like, we're trying to fill into those markets so therefore get the food out to more and more people.

Here, it is important to notice that as a wholesaler, AG is "more interested in keeping the price down and stuff like that." Noah then went on to address the invisible farmer, who, after their sales to the "direct market," for example, farmers markets, were exhausted, would work with Appalachian Growers: "You're getting into the wholesale and then low-impulse sale while the prices are still pretty high, and we're trying to take you from that point so restaurants will pay a higher price all the way up to [organic foods supermarket chain] Earth Fare, who's gonna want the lowest price and stuff like that." This is "a spectrum that we are trying to fill in with that using [a] wholesale, efficient market." In Noah's account of the "spectrum," for the farmer, price increases only up to a point ("restaurants will pay a higher price") when AG moves "way up" to sell to the supermarket chain Earth Fare, "who's gonna want the lowest price"; then the price offered the farmer goes down, but the volume increases. Efficiency for farmers means increasing their volume of production, while being able to lower their prices in order to make their produce more attractive to large buyers and wholesalers.

Later, Noah stated that part of his objective was to have more local farmers become certified as organic food growers, with the result that "along with being certified and keeping up those standards, you can also get a price for it, and it allows you to move things at an institutional level more." Our ethnographer Jen asked whether this was not only to support sustainable agriculture but also to help the farmer get a higher price. Noah replied, hedging carefully, "Yeah a higher price, well, just volume, even. It's not gonna be as high a price as the farmers market, but it's gonna be a more efficient market for them." It will be more efficient, Noah asserts, because with AG marketing their crops, now farmers "don't have to work as hard on the sales side of it, so it's just making things way more efficient. And whether they know it or not, they're not having to work as hard to make these sales, so the money is coming to them easier, at least."

It soon became apparent in this interview that Noah's specifications around the making of more "efficient markets" and for "just making things way more efficient" for farmers had to do with the pressing challenges facing local farmers to "scale up" to meet the demands for local produce of the markets that AG was aspiring to enter.[2] Efficiency is part of the more inclusive set of new transformations required of the small-scale farmer that are part of the "scaling up" process. To start with, the farmer must undergo organic certification.[3] Once the small farmer is certified, then, Noah said, "we're going to carry you with us, and as we start distributing to [produce consolidator] ECO, you're gonna be on our truck, too. So all of a sudden, you can get as big as you can be, and we can still have a place for your [produce]." The farmer will be expected to grow produce in quantity sufficient for "a place" on the back of AG's truck. In addition, as AG learns from its large-scale buyers what produce is in demand, not only "high quantity" produce but also "new specialty produce items" that are "high-end," then AG will transmit this knowledge to farmers, and they will be expected to use it in deciding what to plant and how much.

Moreover, in order to scale up, farmers will have to learn a new financial discipline by calculating their costs better, especially since they will have to charge a lower price for their produce to AG than they would in farmers markets, as Noah described: "I would say calculating your costs and maybe almost accounting of enlightenment on the farm side of it could be an improved area. . . . When Appalachian Growers can sell for somebody, be happy with that 20 percent, and sometimes folks [farmers] don't lower the cost for us to add on that 20 percent, even though we're doing their marketing and delivering for them." In fact, Noah went on, farmers will need a "bookkeeper" or at least "just really [be] tight on their finances." He went on apprehensively to note, "I mean, who knows how worthwhile it would be [for the farmer], because there'd still be farmers doing it at such a mass scale that still you're [AG is] gonna end up setting the prices, so it's still going to be tough."

It thus became increasingly clear through the course of the interview that "scaling up" imposes requirements on the small farmer that have to do with the nature of the large-scale markets that AG seeks to enter and with the demands of the institutional buyers that AG seeks to supply. However, the imperatives of "scaling up" and increased efficiency apply not only to local farmers but also to AG as an aspiring business seeking to expand from being a small farmers' cooperative to a region-wide wholesaler. As Noah reflected on his potential counterparts, the large-scale institutional buyers (supermarket chains, organic foods consolidators, university food services) to which he will seek to sell his produce, he noted that the problem was that buyers always wished to have fewer sellers selling larger quantities of produce with which they do business, and to meet this demand, AG itself would have to scale up. "But buyers [are] always pushing, 'Well, if we're gonna do one [i.e., buy one pallet of produce], we might as well do five, and if we're gonna do'—just to simplify for them, 'cause why should they have another distributor unless they can get a significant, efficient load from that distributor . . . ? So, everything is asking you to scale up, and nobody's really saying do it at your own pace. So that makes it tough." And the challenge for AG to scale up while the small farmers supplying it produce also do so has serious economic risks for farmers, for example, if AG failed to scale up to find institutional buyers. Noah explained, "The farmer needs to increase their production and scale up at the same time that these folks [farmers] need to understand that we are just scaling up. We're not already at that scale, so that creates some difficulties within the market, 'cause it can take a lot longer than just a year to scale up. . . . It would be one thing to just go ahead and plant acres and acres of green beans if you knew you were gonna actually make profit off of that."

The key question that our interview with Noah posed is the following. Can small-scale farmers, using labor-intensive methods of cultivation, such as organic techniques, actually meet the demands of the capitalist food economy to "scale up" while preserving the fundamental characteristics that they and the broader population who desire local and fresh produce, fruit, and livestock both expect—taking care of the earth they work on, preserving the biodiversity of small farming (e.g., heritage varieties), and assuring the nutritional qualities of food grown nearby by small-scale methods? Implied in scaling up, after all, are imperatives to standardize quality, increase quantities (and increase pressure on the limited labor time of farmers), implement the disciplines of economic calculation that many farmers are ill prepared for, commit to branding their produce, and assume the market risk of asking for lower prices while they become increasingly dependent on counterparties such as AG and its institutional food buyers, which have much greater market power than they do. Requiring small farmers to scale up to the standards and practices of the US capitalist food industry may be demanding what is both impossible and undesirable.

Discordant Meanings of "Efficiency": To What End?

The question of the meaning of efficiency for small-scale farmers and local food activists can be looked at from a different perspective, that of Shirley Brown, our interviewee who made the second most references to "efficiency" in her interview (seven of fifty-one). Living outside of Charlotte, Shirley and her husband had a small business running a multifarm CSA, organized farm tours of small farms in the region, and led an organization supporting small farmers, and Shirley had recently been hired as the director of a new nonprofit market. This market's purpose, in addition to selling regionally grown foods, Shirley told Sarah, was "to coordinate efforts across the farmers market and direct marketing framework of [a large city] so that we can have better cooperation and efficiency and hopefully raise the whole . . . local food system rather than cannibalizing one market for another." Later, we come back to this first reference to "efficiency" that she makes because we are convinced that its connection to raising "the whole local food system," rather than setting one market against another, was exceptional and innovative—and very different from the way in which she discussed "efficiency" elsewhere in the interview.

When Sarah asked Shirley what she saw as the most pressing problems of the current food and farming system, she replied that there was "a disconnection or disproportionate scale":

> What I mean by that is that . . . when I go and speak about food, the average person doesn't realize how small changes over time have accumulated to be such a large difference between what their expectation of reality is and what reality is. And because those changes have been happening incrementally and because the food industry is so connected to those changes, the food industry sees them as being fundamental now and . . . imperative. But people don't realize how . . . the scale has changed. . . . When you look at farming that's happening for thousands of people, it's a scale that people can't quite understand. And yet somehow they think that it's happening on a scale of—of one person to one vegetable plant.

Clarifying how people fail to understand what the scale of food for thousands of people entails, Shirley went on to address a quite different definition of "efficiency" than what people think of when they think of personally cooking a vegetable in their own kitchens or growing vegetables in their own gardens: "Based on what I see when I talk with people about how their food's produced and give them the reality, there's . . . a lot of consternation about it. And so I think that the issue of scale is one that has been addressed in a way that people are not really aware of. I also want to add to that . . . that [the] slow pace of incremental changes, incremental efficiencies [is] suddenly putting us somewhere we didn't expect to be." Shirley then gave an example of speaking before the Rotarians Club about school meals programs in the metropolitan area where she lives to illustrate precisely the devastating effects of these "incremental efficiencies":

> One woman said, "Yeah, well, the big issue is schools." And I thought and said, "Well, let's start at the beginning with that and that the Department

of Defense funds the school system nutrition program." And of course the whole room of educated professionals had no idea. I mean . . . you could drop a pin in the room. And they were just stunned that that was the case. And then you say, so that's one issue. And then the next issue is that because the kitchens no longer have knives or ovens to prepare the food, you can't take food in and have it prepared unless it's already been preprocessed, but there's no processing facilities [in schools]. Even if you did give them knives and ovens, the kitchens aren't big enough to house the product and then all the space to prep it, and we've ratcheted down and gained efficiency by hiring less-skilled workers. So we don't have the space, we don't have the equipment, we don't have the skills. It's very hard to go—undo all that's happened over thirty years. And it's just been a slow process of efficiency gaining.

What Shirley so eloquently summarized as "efficiency gaining" has been described repeatedly as a major change that has adversely affected the quality of school meals provided students in public school systems around the nation (Poppendieck 2010, 71–90). Under neoliberal pressures from the federal government to "pay their own way," school meal programs have adopted "efficiency gains" to reduce their labor costs by removing kitchens, refrigerators, and ovens from school cafeterias and by contracting out with private corporations to provide low-cost preprocessed foodstuffs that can be readily brought together in standardized meals by their poorly paid and unskilled workers, who allot them out to students as they pass with their trays, assembly-line fashion, by the workers' serving stations in the cafeteria (Poppendieck 2010, 71–90). Such food, the end result of years of downsizing and restructuring school meal facilities through cost-containment "efficiencies," where a single student's meal (in North Carolina) may cost no more than two dollars, is neither fresh nor particularly nutritious. This is the case despite the recent and valiant campaigns to bring fresh vegetables and fruits into school cafeterias to provide more adequate nourishment—campaigns that have largely failed to entice children to overcome their addictions to high-fat, high-sugar, high-salt processed foods with good "mouth feel" in favor of these far more nutritious fresh foods (Nestle 2013).[4]

It is not surprising, after all, with these changes hidden away in school cafeterias, that Rotarians—themselves socially distant with regard to class and race from the students whom many public schools serve—were, first, "stunned" and then showed a great deal of "consternation" when they realized that what happens in their gardens and kitchens could not be applied whole cloth to the large-scale, "cost-efficient" school lunch programs in a large school district in a major city in the southern United States. It is interesting that Shirley's account of scale and its efficiencies refers to the "downstream" steps of the same food-supply chain that Noah aspires to participate in "upstream" through AG's wholesaling arrangements to buy produce from small-scale farmers to sell to, among others, the buyers for the corporate contractors that service university food facilities. These latter are the upscale versions of public-school meal programs, although in many instances the same food-service corporations (e.g., Aramark) contract out with both. But here again the question needs to be asked: Does the capitalist food industry, with its obsession with large-scale, high-volume, cybernetically monitored supply chains and standardized and often highly processed foodstuffs, provide adequately for the nutritional and psychological needs of schoolchildren at the consumer end of the supply chain, just as we could ask a similar question about farmers at the other end? Is this the best we can hope for?

Here we can return briefly to Shirley's first reference to "efficiency," a different kind of efficiency from what she describes in her history of school meal programs. This is, instead, an efficiency that entails "better cooperation" between farmers markets and between the farmers who sell fresh produce at them, so as to "hopefully raise the whole local food system rather than cannibalizing one market for another." While it is unclear to what extent such efficiency is possible, Shirley in her interview pointed to two other related purposes of the new nonprofit city market she had been hired to manage, to which an efficiency of "cooperation" would be harnessed: to "mitigate food deserts and encourage and promote access to healthy food for all people" and "to bridge the rural-urban boundary by looking at food systems infrastructure and how we support our farmers." Clearly, this is a kind of efficiency that places people's nutritional and social needs first and their capacity to pay for their food second, in food provisioning that by necessity has to redefine the meaning of "large scale."

Finally, both Noah and Shirley used "efficiency" as a term that, through their distinct but complementary reflections on the meaning of "scale" and "scaling up," placed them at a critical distance from econo-mistic discourse. In Noah's case, he spoke of efficiency and scaling up with some ambivalence, as when he feared that farmers might be ruined economically if they "scaled up" but AG failed to match them with its own growth, although it is clear that in the end, he aspired to trans-form both the small-scale farmers and AG itself into more "efficient" and "large-scale" operations. Shirley, in contrast, critically alludes to the tragic outcomes for schoolchildren of "efficiency gaining" within a food system that has scaled up and become "efficient" in ways that save dol-lars but at the cost of children's nutrition and well-being.

Talking about Profits and Profitability: Varying and Dissonant Registers

Perhaps there are no words in the lexicon of capitalism that are more widely used or more highly appreciated than "profit," "profitable," and "profitability." Nonetheless, when we interviewed local food activists, we found their use of those terms to be not that common, with a total of only fifty references to farmers' or, more rarely, farmers markets' profits or profitability among our more than two hundred interviews. And, like other economistic terms, their use was limited to only a few individuals among the local food activists in each of our four study locales. That said, we found multiple meanings of being profitable—but in most such mean-ings, there was little resonance with neoliberal notions that glorified the market, entrepreneurship, and profit as the basis for capital accumulation. Instead, when our food activists did mention profits and profitability, they were most interested in issues that touched on but extended far beyond the limited time horizons of markets, calculation, and cost accounting. Instead, with a few exceptions, most activists we spoke to talked instead about issues around livelihood, scaling up, surviving economically, hold-ing onto farms and managing to operate them and to keep land in farms, over time.

Let us begin with the exceptions. Six out of the fifty references to profits or profitability were encountered in the interviews with a few individuals who had participated in creating a "business-like" space of practice in our rural mountainous site, discussed earlier. This consisted of the married couple George, president of a local farmers market, and

Judy, president of Appalachian Growers, and Noah, AG's wholesaler, and several other activists. As we noted earlier, George himself spoke of the farmers market as a place that was "profitable for people to engage in what they want to do" as vendors, as part of the positive qualities of markets in general.

Like the comments of George, comments from other activists in this rural site referred to people calculating profitability with reference to self-interest as the basis for participating in farming and as local food activists. At one recruiting dinner for farmers thinking of joining Appalachian Growers, Noah noted that the hosts "kept talking about how they were giving it three years (or maybe five) to become 'profitable,' and if they couldn't learn what two young professionals 'should' (or could, if they had their own middle-class jobs back) earn, then they would leave farming, and return to the city." And in one other instance, Tim Franke, another vendor at the Watauga Farmer's Market, recalled that George had insisted, over some resistance by other vendors at the farmers market, to change its vending hours, but what "really bothered people . . . was that George (the president of the market board and the one who made this decision) was the first one to quit selling at the Wednesday market because it wasn't profitable enough"—for George, that is. In contrast, Noah, in discussing the issues around efficiency and scaling up, pointed to the "new farmer" who came into farming because of the high prices of produce sold at farmers markets or "maybe saw this ideal lifestyle or something like that. It's a different reason than they got into it. . . . It never seemed as profitable to anybody as maybe it does now, and still farms are shrinking and disappearing."

It was this theme—being "profitable" defined by whether small farmers could survive at all and prevent their losses from being so high that they had to stop farming or sell their farmland—that appears to come out in yet another, perhaps surprising space of practice. Fully twenty out of fifty mentions of profits or profitability in our interviews came from informants employed as Agricultural Cooperative extension agents— from three agents in one of our rural sites, two in the other rural site. Agricultural Extension offices were important sites for the use of the language of profits. As full-time employees receiving federal funding through land-grant universities in North Carolina, Agricultural Extension agents' mission was to educate farmers with the technical and commercial information they needed to have viable farms, where viability interpreted as "profitable."

Let us take Owen, who mentions profits or profitability seven times (out of twenty). An Agricultural Extension agent in one of our rural sites, Owen put it thus: "Another place where my personal goals and extension goals mesh really nicely is keeping farms viable for the personal profitability of the farmers, but for farmland preservation or for farm preservation more than farmland preservation, you know, just to see . . . private land holders who were stewarding land in a responsible way, you know, being able to continue your year-to-year, and that takes profits." Thus, being "profitable" is only minimally evocative of the cultural figure of the rational, self-interested commercial entrepreneur. Profitability is important but far from the primary goal of most small farmers. Owen went on,

> You know, all the statistics on farmland loss and the age of farmers and the slim margins that farmers have opportunities to make profits, . . . [and] we're at, what, fifty-seven-plus is the average age and . . . where the majority of the farmers market numbers are earning $5,000 a year or less doesn't say it's a nonimportant economic engine, but . . . it sure would be nice to make that sales figure bigger so that—so that people do treat farming as an option of at least a part-time career instead of something that's going to cost them money.

Because of these extension agents' education, their connections to government programs, and their mission to improve agriculture, they played key roles in local food activism.

Yet a third set of activists who discussed profits and profitability (with seven out of fifty mentions) were four activists in Charlotte who worked either part-time or full-time in the regional food economy as small farmers, small restaurant owners, or owners of small businesses supplying produce or poultry. All three were women, and they appeared to know one another. Unlike the prior two spaces of activist practice around profits and profitability, these activists spoke of the serious issues of scale and affordability posed by the commercial food system that confronted not only small farms but also other stages of the local food supply chain, such as restaurants, CSAs, food-processing businesses, and schools. Not only did they actively reflect on these issues, but it was also clear that they saw themselves as having a personal stake in building their local and regional food economies. Maria spoke of the ordeals that she and her three employees faced operating a small café with high prices for the meals it sold, which were "farm to fork," that is, sourced locally: given the high prices of locally grown produce and meats, she said, "for us, as the producers of food, what happens is you have to be really strong, really strong, because it's like I'm going to make my business profitable, you know? I'm going to compete—or not compete—but I'm going to try to survive, doesn't matter how." But here she sees her small-scale café as linked to and part of a broader local food economy in which scale matters: "We don't need to go to California to eat the most healthy food. . . . We also can have it here because we have a lot of farmers, a lot of people growing awesome things. And, for us, why it's important to say it, because I want definitely to say that I'm different and, no, I don't want to belong to any fast-food restaurant. I don't belong to any corporate. We take the decisions in a smaller scale."

Another of the activists, Sonya, whom we met earlier, told our ethnographer Sarah of the challenges she individually felt given the small scale of her farm and restaurant: "One of the roadblocks for small producers is individually we don't have economies of scale that we need to be profitable or to be very present in things like large schools and institutions or even in restaurants that do significant volume." Her solution was to cooperate with other farmers: to "get folks interested in talking about how we can work together instead of being fearful of the competition. . . . So I guess my role in this is to reach out to other farmers to say, 'Hey, let's work together on this or that' or 'Hey, we wanna put your product in our restaurant. How much can you do? When can you do it? How do we work together?'"

Yet a third activist in the set was Shirley, whose views of scale we discussed when she associated a different kind of efficiency (than market allocative efficiency) with cooperation. Speaking to Sarah, along the lines of Sonya, of the need for "synergies" arising from regional cooperation, Shirley discussed how the perishability of fresh foods might be turned to collective advantage through food processing: "If we can deal with it in a way that is holistic, we can solve a ton of other things at the same time. For example, if we're careful and strategic about where we do value-added processing or how we choose to do it, we could be educating people at the same time, providing access to healthier food to those people, creating jobs, and . . . also making it so that we don't repeat the system in which the farmer gets screwed because someone in the middle realizes it's a profit-making opportunity."

The perspectives on profits and profit-making associated with these spaces of practice around profits and profitability were different and distinctive. The first, that of George and Judy, closest to neoliberal discourse, sees farming as primarily an instrument to stimulate the benign functions they see markets perform and views people as acting primarily in relation to their own material self-interest, defined in dollars. Even those who criticize the efforts of this couple see such self-interest as prevalent, inevitable, and a chief determinant of George and Judy's activism. The second, that of our agriculture extension agents, see the profitability of farmers as essential yet not sufficient in defining their capacities to persevere as farmers and to preserve farmland over time: being profitable is no less, but no more, than the means to this end. The third perspective, that of Shirley, Sonya, and Maria, again turns to the challenges of scale for the profitability of local small-scale farmers, restaurant owners, and other small businesses within the local food economy but reflects on these challenges and responds to them through calls for cooperation and synergy among the actors in local food systems. This vision is one at odds with the neoliberal figure of the individualist, self-interested entrepreneur.

If we turn to the three exemplars of sustainable farming activism described earlier, they provide us with other insights into the effects of neoliberalism on food activists. The nonhierarchical organization of the Crop Mob in the Durham area and the avowedly anticapitalist vision of nonmonetized labor exchange connecting its members point to an explicit and self-conscious rejection of neoliberalism itself.

The Lomax Incubator Farm near Charlotte showed the crucial role that the local state can play to empower local organic farmers, when it has the resources, although the farm as a mentoring program provided only a modest stopgap mechanism for a few young farmers to start local farming, given the formidable economic challenges to small-scale farmers in the region due to the scarcity of farmland lost to suburban development, high living costs, and lack of stable incomes or health benefits for young farmers. Yet they were learning how to reconcile the requisites of small-scale organic farming with the skills of making direct sales in the market, while defying neoliberal inducements to "get big or get out."

The third exemplar of the Watauga County Farmers Market board and its conflict with Teen Adventures shows, however, the tensions that farmer-vendors may feel between their attempts at economic sustainability and what some saw as the "unfair competition" they faced from nonprofit organizations for scarce stall space and market patrons' attention. This was a zero-sum moral logic consistent with neoliberal views, probably provoked by the extreme market-fundamentalist views of George Everett, the outgoing president of the farmers market board.

No Neoliberal Subjectivities but No Food Justice Either

To conclude, neoliberalism fails to encompass the subjectivities of sustainable food and farming activists. Although some used economistic language, the contradictions and tensions around small farming and local foods with respect to markets were evident to most activists. As a result, almost all activists disrupted neoliberalism's narrative of the virtues of the market, the entrepreneur, and private capital accumulation. Although largely inarticulate about markets, their views articulated embryonic alternative discourses—those around sustainability and sociality that are excluded from neoliberal thought when they cannot be marketized or capitalized. A few activists' views articulated an explicitly anticapitalist position (Crop Mob) or supported an anti-neoliberal stance that the state subsidies should subsidize vulnerable instead of "competitive" market performers (Lomax Incubator Farm).

We note sadly, however, that most sustainable farming activists we talked to failed to articulate the imperatives for "food justice"—to advocate for more just conditions of livelihood for farm laborers and other workers in the local food economy, not to speak of those working beyond it in the fast-food and meat-processing industries. At this point, most sustainable food and farming activists have not yet sought to invite, much less "bring to the table," those who are most marginalized, exploited, and oppressed by the invidious production relations that characterize some enterprises in the local food sector.

What can be inferred is that two moral logics are at war with each other, uneasily coexisting in the discourses of our activists: the moral logic of neoliberal large-scale "get big or get out" agroindustrial capital-ism, on one side, and the moral logic of sustainable farming and food activism, on the other.

* * * *

In this chapter, we asked the question, Do neoliberal ways of thinking and acting define the practices of sustainable farming and food activists? We then proceeded to answer it by analyzing how these activists actually used key neoliberal concepts like "the market," "efficiency," "profitabil-ity," and "entrepreneur" in the work they undertook. Only rarely did we discover activists who embraced the economistic language that exalted these concepts within the full package of neoliberalism and "free market" ideology. Almost all other sustainable farming and food activists either saw these concepts—with the exception of "the market"—as threaten-ing (as when "efficiency" required farmers to scale up their commercial operations) or creatively reinvented new meanings for these concepts, as when "efficiency" applied to the value of people working collectively to improve their food system or when "profitability" meant only farmers' economic capacity to sustain themselves over time.

A very few activists attacked capitalism as such and its exploitation of labor, and a few took advantage of unusual state support for small-scale farming—a rejection of neoliberalism's "get big or get out" prescription for farmers. However, all activists we interviewed accepted the unregu-lated "market" as the only recognized means for local farmers and other local producers to make their livelihoods, and no sustainable farming activist we interviewed brought up the issue of the exploitation of local farm laborers or other food workers.

5

Food Security Activism in Lean and Mean Times

In July 2011, Patrick Linder asked Len McPhaul, food manager for the community kitchen, food pantry, and homeless shelter at Ministries United, a church-linked nonprofit organization in Northeast Central Durham, what most concerned him about the current food system. Len replied,

> [It's] the terrible waste of food. I have volunteers go to—we have a Food Lion, a Kroger, a Harris Teeter—and pick up their day-old, their stuff that was dated yesterday or produce that's not sellable, and before the food bank set that up, it was thrown away. And even still, if, like today, Charles O'Connor our facilities manager, goes to Costco Mondays and Tuesdays and picks up a ton of bread dated today or tomorrow that they're probably not going to sell. . . . So, they will give us that, and it's still excellent. But if I'm not there by one o'clock to pick it up, it goes in the garbage, because they don't have space to keep it. So, Charley was off today, but he came in and made that run anyway, because he couldn't stand the idea that it would get thrown away. See, I think that's something wrong with that whole food thing, that we throw away so much stuff.

Len's lament about the enormous wastage of food in the US represents a concern shared by many food security activists and, in this era of rising and volatile food prices, many others as well. Some people ask, If so much food is surplus and ends up as trash in landfills, why do so many poor Americans go hungry? Food security activists like Len would respond that the charitable food sector prevents much such waste but is also a solution to hunger and a blessing for poor people in the US. A small number of people will ask, Why are there so many poor and food-insecure people in the US? While we as authors reflect on this last, haunting question, here we ask a third question: What kinds of work do food security activists undertake in this era of advanced neoliberalism, now entering a period of climate chaos?

In this chapter, we discuss the ways in which neoliberalism's effects among charity-based food security activists in our four sites have set the agendas of everyday activism that these activists engage in. In the course of our fieldwork, our ethnographers interviewed, observed, and worked with a large number of activists associated with the charitable food institutions in our four sites. These activists and thousands like them beyond our field sites provided food for millions of food-insecure people as a matter of charity—as the first-person bearers of institutional generosity to those in need. As we have seen, it is impossible to understand the place of food security activists within this system unless we situate their work within the broader arena within which impoverished adults with few exceptions are provided food through the USDA's SNAP benefits program only if they accept (and keep) a wage-labor job—no matter how exploitative or abusive. Otherwise, they go hungry or must resort to the food charity sector and be subjected to its repeated and often humiliating rituals associated with receiving food as a one-way gift. To understand food insecurity and the activism that responds to it, we must inquire about things by their true names: *only the poor—* especially poor working-class people—in the US are food insecure because *only they cannot afford to buy food* within a society where food is a widely available commodity; *only they are expected to be "work ready" as a requirement for receiving government benefits*; and *only they regularly receive food given out as charity*, although frequently it is called "emergency food."

Unpacking this particular conundrum is at the heart of the contradictions in the meanings of food insecurity and hunger in the contemporary United States. This chapter examines these contradictions, which are expressions of the politics of food in an era of neoliberal globalization, by centering our investigation in the one sphere of public life where these contradictions are publicly most visible, even in an obscured way: food charity, as it affected the activists we studied.

Our ethnographers talked at length with staff at regional food banks and with managers at community kitchens, interviewed clergy overseeing church food pantries and backpack programs (of surplus food for schoolchildren), visited managers of Meals on Wheels programs that distributed fresh meals to homebound people, and met with—and assisted—staff and volunteers at food pantries as the latter received surplus food gleaned from farms, collected bread and dairy products at the end of their shelf lives from supermarkets, and prepared produce harvested from church or neighborhood gardens to add to industrial food stuffed in food bags and given out to people in long, biweekly queues.

What our researchers discovered were symptoms of an advanced syndrome of the body politic: how neoliberalism as a discourse that exalts individual market success and stigmatizes poverty as market failure has come to define persons' social worth over the past forty years and has come to penetrate the sites of food charity and the subjectivities of activists providing it. This has taken the form of an economizing discourse that is internalized to some extent by food activists around neoliberal notions of cost-benefit analysis, efficiency, and effectiveness. Over time, this discourse has partially displaced alternative and more humane visions of human needs and their causes, such as the idea of the human right to food, of access to sufficient and culturally appropriate food as essential to human dignity, or of Christian solidarity.

But how have food activists come to be subjected to these now-dominant neoliberal notions? This question has to do with the cultural dimensions of the gross class and racial inequalities that neoliberal politics have brought about over the past forty years. Poor people in the US, unlike people in many countries in the world, have never been accorded access to sufficient and appropriate food as a human right, as in the International Covenant on Economic, Social and Cultural Rights (Poppendieck 2014, 190). Not only in the US but throughout the wealthy OECD countries, neoliberalism's unforgiving exaltation of the capitalist market, profit-making, and private capital accumulation has meant that poor people, no matter how poor they are, do not have the *right* to be provided cash payments or sufficient and appropriate food to meet their needs—even if they incidentally receive such benefits (Riches and Silvasti 2014, 10). The claim of a human right to food is seen by neoliberals as no less than "socialist" anathema that discourages "lazy" people from finding a paid job and from experiencing the bracing moral discipline of the labor market that a job entails.

Instead, the dominant neoliberal assumption, rarely stated overtly nowadays, is that poor people have nothing to offer the market but their cheapened labor and for this reason deservedly have a limited number of choices, none of them desirable, when it comes to food. As set out earlier, they are expected either to "earn their food" through their wages from the job, no matter how exploitative or abusive it is, or to go without and subject themselves (along with their children) to humiliation as inferiors who have to accept food as charity offered by the capricious, socially and morally superior strangers who make up "civil society." Among neoliberal and conservative elites, the deplorable practice of spending so much taxpayers' money on SNAP funding for poor people is compensated partly by the fact that it pushes the deserving "work ready" poor to accept any job as part of the capitalist labor market and punishes those who are recalcitrant by throwing them on the mercy of "civil society." Civil society in the neoliberal imaginary consists of generous citizens and corporations that make up what George H. W. Bush called "a thousand points of light" when they perform spontaneous voluntary acts of charity, from church groups organizing "food drives" to food retailers donating their overstocked foodstuffs to food banks.

We have seen how after forty years of neoliberalism, "food bank nation," composed of institutional food charity organizations organized by and for the private sector, has come to substitute for a state that keeps a large proportion of its residents on the edge of hunger (Riches 2018). It is within this space of mediating between capricious and powerful institutional wardens of food charityng one side and poor people on the other that the food security activist, often a staffer or volunteer at a food charity, finds themselves on an everyday basis.

Despite the rise of the neoliberal discourse of economism—and its harnessing to old class- and race-based humiliations at their worst—it has by no means completely displaced alternative visions of a more just food system among food security activists, as this chapter shows. Through acts of linguistic refusal, subterfuge, and redefinition, the food activists we interviewed demonstrated a capacity to redirect neoliberalism's economizing discourse without completely accepting its corrosive effects on the humanity of poor food recipients. Food security activists focused on challenges they faced in providing large quantities of food on a daily basis to low-income, food-insecure populations. While some activists provided food aid to recipients across lines of class and often racial difference, others, especially volunteers, were themselves poor, working class, often people of color, and themselves food insecure. Most performed their labor at food charities while trying to respect the dignity of those to whom they gave food.

A Brief History: From Advocacy for Those Who Are Hungry to Administering Hunger

Over the past forty years, there has been a shift among food activists and the nongovernmental organizations they have worked for away from advocating for economic justice *for* those who are food insecure toward administering food insecurity *to* the poor. When we write that food activists now "administer hunger" and "administer food insecurity," our use of these words is deliberate. As a generalization, charitable food recipients have long been subjected to being doled out food whose quality they at times found objectionable (e.g., culturally inappropriate, nutritionally inadequate) and questionably sufficient in quantity. But since the 1980s, the food activists who were previously their allies have been converted into those who administer food to them, while food recipients have been subjected to new humiliations. As we saw earlier, from the 1980s onward, culminating in President Clinton's 1996 "welfare reform" law (the Personal Responsibility Work and Opportunity Reconciliation Act, or PRWORA), the SNAP federal food benefit has shifted from being a *right* of citizens and residents to being a *privilege* that only formally employed citizens and a relatively few non-"able-bodied adults" can petition the state to provide them with. This benefit has become tied closely to working-class people being "work ready" to take any formal "job," irrespective of how abusive it is, while the SNAP program disregards other contributions they make to the broader economy or to the common good, such as from self-employment, self provisioning, unpaid caring labor, or volunteer work for their communities. At the same time, however, government benefits provided (e.g., SNAP, WIC) have not been sufficient to meet the food needs of the poor (National Research Council 2013, 177–179). What does it mean that the charitable food sector is at the tail end of a system that maintains food scarcity within the industrial food economy by "enclosing food surplus" (Lohnes 2021) yet often leaves those who are food recipients falling far short in the quantity of the food they receive, its nutritional adequacy, and its cultural appropriateness, within a culture where not having sufficient food to feed oneself or, worse yet, one's children is considered shameful?

Therefore, food activists working for food charities administer to people who are food insecure and fall under a large penumbra of uncertainty and ambiguity of just how food insecure they are—within a broader food bank nation that, as we will see, accounts for every pound of surplus food it provides food-insecure people. Activist staffers and volunteers are thus cast in the position of determining who is food insecure and hungry and who is not and making decisions to measure out food to poor people in food pantries, community kitchens, and even school meal programs. Irrespective of their own personal sense of injustice, class exploitation, and so on, food activists on the front lines of charity are those who reinforce the social realities of scarcity that the whole system of surplus food pushes everyone toward. How did this come about?

Under the conditions of neoliberalism, advocacy for economic justice for the poor population with regard not only to food but to all the necessities of life has over the past four decades come at its best to be seen as naïve and at its worse as not only politically but ethically inappropriate. It was not always so. Postwar advocacy around economic justice as revealed by the injustices of hunger was originally associated in the 1960s with Edgar R. Murrow's documentary *Harvest of Shame*, Michael Harrington's classic exposé *The Other America: Poverty in the United States*, and Marion Wright Edelman's congressional tour of the Mississippi Delta and its discovery of hungry Black sharecroppers there (Poppendieck 1998, 9–12). When it came to claiming the *right* to food by the working-class population, the Black Panther Party's community children's breakfast program was so effective in making the argument for this by alleviating child hunger in inner city areas in the early 1970s that its successes scandalized US Cold War political elites and in part provoked the persecution and killing of the party's leaders by the FBI's J. Edgar Hoover (Pien 2010; Patel 2011; Pellizzari 2020).

All that began to change with the election of Ronald Reagan in 1980s and the advent of a dominant neoliberal political program in which, in Reagan's words, "government is not the solution to our problem; government *is* the problem" (Reagan 1981). Capitalist markets were promoted by Reagan and his supporters as the solution to all social problems—well, at least those worth solving. "Deregulation," "freeing the market," the need to achieve "efficiencies" particularly in labor markets burdened by recalcitrant labor unions, getting rid of "welfare cheats," and eliminating taxes for corporations and wealthy "entrepreneurs" became cries of the decade—and have continued to resonate with US political and economic elites up to the present.

One of the most significant and pernicious aspects of neoliberalism, especially since the passage of PRWORA in 1996 but even before, has been the pressure that neoliberalism has placed on activists in nonprofit organizations to shift away from the work of public advocacy on behalf of *citizens (and noncitizen residents)* with rights who seek legal protection of these rights toward activists' intensified labor to provide privatized and devolved services (like food charity) to an impoverished population treated as pro forma *customers* (Hasenfeld and Garrow 2012, 317). The access of the working- and middle-class majority of the population to state benefits, especially cash transfers, has been cut back precisely on the pretext of the neoliberal imperative to shrink government and the services it provides the population so as to reduce "onerous taxation" on private entrepreneurs (Hasenfeld and Garrow 2012, 301–303). Neoliberal austerity policies have pressed activists to turn away from advocacy based on the rights of people to food from government toward the privatized nonprofit provision of food to fill the void in benefits, often literally empty stomachs, created by these policies (Hasenfeld and Garrow 2012). This *shift away from advocacy to protect the rights of the poor population to state benefits toward administering privatized services to meet the urgent needs of the poor* has applied to food activism as it has other areas of service (e.g., housing) within the nonprofit sector.

In this chapter, we cannot trace out the history of this poignant shift among food activists, which has been documented in certain locales (Povitz 2019; Poppendieck 1998) but not in our North Carolina field sites. In these accounts, this shift was evident not only in the imposed changes in value orientations of food activists but also in their work and everyday lives in response to the pressures of neoliberal policies and rhetoric. The oppressive exercise by political elites of their influence to attack poor people and their rights has characterized the past four decades of US politics. This assault culminated in the drastic increase in misery caused by PRWORA "welfare reform" in the late 1990s, due to the law's restrictions on poor people's capacity to receive SNAP benefits, and has continued since then (Collins and Mayer 2010; Dickenson 2020). The advent of food bank nation described earlier has led to the build-out of organizations to administer projects to distribute charitable food donated by corporations or individuals to prevent or at least lessen food insecurity under these new political conditions.

An example will have to suffice to illustrate the shock imposed on food activists by the shift in the late 1980s and early 1990s to neoliberal policies as these became ascendant. The wrenching changes away from advocacy toward administering charitable organizations daily—food banks, food pantries, community kitchens, and the like—to feed those who were food insecure could be agonizing. The historian Lana Dee Povitz records the transformation that occurred for Kathy Goldman, executive director of Community Food Resource Center in New York, and one of the foremost national antihunger advocates at the time, as Goldman moved toward managing the All Souls Church Community Kitchen in Harlem in the early 1990s, when the kitchen was pressed to provide meals to an increasing number of hungry and destitute people.

Povitz (2019, 210) describes Jan Poppendieck, witnessing this change in Goldman, a close friend:

> [Poppendieck] recalled that in the early 1990s the Community Kitchen was plagued by a series of break-ins during the Christmas holidays, forcing it to close for two weeks. In response, Goldman was required to get burglary insurance, which came with an alarm system that personally notified her in case of any incident. Poppendieck recounted, "I commented to Kathy that she was looking awfully tired. She said, 'Oh, the alarm at the kitchen has been going on every morning this week at about 3 a.m.' [Eventually it was determined that] rats were setting off the burglar alarm. I remember sitting there thinking, 'This is the most skilled, most effective advocate for the right to food that I know, and she is being debilitated [by the situation]' and that was the point at which I think I really *got* it: [where] people who care most about poor people and poverty will put their effort. And if you are putting it into running a soup kitchen, you won't be wide awake at the legislative hearing the next day.

As we saw earlier, the food security activists we interviewed were highly diverse, and the burdens of administering hunger by measuring out charitable food to recipients affected them differently depending on their own situations. Upper-level paid staff of nonprofit charities, such as executive directors and managers of food banks, food pantries, and homeless shelters, were relatively distanced from direct interactions with poor people who requested aid, yet some found the work satisfying, if logistically challenging, as we show in this chapter. For other staff, their paid positions lasted for a year or two years, dependent as these were on the vicissitudes of funding received by the nonprofit organizations they worked for. Thus, few were able to make careers from these positions in nonprofit management. A few were clergy for whom professional training for charitable administration was expected. A minority of staffers came from poor backgrounds or were people of color.

In contrast, volunteers who labored in the food pantries and community kitchens at the terminal end of the charitable food system had more direct contact than staff with people who qualified as food insecure and differed from employed staff in having to exert the emotional labor entailed in confronting people seeking food in weekly food giveaways. This was no doubt difficult for many. Volunteers tended to occupy liminal positions as students, retirees, and part-timers between jobs, and their tenure was not a long one. Generic to the charitable food sector was a considerable "churning" of relatively large numbers of volunteers administering food through charitable organizations—as volunteers might work for a few weeks or months, then leave, with new volunteers coming in to replace them.

The Food Bank: "It's Not a Democracy. It's Not a Republic. It's Really a Business."

How have the business logic and discourse of economism that are present in contemporary neoliberal thinking affected antihunger activism in North Carolina? A good place to start is the food bank: the food bank was not only the apex node of the food infrastructure network but also a space of local practice, in this case an institutionalized space of practice, that stood between food suppliers (such as retail grocery chains and large farmers) and those who directly provided food to the hungry "on the ground." Given the sheer scale at which food banks operated—regionally, across several counties, and each servicing scores of "agencies"—food pantries, community kitchens, and shelters for the homeless—it would not be surprising but should not be taken for granted to find them display the business logics of the modern corporation. For example, the food bank in Mecklenburg County within our research area served nineteen counties and transported food to and from the food bank in thirteen semitrailer trucks every day. Challenges of scale—the sheer extent of hungry people to be served, the large amount of food they required on a daily basis, and the many miles and complex logistics of transporting food within the US economy—faced food banks at every turn.

And indeed this is what we found—up to a point. Rationalities of cost minimization, optimal use of limited resources, and the maximization of people served (translated as tons of food given out per day) clearly characterized the operations of these food banks. But this leaves unanswered the question, How was value constructed in these spaces of practice? Unlike the modern private corporation, the food bank has a logic of maximization devoted not to maximizing the profits of private corporate shareholders but to maximizing the amount of food delivered to the largest number of people in hunger. Liz, a staff member of one of the larger food banks in Charlotte that we learned about, when asked by our site ethnographer Sarah what the goals and the values of her organization were, replied, "As a food bank, we have a mission: to distribute as much food as possible to people at risk of hunger." Amplifying on that at another point in her interview, Liz stated, "It's not a democracy. It's not a republic. It's really a business. And it's a business that operates with different goals. . . . I think we're primarily guided by how we can distribute the most food to the most people."

The effort to maximize the amount of food, in terms of the sheer number of pounds and tons of food distributed to agencies, is central to the driving ethos of the food bank, which occupies a critical logistics role within the emergency food economy. But in the US capitalist economy, in a neoliberal setting within which, after all, food banks operate, not only must the numbers of clients served be maximized, but these numbers must also be compiled, analyzed, and presented to corporations, philanthropies, and the public as measures of performance. The kind of public-private partnerships that constitute the bulk of the contemporary US "emergency" food system are a neoliberal form of delivery of charitable services but, as such, are to be operated as close to a business model as possible (Holland et al. 2007). Collecting, keeping, analyzing, and reporting the "numbers"—enumerating the amount of food collected and allocated, the numbers of agencies served, and ultimately the number of hungry people provided with food—is crucial to the assessment of work performed by the food bank's staff and thus by the food bank itself. For example, when Liz was asked how the leaders of her food bank should have their performance assessed, she replied,

> I think the food bank needs to be its own best critic and administrator, and I think we do that. We generate documents at the end of each fiscal year. We set goals for the next fiscal year, at the end of the year. It says, "Did we achieve those goals? At what rate did we achieve them—100 percent, 105 percent?" It includes everything from how many outreach efforts we're gonna conduct, how many speeches we're gonna make, to how many volunteer hours we're gonna have to how many pounds of produce we distribute. So we measure and check ourselves pretty well.

The obsession of food bank staff with collecting and keeping the numbers is in turn driven by the processes through which the work of food banks is itself supported economically: donations and grants from institutions—private corporations, local governments, and individual donors. Thus, collecting, processing, analyzing, and publishing the numbers is part of a broader system of reporting—of "accountability." When Liz was asked what kind of data were important to her for grant writing, she replied that they consist of "number of people served, number of people in poverty (from census), should [be] number of pounds distributed to children at risk of hunger. . . . There might be a grant that asked, 'How many pounds did you distribute to children at risk of hunger last year? How many pounds did you distribute through Backpacks? How many pounds of produce?'"

This logic of reporting and accumulation of numbers arises, we can readily see, when we consider the partners, that is, the donors and funders, of food banks. Some donors provided direct funding, such as the Critical Response Fund of the huge Charlotte Philanthropy Foundation for the Carolinas, largely funded by corporations and individual philanthropists, with more than $1 billion in assets. So how were their funds being used by food banks to feed the hungry? But equally important were those donors who provided either surplus food products to the food bank or organized volunteer efforts by their employees to assist the food bank in its daily work. For example, in the case of the food bank in Charlotte, corporate food donors included the major supermarket chains Food Lion, Harris Teeter, and Walmart, as well as Bi-Lo, MDI Lowe's, and US Foods, while employees of corporations such as Bank of America, insurance companies, utility companies, and others were occasional volunteers.

In addition to compassion and a sense of professional obligation, food bank staff had another reason not only to maximize the contribution that food banks made to alleviate hunger in terms of sheer quantity of food provided directly to agencies and indirectly to large numbers of people in hunger but also to quantify, enumerate, and publish their contributions. Food banks were expected by corporate funders and donors to provide quantitative measures—such as the number of meals or pounds of food they gave out annually—that would legitimate the charitable donations of corporations.

In addition to avoiding the high costs of food-waste disposal and managing food scarcity (Lohnes 2021), tax-write-off values of corporate food donations are significant. A corporate food donor under the IRS Code is allowed to receive a deduction from its taxable income of the sum of the cost of an item donated, plus an amount of either up to half of the difference between its cost and its "fair market value" or up to twice its cost, whichever is less (Feeding America 2013; USDA 2021a). Thus, if the cost to the donor of producing an item is $1 and its selling price (fair market value) is $4, then it can deduct up to $2, or twice the item's cost, from its taxable income.[1]

The enumeration of the amounts of charitable donations by donors in the annual reports of food banks thus provided corporations public support for the tax deductions of charitable food that they claimed. An estimate of what this represented in corporate tax deductions can be gained by examining the claimed monetary value of corporate food donations to the three largest food banks in North Carolina over the years 2015–2017: it totaled more than $200 million per year for corporations like Kroger, Food Lion, Costco, Walmart, and other food retailers and manufacturers (calculated from the annual reports or Form 990s of these food banks).[2]

So, after all, it turns out that the profits of corporate donors *did* materially matter in assessing food banks' performance, given the place of tax deductions and credits and offloading waste-disposal costs onto the public domain within the corporate welfare system in the United States (Zepezauer 2004). Thus, for both financial and reputational reasons, the raisons d'être of the food bank and the good reputation of the brand names of the corporations that supported it were closely interlinked both in everyday practice and in the annual reports of the food banks. Such links were embodied in the prominent presence of representatives of food retailers and dealers and other business owners and managers on food banks' boards of directors, which set the policies and procedures that the staff of food banks followed (Riches 2018, 51–54). This was also true for the regional food banks serving our two urban field sites: the Charlotte metropolitan area and Durham.

Therefore, while considerations of profit for food corporate donors and their shareholders did not visibly drive the operations of the food banks we studied, the application of accountability measures required by donors to food banks and their staff and the discipline of enumerating and publicizing donations were internalized by food bank managers. As Liz put it, "The food bank needs to be its own best critic." Toward that end, the standards of cost-benefit accounting from the corporate world were quite clearly expected, and they were applied. Liz told Sarah, referring to her food bank's operating budget, "I think we need to continue to operate—we pinch every penny until it screams, and I think that your ratio of administrative and fundraising to your bottom line presses 3 percent—that's low."

In Liz's work, however, there were limits she drew to applying economizing business logics. In the case of food banks, these limits led to criticisms of food bank practices by corporate donors and employees about food banks' lack of "efficiency." At one point, Liz told Sarah, "Some people would want us to always go for the bigger bang for the buck." By this, she meant that "some people"—perhaps corporate donors—believed that to be maximally efficient, food bank trucks should preferentially deliver donated food to agencies (food pantries, community kitchens, etc.) that served the largest numbers of hungry families and bypass smaller ones, "since it takes the same amount of time to shop and check out a small agency as it is a big agency, because they all get thirty minutes." Responding to such criticism, which would more highly value an agency such as a food pantry that serves four hundred families a month over one that serves only fifty families per month, Liz observed, "There was a time when that little agency serving fifty people might've been the only agency in that area. So what, just because that need is still there, it's a mom-and-pop shop. A lot of places are."

The issue of "efficiency," which arises as a corporate logic to increase productivity within the private business sector, continued to trouble the public-private partnerships of the emergency food economy and constrained the staff of antihunger organizations situated elsewhere than food banks in that economy.

Whose Efficiency in Meals on Wheels? "We Just Can't Get These Poor People to Live in Close Proximity to One Another"

Economistic discourses associated with market rule often appeared to penetrate deeply into the operational aspects of major organizations within the emergency food economy—far more so than we have seen up this point. That is, the power of the discourse of efficiency and cost-benefit analysis constrained the operation of these organizations and limited their performance, with the result that organization staff simultaneously acted as if they should apply this discourse even as they rejected its specific application to the work they did. Let us take for example the Meals on Wheels program of one of our urban sites. Metropolitan Meals on Wheels (M2W), in the words of Amanda Spencer, its director, had as its objective to "deliver healthy nutritious food to people who don't have access to it on their own, whether by age, infirmity, health, chronic illness, temporary physical situations that restrict their ability to get healthy food, and don't have anybody [else] to do it for them." In the course of a day, Metropolitan Meals on Wheels delivered nine hundred meals to several hundred people within the large city and county it served and relied on thirteen hundred volunteers. Amanda's job was to coordinate this enormous logistical task of preparing and distributing hundreds of hot meals on a daily basis to a large number of needy people who were spatially dispersed across the county and beyond, by means of a workforce made up by a large percentage of volunteers, who concentrated on the actual delivery of meals to people, while M2W's full-time staff worked with local culinary school students to undertake the large-scale meal preparation and food purchasing and storage entailed.

The demands of coordinating this complex task presented Amanda with structural problems that she talked about in terms of efficiency. When our ethnographer Sarah asked a question about the meetings that M2W held to carry out its work, Amanda replied, "We have—we are implementing a weekly sort of strategy. We had our first one this Monday. And that's new. We will just try to pay attention to some of the details that are so vital to keeping an efficient [route] for the volunteers system going—making sure that the directions are right on the route sheets. We need the volunteers, because they control costs. So we don't want them frustrated." The necessary reliance on volunteer labor, as we see throughout this book, was one of the major institutional obstacles to providing fresh and nutritious food to poor people, food that many activists now expected the emergency food system in Charlotte to provide—but this reliance took a specific form.

What precisely was the problem that Amanda confronted? When Sarah asked Amanda about the goals that M2W ought to work toward, she replied, "[They] are about the relationships with the people that we serve. I think sometimes they [the volunteers] think we should be more efficient with our routes—you know, that if, for whatever reason, a route dwindles and gets so small, then we've had volunteers say, 'You know, it's not worth it for me to drive all the way over there to deliver just two meals.'"

Amanda then showed what she sees as being at stake when she replies to the imagined complaining volunteer: "And sometimes my response to that is, 'Well, we'll tell those two people that on this day they won't get their meals,' to really just try to jolt them [the volunteer] into realizing, well, you know, this isn't Bank of America. And, you know, we're serving people in a unique way. And maybe on some level that's not . . ." Her exasperation, evident in her remembered or fantasized recollection of dealing with volunteers and in her incomplete sentence that hints at their moral blindness, arose precisely because she sought to order the work of volunteers along different lines of "efficient" distribution of meals over space than many volunteers expected. She added sarcastically,

> Another thing is we just can't get these poor people to live in close proximity to one another so that we can keep these routes efficient. Imagine that. You know. We need ten poor people to live over here who need these meals. And there's only two. . . . Well, and then sometimes we'll say, you know, on the way we'll structure a route that isn't all in the [lower-income] Belmont neighborhood. We'll structure a route where there are two or three on the way there, where you can deliver two or three close and then drive all the way out there.

She tried to optimize the travel time of volunteers given the number and locations of hot meals they need to deliver, because she realized that one of the goals of volunteers "is to not waste their time." However, the tyranny of efficiency discourse meant that while such goals set the terms for efficiency, the very nature of volunteers' time as "freely" given meant that whose objectives were to take priority was always the subject of negotiation and dispute: "But they still don't agree that that's—you know, they think that we . . . don't know what we're doing. They want to say, 'Well, you know, you just don't know what you're doing. If I was managing this, then I'd . . . do it this way.' "

Economizing arguments around "efficiency" work best, of course, when those who set goals have the power to enforce them on all participants—characteristic of the corporate hierarchies in the city—but this was precisely what Amanda's dependence on volunteers lacked. And any concerns about the adequacy of the M2W program that the actual recipients of meals might have, such as about the quality of the meals they received (and when etc.), disappeared completely from the accounting. Economizing arguments about "efficiency" may not have met the needs of poor and house-bound meal recipients, even if they were relevant to how corporate volunteers tried to optimize volunteers' time as they saw it.

Caught within the confines of a corporate efficiency logic but absent its compulsion, with little incentive for volunteers other than "doing good for the community" and, for some, getting a day off at their employer's expense, Amanda justified the performance of M2W within the discourse of efficiency, even as she fundamentally questioned its assumptions: "I think most people who come and experience what we do come away thinking, 'Wow, you run an efficient business, and you are not wasting my time.' The majority of people are pleased with the way that we have structured the work for them."

"Efficiency" and "Effectiveness" in Providing for People Seeking Food: "Church-Shopping" Tales in Two Locales in North Carolina

At this point, we turn from the large-scale institutions (the food banks, coalitions of churches, and large-scale Meals on Wheels programs) to those who managed the institutions that directly provided food to poor people (the food pantries, small-scale Meals on Wheels programs, the operators of soup kitchens, and creators of gardens of produce raised for the hungry) and to how they came to terms with the exigencies of the emergency food economy.

We start with the interesting practice of "church-shopping" by charitable food recipients in our two rural sites in North Carolina. We found that food charity in North Carolina in many rural areas took place largely through Christian churches and other faith-based organizations, but all too often, those who staffed and volunteered to operate church-sponsored food banks, food drives, food pantries, and community kitchens told us of the distrust they felt toward those who presented themselves as food recipients—those who might be the undeserving or even dangerous poor—and this distrust was distilled in concerns about the "efficient" or "effective" provision of food charity to them.

In our research locale in two counties in mountainous western North Carolina, our ethnographer Jen asked Judith Ravenau, a board member of Welcome House, about its history. Welcome House was a facility built to serve poor people who were hungry and homeless, or at risk of becoming homeless, by providing them with daily hot meals and distributing food boxes. It served a multicounty region, providing a suite of services to poor people, including a community kitchen with daily meals, food boxes on a periodic basis, shelter for those who had been evicted, and a referral service for financial assistance (e.g., heating subsidies) for those who were threatened with eviction. In one interview, Jen asked about the origins of Welcome House and about its program to assist people in danger of eviction, High Country Assistance Center (HICAN), founded in the late 1990s. Judith replied, "I almost would have to say that the most important thing was HICAN." This program was begun by a coalition of churches coming together to plan for the building of Welcome House. Judith later commented, "The whole theory being—and it's more than a theory; it's an actuality—that it is much more efficient and less expensive to keep people in their home, in a home situation, rather than for them to lose that. . . . The way back is way more expensive than it is to [keep them in their home]."

While most readers would agree that the overall economic cost of keeping people in their rented or owned residences would be lower than providing for their needs after they are evicted, Judith's description of the formation of HICAN points to a different set of issues around administering to hunger, homelessness, and poverty in the US than just the cost of homelessness. Going back to her telling the history of formation of HICAN, Judith described its history as arising from the experiences of encounter between church clergy and volunteers, on the one hand, and those seeking food and other assistance, on the other:

> Because it was because these church offices were having knocks at their door all the time. . . . And not only that, but people would go from one church to another so . . . um . . . you know, then maybe fewer people were sapping all the funds from the churches and so forth. So but then I don't know, but I think very soon after the establishment of HICAN, it being under the umbrella of Welcome House. And then at that point, then . . . Welcome House collects the funds from the donors [for] HICAN, and they disburse the funds. . . . Rarely—I'd say never—is there a handout of cash; it's always a voucher, for that could be taken to a utility company to pay a utility bill or to be taken to a pharmacy.

What we found, then, was that Judith recounted a history of distrust between better-off church members and poor people seeking assistance from local churches: church offices were disturbed by "having knocks at their door all the time," the implication being that the same set of cagey poor people would "shop around" by going "from one church to another" seeking multiple donations and "sapping all the funds from the churches." This salient image or cultural figure of poor people is disquieting: that their uncontrolled access to and importuning of individual church offices was the source of waste, inefficiency, and even fraud (e.g., their claim that they needed food but, if given cash, would actually spend money on other, less necessary expenses). HICAN's policies were thus designed by its founders to allay their distrust of the poor while also allowing themselves and HICAN staff members to distance themselves socially from the poor, for instance, in the policy that "never is there a handout of cash; it's always a voucher." Thus, the needs of the poor must be regulated and controlled, if the "efficiency" of keeping people in their homes is to be achieved. As Judith put it in the interview, Welcome House "gives a hand up, not a handout."

In our other rural field site in eastern North Carolina, we discovered a similar coalition of churches and nonprofit organizations that had come together to offer services to poor people in need of food and other assistance but that also acted to regulate poor people's behavior and reinforce social distance between their largely white middle-class clergy and members and poor people, most of whom were African Americans. When poor people come to depend for food on the emergency food economy, then class and race suspicion, antagonism, and fears of the poor and racial "others" held by these constituencies all too often became institutionalized in programs, discourses, and imaginaries that sought to regulate the poor and moralize about them.

Food activists we interviewed justified such coalitions not only in economistic terms like "efficiency" but also in more general terms like "effectiveness," as used by one minister directing this coalition in eastern North Carolina. When our ethnographer Willie Jamaal Wright asked about the similarities and differences among the different churches in the coalition, the minister replied,

> I would say their goals are all—I mean, we have very similar goals. We want to help the people who come to our doors in an effective way, and I think you would hear from all pastors the same challenges I've just described to you. It doesn't matter whether they come to a Pentecostal church or an Episcopal church or a Catholic church, people come to the door, and they knock on the door, they say they need $100 for their light

bill. And if you're convincing, you'll get the hundred bucks; if you're not convincing, you know, you won't get the hundred bucks. And we know there are lots of people who aren't convincing who really need it, and vice versa.

In the book *Sweet Charity?*, Janet Poppendieck (1998, 238) describes "professional pantry shoppers" in what may be a recurring pattern for many poor people in North Carolina whose SNAP or other government benefits (if they receive them), supplemented by pantry give-outs, fail to meet their monthly food needs:

> Most pantries will supply a maximum of three days food, . . . [while] most forbid giving to an individual household more often than once per month or even once every three months. For clients who come in the aftermath of an actual emergency or an exceptional occurrence, such limitations may make sense. But increasingly, poor families use food pantries as a chronic supplement to low wages or public assistance that does not last the month. Food stamps typically run out after two and a half to three weeks. The three day supply will not fill the gap, so of course those who have access to multiple pantries will seek help from several.

If this is so, then it is worth asking about efficiency: Efficiency for whom? In meeting whose goals? And how might staffs' or volunteers' class- or race-specific apprehensions about food recipients play a role in defining it?

"We Made That Decision Based on the Economy of Food Scale": Economizing Language and the Challenges of Being a Front-Line Food Provider

How might the very language of the "economic" be fruitfully enlisted in a moral logic of food toward the end of ensuring the long-term food sufficiency of poor people? This was a question we encountered in our site in the Appalachian western region of North Carolina. Economistic discourse—the language of profitability, efficiency, and cost-benefit analysis—was, as should be clear at this point, by no means the self-evident terms in which the charitable food economy should be described. The language and logics of business often proved difficult to apply for those who thought of hungry people as people with feelings and needs. At the same time, however, this discourse could become a useful tool, particularly among those who were experienced in its use in a previous career or line of work. Such was the case for Rev. Jon Fisher, who, prior to becoming a Methodist minister and being assigned a rotating circuit of churches to serve in one of our rural sites, took his degree from a regional university majoring in advertising and marketing management and worked in advertising and cellphone sales before entering the ministry. His experience was by no means unique.

Rev. Fisher was executive director of Mountain Service Ministry (MSM), which served an estimated twelve hundred people in mountainous western North Carolina with a food pantry, a soup kitchen, a small Meals on Wheels program, a backpack program, a community garden, and a gleaning program. MSM had been in operation for the past decade, and Rev. Fisher had worked energetically over many years to achieve MSM's stability and to increase capacity by finding funding from statewide philanthropies, establishing locally respected programs, and holding successful fundraisers in an economically distressed region of North Carolina, such as a popular twice-monthly fish fry. Through his efforts and those of volunteers, MSM met the food needs of hundreds of poor people who were widely dispersed in small settlements and towns over several counties in a mountainous region, one in which people had to travel over long distances to purchase any food they needed that could not be grown in their own gardens and farms.

At one point, our ethnographer Jen asked Rev. Fisher what the objectives of MSM were. He replied that his first decision was to specialize in providing only food to poor people, instead of several other services (e.g., rent and heating fuel subsidies), because MSM's "dollar could be stretched much further":

> We do have a mission statement. The goal is to feed. Our new tag line is, "Giving Food, Growing Hope." Everything centers around food. We decided—it's not a mission statement, but we decided early that, you know, the dollar could be stretched much further purchasing food, especially through Second Harvest Food Bank. We could expand that dollar to equal a lot more benefit than we could putting a gallon into

someone's tank, to provide heating assistance. So, we narrowed our focus, but we expanded the capacity of what we're able to do, and that is to feed people.

He went on, "The second [goal] . . . is to teach food responsibility. And that not only am I responsible for going up to the food bank and getting my food—nobody's going to deliver it to me—I'm also able to grow some of my own food. And through a program, a little program called Seeds to Feed, and another one called Field to Table, we're helping to nudge folk along the way." Explaining, he indicated how MSM had tried to "nudge folk." Many poor people in the region, he claimed, "have lost the art and/or desire to garden," but worse than that, as poor people, they were "recipients [who] have a little more time than others might." Thus, MSM sought to "incent" these folks to participate in growing their own food by providing them with a container with vermicultured potting soil and vegetable "starts." Another inducement was to mobilize them to participate in gleaning operations when farmers had surplus food that could be gleaned from their fields in return for part of the harvest that they gleaned. His assumption was that poor recipients would have "a little more time" than others to participate in such unpaid labor.

As it turned out, Rev. Fisher harnessed the theme of personal responsibility to his broader argument about "economies" in the food industry and the increased cost of fuel in the future. When Jen asked Rev. Fisher about his personal reasons for founding MSM, he returned to the theme of "stretching the dollar": "So we made that decision based on the economy of food scale, I guess you would call it. But eventually, those economies are going to diminish because of transportation costs and infrastructure stuff and total cutbacks to the very leanest possible. We've seen it dramatically over the past three years." Previously, in contrast, many kinds of processed food had come into the regional food bank for redistribution by MSM due to "overproduction." However, "now, those production numbers have been cut to equal the economy and sales figures and actual demand for . . . And, that just doesn't really exist." As a result, this has decreased the "volume of stuff that [MSM] can get, but not really . . . the essentials of stuff." That is, the "volume" of processed foods donated to the food bank by large food distributors and retailers had decreased in quantity due to inventory controls, and what remained in abundance were the fresh staple crops that had to be quickly sold or distributed because they would soon spoil.

This mattered, Rev. Fisher made clear, because of the increased cost of the fuel for the trucks that carried the food from the regional food bank to MSM's headquarters for redistribution: "Sooner or later there will be a moment when [because of rising fuel prices] it's almost cost-prohibitive to drive and get prepackaged food." And this was why poor recipients of food donated by MSM had to acquire "food responsibility." Rev. Fisher explained, "Thus, we try to help teach people, along the way, best we can, how to grow their own food. Because if they can take a certain percentage out of the garden, then that didn't have to come from MSM or fuel assistance or anybody else; it was just the value of seeds, amendments [e.g., vermicultured soil], and time growing."

Rev. Fisher's explanation for the necessity for "food responsibility" on the part of poor people especially in the face of coming challenges to transportation, then, was remarkably laden with economistic and technocratic language: he spoke about "the "economies of food scale," "transportation costs," "overproduction" being reduced to "the economy and sales figures and actual demand," "total cutbacks to the very leanest possible," and "cost-prohibitive" fuel prices. Rev. Fisher's prior education and experience in marketing and sales were quite clearly evident. At the same time, his awareness of how "the economy" worked led him to think of a future constrained by "peak oil" and rising energy costs, and he had pointed to a new self-discipline that would be required of many poor people (and others) in a future of scarcity.

The idea that the individual should take personal responsibility for their own situation is one associated not only with neoliberalism/market rule ("people need to take responsibility" for their performance in markets) but also with Christian Protestant religious tradition in North Carolina (people have to seek "salvation" for themselves). It is therefore not surprising to find the idea of personal responsibility integrated thoroughly into the discourse of a Methodist minister with a business background. But one must be cautious in drawing the conclusion that this is a sufficient explanation for Rev. Fisher's concern to "incent" poor people's "food responsibility" and instead see it as part of a broader set of concerns around the looming threats to future provision of charitable food to poor people.

Rev. Fisher's argument was based on an economic analysis—one in which a new food responsibility becomes a future imperative due to a predictable decline in corporate food donations, arising from increased systemic efficiencies in the food wholesaling and retail industries. Thus, his injunction that poor people take responsibility by growing their own was what the philosopher Immanuel Kant called a "hypothetical imperative," that is, a moral imperative that exists because certain conditions exist that require it now or in the near future. In this case, this is the imperative that the declining provision of food from the corporate sector, combined with the increased cost of fuel for transport—not negligible in a mountainous region with a dispersed population—would sooner or later throw people back onto their own capacities and require them to develop food responsibility. Rev. Fisher was not primarily demanding food responsibility from poor people in the region as a moral absolute, or what Kant called a "categorical imperative," arising from the tenets either of neoliberalism or Protestant Christianity. Instead, his economizing language combined with a bleak but perhaps realistic vision of a near-term future of increased scarcity of emergency food, or expensive fuel, and an increased incapacity of organizations like MSM to meet the food needs of the poor population in this region.

What Did the Community Food Security Activist Exemplars Reveal?

The exemplars of community food security that we presented earlier have other lessons than those discussed to offer about the influence of neoliberalism among food security activists. We see, for example, the deep ethical concern by activists from Durham's West End neighborhood for neighborhood residents' food needs when the Interfaith Food Shuttle's mobile market's deliveries to neighborhood residents were cut off and their eagerness to work with IFFS staff in order to restore it. Here, we see little uptake of neoliberal notions of efficiency or of a supposed need to enumerate pounds of food delivered.

On the other hand, Gerald Payton, IFFS's food delivery manager, was preoccupied with meeting the organization's goals of rationing food to a population, some of whom he believed might make unreasonable demands, as when recipients took advantage of the "open market" he described to take more of the kinds of food they desired than he thought proper. This suggests that the economizing "business-like" performance measures characteristic of other food banks might apply to IFFS as well, despite its reputation as an innovator in bringing fresh locally farmed foods to charitable food recipients across a multicounty region. Clearly, the polite argument between IFFS staff and Durham organization volunteers about how IFFS should provide charitable food to food-insecure neighborhood residents suggests how racial and class differences could place activists at odds with one another. Put slightly differently, IFFS's charity-based food security activism tied to the corporate-driven food bank nation that backstopped it was in tension with the community food security concerns of neighborhood activists.

A similar deep commitment to providing fresh food to the neighborhood was evident in the ingenious efforts of Dorothy James (of Charlotte's Black Women's Health Network) to bring mini farmers markets to African American churches in order to bring fresh produce to food-insecure residents with health and disability issues. Again, there was little talk of efficiency, profit, or productivity: people had the right to fresh food, and the question was how to bring it to them.

In the Happy Hills Community Garden of Rocky Mount, we saw the way that Lawrence Farmer, a compassionate retiree, used his own money to provision a community garden and welcomed in impoverished neighborhood residents to grow food there, thus allowing them to get by economically, while providing the setting for a rich shared experience among the gardeners that crossed the racial boundaries that divided the city. The principle of *community* food security—as distinct from the food security of individuals alone—was manifest in this case.

To conclude, what we observe in this chapter, as in chapter 4, is a war between two moral logics—in this case, between a neoliberal moral logic deeply ensconced in the corporate dominance of the charitable food sector, that is, the imaginary of "food bank nation," on the one hand, and a community food security moral logic based on social justice concerns that was manifested by so many on-the-ground activists, on the other—taking the form of an uneasy and tense coexistence between the two within the daily practices of activists.

* * * *

This chapter examined the effects of neoliberal discourse on food activists who worked in the charitable food sector in our four sites. First, we recounted the shift for food security activists that began in the 1980s from advocating on behalf of people who were poor and food insecure toward administering to their needs through the nonprofit charitable food bureaucracies set up within the nonprofit sector. In making this shift, food activists met the pressing material needs of food-insecure citizens and residents that the US corporate state itself had abandoned. This transformation took place within a broader shift in the corporate state that required poor people to seek, find, and keep wage-labor jobs within the capitalist economy as a precondition for federal SNAP food benefits, relegating those in the working-class population who are unable or unwilling to hold such jobs, as well as other poor people, to being the recipients of the charitable food sector.

Second, we examined the ways in which food security activists who worked in food banks and Meals on Wheels programs were simultaneously constrained to follow the cost-benefit accounting practices of private corporations in line with neoliberal "performance" measures, while they also sought to subvert neoliberal concepts that defined "efficient" food delivery in ways that deprived recipients who genuinely needed food from receiving it.

Third, we also looked at the ways in which food security activists who managed food pantries and community kitchens in rural North Carolina churches employed the neoliberal concepts of "efficiency" and "effectiveness" to reinforce racial- and class-based stereotypes that stigmatized the behaviors of food recipients from whom they were socially distant.

Fourth, we analyzed the ways in which economizing language—of "efficiency" and "scaling up"—might be detached from neoliberal objectives of exalting markets, individual self-interest around capital accumulation, and making working-class bodies "work ready" and instead be redirected toward meeting the collective needs of food recipients. This

Apple Day: apple expert gives talk, signs book, receives award, at Ashe County Farmers Market, West Jefferson (Photo: Jennifer Walker)

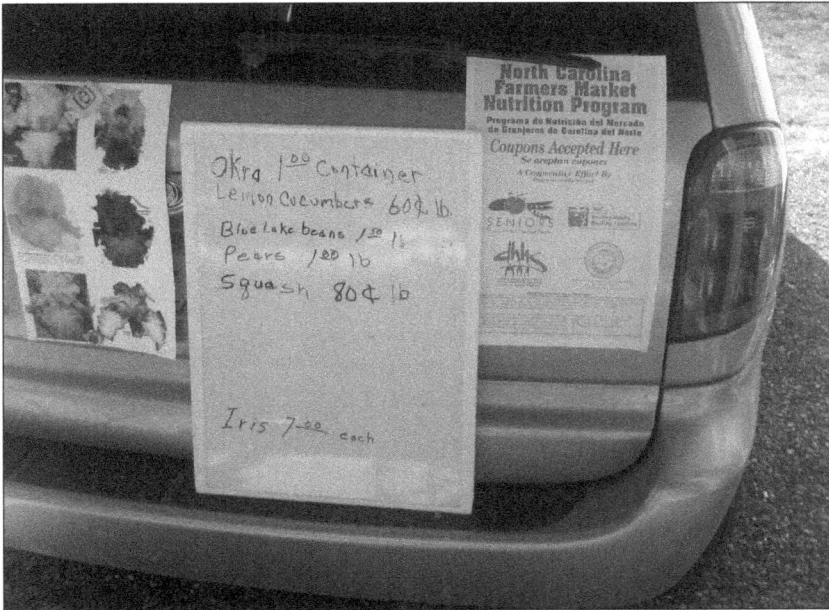

Back of truck price list and WIC notice, Ashe County Farmers Market, West Jefferson (Photo: Jennifer Walker)

"No Pets Allowed," Ashe County Farmers Market, West Jefferson (Photo: Jennifer Walker)

Vendor, Down East Partnership Market, Rocky Mount, Edgecombe County
(Photo: Willie Jamaal Wright)

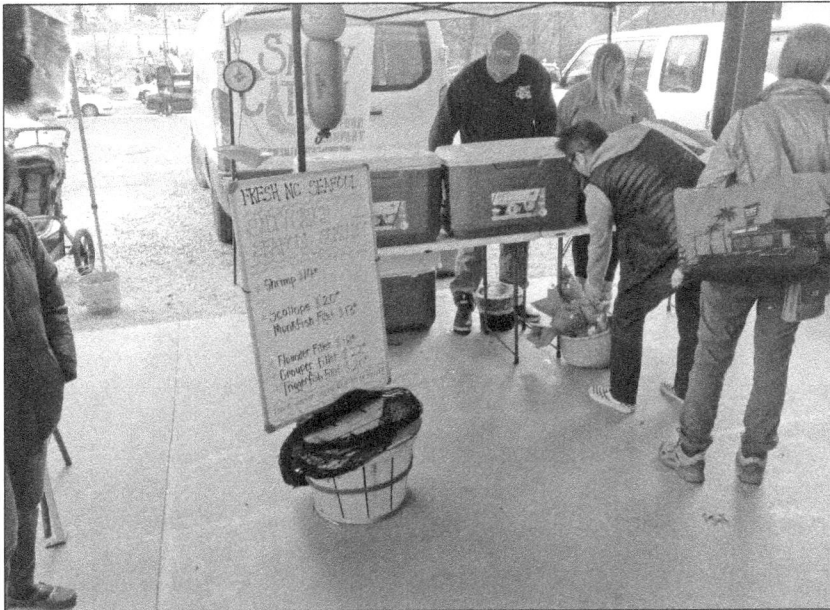

Price list, fresh seafood vendor, Durham Farmers Market, Durham (Photo: Don Nonini)

Market stalls, Farmers Market, Downtown Charlotte (Photo: Sarah Johnson)

Produce section, supermarket, East Charlotte (Photo: Sarah Johnson)

Produce section, supermarket, South Charlotte (Photo: Sarah Johnson)

Genesis Park Garden, Charlotte Green Project, Charlotte, serving "inner-city neighborhoods that do not have access or have limited access to a grocery store" (Sarah Johnson, field note, 2012; photo: Sarah Johnson)

University students build soil at school garden, Old West Durham, Durham (Photo: Patrick Linder)

Sign, Hope Haven Garden, Hope Haven, Charlotte. The text in the upper-right corner reads, "rehabilitation center made up of 200 men, women & families w/ children" (Photo: Sarah Johnson)

Signs, the Males' Place Garden, Charlotte. The sign includes the text, "Helping males develop into manhood" (Photo: Sarah Johnson)

Potato harvest, Clarketon Community Garden, Clarketon, Edgecombe County
(Photo: Willie Jamaal Wright)

Corporate volunteers at FoodBase workday on urban farm, Durham
(Photo: Patrick Linder)

Damaged goods to be picked through by volunteers, Food Bank of Central and Eastern North Carolina, Durham (Photo: Patrick Linder)

Chart explaining to volunteers how to fill food bags, Urban Ministries Food Pantry, Durham (Photo: Patrick Linder)

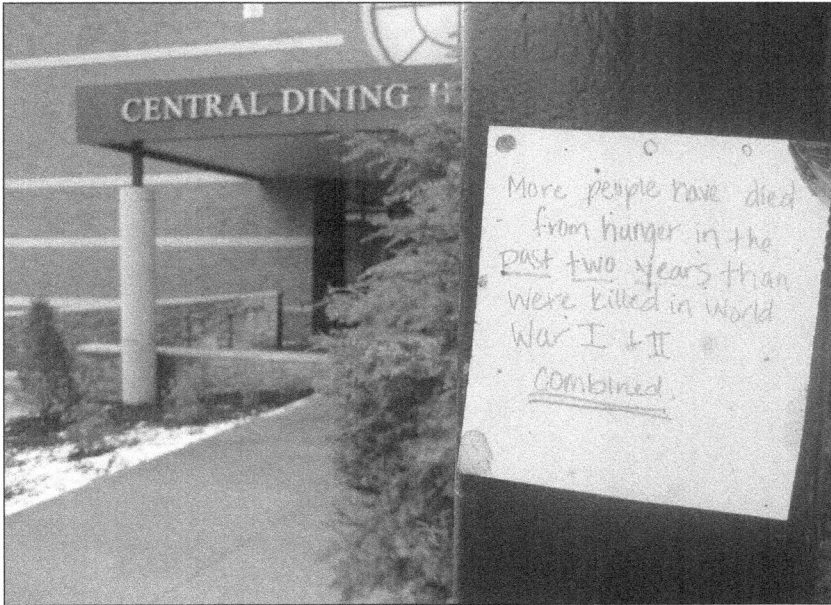

Sign, local university campus, Boone, Watauga County (Photo: Jennifer Walker)

was evident in Rev. Fisher's efforts to provide rural food recipients in a mountainous region of North Carolina with the initial tools and skills to enhance their food self-sufficiency, given the future limits he envisioned in the capacity of the fossil-fuel-based transport systems on which charitable food deliveries depend to provide remote rural populations with the food they need.

Recurrently, we saw a dialectical "war" between two different moral logics of food: that of the corporate food sector and corporate state, on one side, and that of food security activists committed to socially equitable food distribution to needy, working-class populations in our four North Carolina sites, on the other.

PART III

Cross-Over Experiments in Sustainable and Socially Just Food Systems

6

Slow Food Charlotte, Farmers, and Food Insecurity

I'm definitely committed to serving farmers and creating access [to markets] and doing whatever we can to help them thrive, . . . do better, make a go of it. I think education needs to be a big part of it, . . . [like] writing articles or being interviewed by the newspaper and attempting to get the word across that most of the markets in town are resellers and you're not really buying from a farmer. I think it's also our mission to keep it fun and not make everything this dreary chore, make everything duty. So what is that, education, awareness, help farmers, make it fun.
—Fred Pearson, leader of Slow Food Charlotte, when asked by Sarah Johnson what goals it ought to serve

Slow Food Charlotte: Querying the Politics of Pleasure

Over the past thirty years, Charlotte, North Carolina, has become the third-largest banking center in the US (after New York and San Francisco) with a flourishing financial and high-tech sector (Burns 2017). Bank of America has its national headquarters there, as do several regional banks. The Charlotte metropolitan area has been one of the fastest growing areas nationally in the past several years, with large numbers of people moving into the region seeking desirable managerial and professional positions. As a result, a numerically large upper middle class of elite consumers has come to patronize Charlotte's burgeoning upscale restaurant industry.

In 2004, Fred Pearson founded Slow Food Charlotte (SFC), which became a chapter (or "convivium") of Slow Food USA and the international Slow Food movement. Committed to the Slow Food movement's "eco-gastronomic" vision that the pleasures of locally grown, prepared, and cooked meals and the sociality such meals create could be coupled to a progressive commitment to justice and sustainability of land, farming, and small farmers (Andrews 2008; Slow Food 2022a), Slow Food Charlotte has in subsequent years become a major player in the food politics of metropolitan Charlotte. Its founder's timing was crucial. Soon after its founding, Slow Food Charlotte became well recognized within the elite circles of Charlotte's financial, high-tech, and entertainment industries for its epicurean interests in local food and fine dining. By 2004, the reputation of the Slow Food movement, with its sensual ambience centered on the enjoyment of food and its exotic roots in the imagined gourmets' paradise of Northern Italy, drew Charlotte's elite diners to the Slow Food Charlotte chapter. SFC's founding must be situated, therefore, within the broader rise of the tourism, entertainment, and dining industries in this metropolis of almost one million people.

SFC's founder, Fred Pearson, members of the board, and almost all members of SFC were themselves white, highly educated professionals and business people and far more affluent than the vast majority of Charlotte residents—and were therefore readily accepted as players within the new dining cultural milieu of the city. Fred noted that one of the goals of SFC was to infiltrate a broader spectrum of society that needed to be included and that combating elitism needed be led by rich white people, since, as he said, "we're the ones who got us into this mess." Fred was referred to by one SFC board member as "somebody who . . . has means of support that doesn't require lots of hours," while other Slow Food board members were small business owners, executive chefs, retired corporate managers, and landowners.

Slow Food Charlotte's mission, in the words of Fred Pearson, was to "make it fun," "keep it fun," while it supported small farmers through public education. This resonated with the mission of the international Slow Food movement. Founded by the charismatic "eco-gastronome" Carlo Petrini and several others in the small Piedmont town of Bra in northern Italy in the mid-1980s, the international Slow Food movement had grown to eighty-four thousand members in 120 different countries around the world by 2008 (Andrews 2008, 12). By the time of this writing, "Slow Food has grown into a global movement involving millions of people in over 160 countries" (Slow Food 2022a). The movement has come to stand, for both its supporters and critics, as a landmark in the history of the local food movement across the US. Combining the cultivation of pleasure in dining on locally grown, fresh food with support for local or regional small-scale farmers, Slow Food has become a lightning rod for one of the long-lasting and interesting debates around local food. It goes as follows.

To the supporters of the Slow Food movement, invoking the importance of pleasure and the cultivation of taste, it has successfully resisted the standardization and lack of nourishment and taste of "fast," globally sourced industrial food by promoting its alternative—"slow" food. Many recalled the subtitle of the *Slow Food Manifesto* put out by Petrini in 1989, which announced the founding of an "International Movement for the Defense of and the Right to Pleasure" (Slow Food 1989). Slow food was described as food that was a delight to the senses because it was fresh, nutritious, and thoughtfully grown by farmers who were knowledgeable about the crops and livestock that were best adapted to the local terrain and its ecologies; it was food prepared by cooks using long-standing ("traditional") nonindustrial culinary practices (Slow Food 2016; Andrews 2008, 86–102). Moreover, slow food was resistant to scaling up to the mass production and standardization that characterized the "efficiencies" and "profitability" of the corporate food industries, a phenomenon discussed earlier.

To detractors, the Slow Food movement merely provided the pretext for local elites to seek to distinguish themselves from the masses by engaging in self-indulgent feasting, while the justifications they offered, invoking knowledge and taste, merely served as a rationale for prioritizing their own pleasure relative to the labor of the nonelites who made their consumption possible (Chrzan 2004; Gaytan 2007; Laudan 2004). Nonindustrial methods meant that the quantities of food grown and prepared were small and expensive to produce and their prices high, and thus the number of people who could afford to consume them was limited to the middle and upper classes. During the period of our ethnographic fieldwork in Charlotte, from late 2010 through early 2012, we discovered that the activism of the leaders and members of the Slow Food Charlotte chapter came to be assessed within the terms of this debate and that the debate itself began to reshape the mission of the chapter.

The broader question that this perennial debate brings up is, What is the place of pleasure in the politics of food activism? Must those who support the nutritional, culinary, and environmental virtues of slow food through their consumption be upper-middle-class epicures committed to their own selfish bodily pleasures of "fine," that is, expensive, dining at the expense of the majority of the population and especially working people and people of color? That is, does local food activism oriented around the pleasures of consumption hide class and race privilege that necessarily limits its value? Given the extreme class and racial inequalities that distinguish contemporary Charlotte (Semuels 2017), what might lead the leaders of Slow Food Charlotte to move beyond their own pleasures not only to acknowledge but also to work to improve the life situations of two distinct but greatly disadvantaged groups within the US food system: on the one hand, the impoverished inner-city populations of Charlotte and, on the other, the dwindling population of small-scale farmers of the region?

Meet the Slow Food Movement: A Movement of Political Consumerism

The Slow Food movement at the time of our fieldwork in 2010 coupled a commitment to the pleasures of dining with a critique of and consumer-based resistance to "fast foods" and the "fast" way of life associated with the globalized industrial food system; its destruction of biological species and poisoning of rural soils, air, and waters; and its degrading treatment of farmers, farmworkers, and Indigenous groups in both global North and South nation-states. Through the international Slow Food movement, thousands of activists, farmers, cooks, farmworkers, artisanal processors, academics, consumer advocates, and others affiliated with scores of culinary traditions and with hundreds of food professional associations, peasant organizations, Indigenous peoples groups, and farm laborers' and rural women's organizations have undertaken diverse initiatives of critique and resistance. Throughout, the Slow Food movement has consistently insisted on the right of the majority of the population to their most endangered sensations: the pleasures they share in their own cuisines and the social relations that shape and are shaped by these cuisines.

The international Slow Food movement represents a social movement based on "political consumerism" (Bossy 2014), a movement "in which a network of individual and collective actors criticize and try to differentiate themselves from traditional consumerism by politicizing the act of buying in order to search [for] and promote other types of consumption" (Bossy 2014, 182), in particular, the consumption of local foods in the company of others and the pleasures it connotes. We must qualify Bossy's definition of political consumerism by noting that it is not what people buy in markets that matters as much as how they sustain themselves and the people dependent on them by what and how they produce and consume—all of which *determine* what they buy in markets—and that the rhythms of production and consumption (and taking of pleasure) in everyday life also structure their use of time.

The Slow Food movement insists that culinary practices have a utopian dimension in that they represent efforts "to go beyond traditional activism by encouraging the promotion of alternatives" (Bossy 2014, 188). In the history of the international Slow Food movement since its founding in 1989, it has become globally renowned for its innovative experiments and alternative projects grounded in transnational networks of producers and consumers, and many of these were already under way at the time our fieldwork began in 2010. These include Terra Madre, the movement's biennial convening of thousands of food activists (farmers, cooks, youths, educators, et al.) from both the global North and South who interact with and learn from one another in Turin, Italy; its Presidia, "Slow Food communities that work every day to save native livestock breeds, local fruit and vegetable varieties, bread, cheeses, cured meats, sweets, and more" (Slow Food 2022d); its Slow Food Foundation for Biodiversity (Slow Food 2022c); and its many innovative partnerships and funded projects, for example, "Food Heroes," "Slow Plate: Knowledge of and Healthy Access to School Meals," "Food Trails," "Agroecology and Alternative Food Systems in Kenya and Uganda" (Slow Food 2022b).

Slow Food movement actors "prefer to concentrate their energy in saying yes to something rather than saying no to the existing world" (Bossy 2014, 188). For the Slow Food movement, this "saying yes" begins with a transformation of present institutions and practices around consumption and consumers to connect them to the worlds of food producers and production. This utopian aspiration, scaled up globally, prefigures a collective and structural change in people's use of time, so that they will spend more time doing what is necessary to sustain their local ecologies and what they enjoy, especially the pleasures of commensality while eating food that is fresh, nutritious, flavorful, and sustainably and ethically produced (Petrini 2007). The Slow Food movement is thus at the cutting edge of a global shift by a large number of people seeking to transform their consumption, sustain their environments, reduce their work, and free up their time to be more sociable and enjoyable—around eating together and other everyday pleasures. Even in the United States, the country with arguably the most harried middle class, many "down-shifters" have already sought to make changes in their own lives by reducing their consumption, the incomes required for it, and thus their working and commuting times, while learning new skills and habits to support and simplify their lives and to become more in control of their time (Schor 1998, 111–142; 2011).

Above all, to come to terms with this debate about the relations between the pleasures of eating well together and food politics in Charlotte, one must investigate the meaning of pleasure and of eco-gastronomy in their contexts—something our ethnography was well suited to do.

Slow Food Charlotte: Supporting Farmers through Dining and Educating Consumers about Farming

At the time of SFC's founding in 2004 but before the international Slow Food movement turned in 2007 toward a more radical food politics (Petrini 2007; Andrews 2008, 48–66, 148–164), Fred Pearson had a distinctive vision of what it meant to be a slow food consumer. He associated the "community" of epicurean elite diners with an attachment to "tradition" and culinary "culture" centered around the gastronomic practices of Charlotte's diners, chefs, and gourmets, who increasingly sought out locally grown, fresh produce and other foodstuffs in line with the consumer trends in the US from the 1990s onward favoring local produce and foodstuffs, as described earlier. Fred saw these connections between food and community as the basis for effecting incremental changes to Charlotte's food system.

Key to these connections were the sociality of people coming together and dining with one another. As Fred put it,

Culture is this entity that is carried by community, and if the community is fractured, it can't carry a culture forward. And culture is made up of tradition, all the stuff that's worked. And so with the community . . . fractured and culture . . . not carried forward, then you begin to lose things. And not only do you lose the traditions and all this great knowledge, but you also lose the opportunity to add your voice to the great conversation of life. *To me, by growing a community, which is having friends and gathering and getting to know people,* you are ultimately preserving everything collectively that we all carry with us. And it goes forward in time and we can add our voice to it, so it gives a place the "there," as you say. It creates place. It differentiates. It makes it feel like home. (emphasis added)

From SFC's founding in 2004 as a convivium up through the start in 2010 of Sarah Johnson's fieldwork in Charlotte, much of SFC's recruitment efforts and its impetus of "growing a community, . . . having friends and gathering and getting to know people" took the form of periodically holding potluck dinners to attract potential SFC members. The requirement for what kind of food to bring to the potlucks set a fairly low bar, asking invitees only to bring a dish that they had made containing one or more local ingredients in it. These potluck dinners were occasions for casual socializing and taking pleasure in tasty food in the company of like-minded others, and most—but not all—Slow Food members enjoyed them and approved of them. And therein hangs a tale we return to shortly.

Another event that brought together SFC members was the convivium's annual meeting—held at a local restaurant over courses of locally and regionally sourced dishes combined with fine wines. Our ethnographer Sarah Johnson described the annual meeting of 2011 as follows:

The meeting takes place at FABO Café. It's a small café in the Myers Park area (upscale) that has local paintings & crafts for sale along the walls. The meeting is scheduled to start at 6:30 p.m.; when I get there, a few minutes late, none of the board except Fred has arrived yet. There are about thirty-five people there, mingling and eating and drinking wine. The food is provided by Beverly's Gourmet Foods; the wine was purchased from FABO at wholesale rate. . . . It's relaxed, social, though most of the people don't seem to know each other. Mostly white—one African American couple there. Varied in age, mostly middle class.

According to Fred, Slow Food Charlotte was founded to provide support for small-scale farmers on the part of consumers of local produce and foodstuffs. As quoted in the epigraph to this chapter, when Sarah asked the founder of SFC about its values and goals, he replied, "I'm definitely committed to serving farmers and creating access and doing whatever we can to help them thrive as much as—if 'thrive' is even a word you can use in this environment—do better, make a go of it. I think education needs to be a big part of it."

In SFC's first six years prior to Sarah's fieldwork in 2011, its education of consumers about the achievements of the small farmers in the region of the fourteen "necklace" counties of the Carolina Piedmont that surrounded the "Queen city" of Charlotte, whose foodstuffs appeared weekly in the farmers markets of Charlotte's Mecklenburg County, took only a few forms. One important recurring activity was SFC's organizing of "farm-to-table" dinners featuring the produce and livestock of regional small farmers as the basis for gourmet meals prepared by outstanding local chefs. One Slow Food member, who worked as a professional chef, commented when interviewed by Sarah during a tasting event held at one of Charlotte's upscale farmers markets, where he joined a local winemaker by cooking regionally grown artisanal meats served with wine, "Chefs feature in support of the farmers. . . . It's a good relationship. . . . Chefs and farmers work together to restore the food system. . . . Chefs can educate the public." This was very much in accordance with Slow Food's international focus on connecting farmers, cooks, and consumers through the interpersonal connections formed by such an event, although it was also consistent with the branding and celebritization processes characteristic of the flourishing Charlotte restaurant industry.

A leader of SFC referred to what happened when one farm, Far Village Farms, asked SFC about doing a farm-to-table dinner: "We set up a dinner out there with judges and press to go eat, and they had like fifty people and were completely blown out of the water. They had no idea that food could be that good, and it's all local." Thus, consistent with a "classic" (Euro-centered) gastronomic practice, local gourmets, certainly only those who could afford the high price of the farm-to-table, found themselves occasionally occupied with learning about a "culture . . . made up of tradition," that is, a culinary regime—contemporary neotraditional southern farm cooking and preparation of local food complemented by wine vintages—that was more in a process of invention by Charlotte-based chefs than it was of recovering cuisines from the pastoral spaces of the western Carolina Piedmont and its farms.

In a second, far more modest effort, SFC leaders recognized that shoppers at Charlotte's farmers markets were not aware that many farmer vendors whom they were buying "local" food from were not selling produce that they themselves had grown but were resellers of produce grown by someone else. As a project, SFC provided farmer vendors who sold only their own produce with small signs to authenticate the fact. These they could display at their vending stalls, so that they might have some edge against unfair competition from the lower prices that volume-based resellers offered. As Bill Fletcher, an SFC board member, put it, "We printed up a flag that says 'local farmer.' And there's always this definition of who's a local farmer. Well, we know a local farmer when we see one. I mean, we know who they are. They're our friends. We know them. So at the farmers market, where there are both, we give them this cool flag from Slow Food, and we tell people all the time, 'Just go to those farmers.'"

It is interesting that this experiment in distinguishing the local farmers who sold their own produce from others who did not contrasted sharply with the approach of a farmers market manager in an outlying necklace county. The market manager stated to Sarah Johnson that "she doesn't have a problem with [reselling] at her markets because it does help farmers. All her vendors are growers, but many are not able to grow enough year-round to support themselves. By reselling during the slow months or when their crops are not yet in season, they are able to keep themselves afloat and ensure that the market has enough to offer to keep people coming back." This juxtaposition contrasts SFC's "at-arms-length" relationship with farmers—sponsor and turn a few select farmers into celebrities through farm-to-fork dining events, provide signage for "real" local farmers, and educate consumers a bit—with that of this farmers market manager, who spoke knowledgeably about farmers selling at her market and about their precarious economic condition on a week-by-week basis.

What connects each of these efforts was, as Pearson put it earlier, the "education [that] needs to be a big part of it." But educating consumers was no easy task, and beyond publicizing the food products of a few farmers through farm-to-table dinners and authenticating local farmer-vendors with signage for their stalls, these SFC initiatives were dwarfed by the challenges of educating consumers about the threats posed by the industrial food regime to both consumers and small farmers. When asked by Sarah whether SFC in its work ever found that it could only serve the small farmer, on the one hand, or the consumer, on the other, Fred Pearson replied, "I would say my biggest hurdle is a frustration with the deaf ears that you're attempting to reach by pointing out to people that this is what's going on and how they're all too busy and caught up in their lives to even register the fact that the industrial food system is making them sick, not to mention destroying the planet, not to mention undermining culture and community. So that's my biggest frustration, which is huge. I mean, to me, 2 percent of the population is driving this whole thing, maybe 1 percent." The "2 percent, . . . maybe 1 percent" whom Pearson referred to, that is, were the elite of SFC members who were committed to supporting local small farms and the regional cuisines they made possible. The challenge of seeking to inform most people of the imperative to break with the industrial food system so as to not make themselves sick, destroy the planet, or undermine culture and community, and instead support local farmers, was a formidable one.

A Schism within Slow Food Charlotte's Board: Did SFC Really Support the Region's Farmers?

In the year from 2008–2009, changes to SFC occurred that were to affect its work crucially in the years to come. A conflict broke out in mid-2008 within the board of directors of Slow Food Charlotte that revealed the contested meanings around SFC's mission. Because SFC board members whom Sarah Johnson interviewed did not directly mention the conflict except in passing and long after it had led to a schism in the chapter, she recognized it only late in her fieldwork in late 2011, as evidence mounted for it in her interviews.

The conflict became overt when two SFC board members expressed their displeasure with the periodic potluck dinners that accompanied the meetings of SFC for potential members and, on other occasions, by referring to SFC as being no more than a "supper club" committed only to the dining pleasure of its members. Bill Fletcher, a friend of Fred Pearson's, who allowed Fred to recruit him to the board at the onset of this dispute to support Fred, gave his own reconstruction of it: "There was a certain faction of Slow Food who were like—who didn't want to do anything fun. They were like, 'This is not a dining club.' And so they didn't like the potlucks. I don't know if they disliked them, but they were always scoffing at them. And they did split off for a while and form . . . [another] chapter."

Bill went on to elaborate his interpretation of what the dispute was about:

> The girls at Felicity Farms . . . went on a rage that Fred was just running a dinner party—a "supper club" was the word that they used—and that they were going to make more change and that they were going to strive a little harder. . . . And also, they grow up north, so it was kind of hard for them to come down, and they were the ones that wanted—they set up another chapter. And then, frankly, we kind of raised our eyebrows, going, "Should a chapter really be run by the people who also profit from the education that they're putting out?" I mean, it seemed a little too tied up to a for-profit business.

What did it mean to imply, as Bill did here, that leading food activists should not "also profit from the education that they're putting out" or that Felicity Farms was "a little too tied up for a for-profit business"? This was the claim that a liberal disinterestedness independent of commerce was essential for ethical food activism. We begin by noting the discordance when Bill referred to the two partners of Felicity Farms as "girls," despite the fact that they were both middle-aged. Leaving this aside, it is important to state that the owners of Felicity Farms, Sonya James and Jessie Newman, were working small farmers who combined their farm's business raising about one hundred heritage Tamworth hogs and a few chickens with keeping open their own restaurant in a small town thirty-five miles north of Charlotte, which served exclusively local heritage produce and meats and for which Jessie was chef.[1] What, if Slow Food was to be more than a "supper club," did they hope it would become?

When Sarah Johnson interviewed Sonya in early 2011, Sarah was not yet aware of the schism in the SFC chapter. Nonetheless, what Sonya spoke of as most important in her work as farmer and food activist is relevant to the dispute. Instead of referring to Felicity Farms by itself, Sonya spoke of leading "an informal group, . . . more of a loose collection of like-minded farmers, essentially, and some restaurateurs and friends of ours who we know all think that we need to do some things differently with our food system in the Charlotte area." She and her partner, Jessie, in Felicity Farms were "leaders in getting a lot of farmers . . . some visibility": "We have a restaurant in uptown, and we have the food cart, and that tool has really helped us publicize the farmers that we use whose products we use for those purposes but also to get folks interested in talking about how we can work together instead of being fearful of the competition." Sonya, as an established livestock rearer, also helped lead an organization of independent small animal producers committed to establishing a regional abattoir for slaughtering small batches of poultry and rabbits, since the closest one for local farmers was 120 miles to the east. Felicity Farms was also active in organizing farm internships for students, holding an annual farm dinner, participating in farm tours, working with the county's Agricultural Extension office, hosting tours, and giving talks on farming to local groups.

When asked about the goals of Felicity Farms, Sonya replied, "Well, we want a working farm that is profitable and positively impacts our community. And you know, that really sums up our mission. . . . It's not just about producing food in a vacuum. It's about connecting people and getting people to care about where food is produced [and] how food is produced." When Sarah went on to ask Sonya what her personal reasons were for farming, she replied that a few years previously, she and Jessie decided that they needed to contribute to changing the current food system by farming. She explained, "We can produce without all the petrochemicals and without all of the injustice to animals and injustice to people and to supply a wholesome and healthy product that is raised sustainably." She went on,

> So that's why we started farming. It is very much a value-based decision. I mean, anybody who starts farming doesn't sit down and go, "Wow, I'm gonna get rich doing this." . . . We're doing this not because it's an altruistic endeavor but because it's the right thing to do, and we do need to

make it economically sustainable. Farming is the one profession that we cannot live without. And it's a noble profession, and I want to be part of encouraging people to start farming and to get back to farming because we are losing small farms and farmers at a ridiculously fast rate. And I don't know about you, but when I'm old, I don't want to have to rely on other countries for my food. That's the place we will be [in] if we don't grow farmers, if we don't protect spaces to farm in, if we don't protect genetic diversity in our plants and animals.

Responding later to Sarah about why genetic diversity livestock was important, not just to her farm's operation but to farmers and to consumers as a whole, Sonya mentioned that livestock species diversity—including the preservation of the Tamworth heritage hog variety—was vital because of the "meat quality, genetic diversity and disease resistance, and connection to our agricultural and culinary heritage."

Sonya's strong convictions about the serious losses of farmland, farmers, and livestock species diversity were ones that are consistent with an increasing body of findings from geographers, biologists, agronomists, and planners about the threats to the continued existence of small-scale farms, their farmers, and rural biodiversity in the Charlotte region and elsewhere in rural North Carolina (Dorning et al. 2015; Israel 2019; Griffin 2019; Outz 2007; Carolina Demography 2018).

However, we found no ethnographic evidence that the destruction of the small-scale farming sector in the Charlotte metropolitan region due to suburban development was of specific concern to, much less served as a goad to action for, Slow Food Charlotte leaders—other than Sonya and another dissident SFC board member—in the years leading up to schism in 2008. For instance, when we searched our entire Charlotte Mecklenburg ethnographic database of interviews, we found only seventeen references to the loss of farmland, made by six people and two organizations—none of whom were affiliated with Slow Food Charlotte.[2] Within the metropolitan region, the development of land has occurred both within the city limits of Charlotte, where it has taken the form of gentrification of African American neighborhoods and their displacement (Smith and Graves 2005; Yonto and Thill 2020; Hanchett 2020, xiii–xxxvi), and in the surrounding areas, where it has been manifest as the construction of new large-scale suburbs and the resultant grave loss of farmland (Griffin 2019; Israel 2019). The development of land and its financing constituted and still constitute a major source of profit and capital accumulation for the finance and construction industries that dominate Charlotte's economy. This probably represents the working of a local hegemony: where the limits of what might be thought come up against the unthinkable—that is, taking the radical steps needed to preserve the small farming sector in the Charlotte region by protecting farmers' access to land, their very means of production. It would be perfectly understandable that the thinking of SFC's relatively affluent leaders in the early 2000s just "didn't go there."

We also believe that class differences had much to do with the schism within the Slow Food Charlotte chapter. On the one hand, those who remained on the SFC board depended on business, corporate, or investment incomes, which left them a certain amount of leisure time for their Slow Food activities. On the other hand, the frenzied day-in, day-out physical and mental labor of maintaining a small farm and a working restaurant was the source of the precarious incomes of Sonya and Jessie, similar to the livelihoods of other rural small farmers, food provisioners, and restaurant owners. These latter, the property they owned, and the labor they expended were at the center of Charlotte's local/regional small-scale farming economy.

In an unhappy but perhaps not unexpected coda, in September 2020, at the height of the COVID-19 pandemic, with a regional population of millions facing lockdown, Sonya and Jessie closed their restaurant. Jessie, its chef, wrote, "Due to covid restriction, the state of our economy and all this uncertainty, we are closing our doors for good. Anyone who truly knows me, knows that my craft here at [the restaurant] was my passion. I came . . . [here] to show how amazing it can be, to buy products from my neighbor farmers and that freshness and quality is utmost important" (Harvest Moon Grille 2020).

What should be clear from Sonya's reflections about "the one profession that we cannot live without" is that her and her partner's occupations as farmers and producers of food, and not merely as food consumers, had everything to do with their dissatisfaction with the "supper club" quality of the periodic, casual SFC potluck dinners used as a recruitment device to attract epicurean diners from among Charlotte's cosmopolitans. Moreover, following the broader trend of the Slow Food international movement and the Slow Food USA movement, Sonya and Jessie actively worked to form personal bonds of mutual learning and support among farmers, chefs, and consumers. Sonya saw the changes she mentioned—"losing small farms and farmers . . . [and the loss of] spaces to farm . . . [and] genetic diversity in our plants and animals"—as urgent issues at regional and global scales that would shape the future of the human food system. It was not at all clear, the women of Felicity Farms were implying, that such priorities critical to the sustainability of the small farming sector and the economic profitability of small farmers mattered fundamentally to SFC leaders.

In November 2008, Sonya and her partner at Felicity Farms and several other members of SFC left the Charlotte chapter to form their own chapter in an area north of Charlotte, the Slow Food Carolina Piedmont Region chapter. Their mission statement of February 2009 spoke of the new chapter as a group "focused on cultivating a sustainable food economy for our region of the Carolina Piedmont. . . . Slow Food is an idea, a way of living and a way of eating. It is a global, grassroots movement with thousands of members around the world that links the pleasure of food with a commitment to community and the environment" (Slow Food Carolina Piedmont 2009). Within two years, this separate chapter of Slow Food closed down, with some of its leaders and members returning to SFC, while others, including Sonya and Jessie, remained outside it.

Those board members and others who had remained in SFC since the schism felt for the most part unfairly criticized by those who had left the chapter but also, it is clear, took the criticism to heart. Were they self-indulgent epicures interested only in their own pleasure—albeit shared with others of like mind? Were the members of Slow Food Charlotte, in Petrini's words, which applied to some older Slow Food convivia in Europe, "reducing themselves to an exclusive club" (Petrini 2007, 8) with no other purpose than to cultivate their own individual palates?

Slow Food Charlotte and Friendship Gardens: "As All the Relationships Come Together, They Are Really Kind of Daisy Chaining"

After this schism, the members of the board who remained with SFC began a period of several months' self-reflection. Bill Fletcher had become particularly interested in gardening, and he recalls a conversation with SFC's founder: "And I was like, 'Fred, this is the food deserts and community gardens is the . . . [pauses]—we're not a freaking dining club issue. You cannot beat that issue.' It's awesome, I mean, I just wanted to do it, and Fred was 100 percent on board. He jumped in completely." Subsequently, SFC found three different locations for community gardens and began to fund them. Fletcher described what happened next:

> Around that time was when Amanda and the folks over at Friendship Trays [Charlotte's Meals on Wheels program] kind of reached out. They were doing—they built three little beds there, and they wanted to do a garden. They didn't know exactly what they wanted to do, and they just— Fred and I went over there, and it just clicked. I mean, this is perfect—the garden, the food goes into Friendship Trays. I mean, that rings like every bell. I mean, that's not just a food desert. That's like every food desert in town [laughs]. Because they have this wonderful Meals on Wheels program, but they necessarily use institutional food. And because of so many, nine hundred meals a day, they couldn't help it. But anyway, for what we wanted to do with the gardens and what they were doing and thinking, it just clicked completely.

As a result of this encounter, SFC started an "education and demonstration" garden associated with the Community Culinary School as a means to educate the culinary workers trained there about the virtues of locally grown, fresh food. The garden was located near the culinary school, behind the main office of Friendship Trays, which the school supplied meals to. In the words of Fred Pearson, the garden was started "not so much to grow food but enough food to train them [culinary school students] in what food looks like, where it comes from": "They tend to it, and then we can also use it to educate schools, kids, gardeners in any way that comes up, which we have [as] an ongoing series of workshops."

The professional connections that SFC leaders and members had as affluent whites to the broader food and culture industries may well explain what happened next. In the words of Fred Pearson, "[The education demonstration garden] got the attention of people in the Women's Impact Fund, and they literally contacted us, saying, 'We really like the work you're doing. We wish you would put in a letter of inquiry.' And so we did, and we had this vision but didn't have the money to do it. And so they liked it, and we got the grant. It's really as simple as that." Simple or not, SFC received a grant of $70,000 to coordinate the expansion of a new network of gardens within Charlotte that would supply local, fresh produce to Friendship Trays to include in its meals sent out to Meals on Wheels recipients; central to SFC's role was its hiring of two garden managers through the grant.

By spring 2011, SFC had helped start or had funded twelve community, school, or faith-based gardens, and three more were under preparation in the city of Charlotte. These gardens were all known as "Friendship Gardens" because each at harvest provided produce for Friendship Trays. Friendship Trays, through the Community Culinary School, supplemented by many volunteers, prepared and delivered daily cooked meals that featured freshly harvested local produce to nine hundred house-bound and other mobility-impaired residents within the Charlotte metropolitan area.

This was indeed remarkable, as most urban Meals on Wheels programs elsewhere in the US depend on institutionally streamed foods from food banks, which receive produce from supermarket and food-distribution corporations—foods that are most often processed but, even if not, are not at all fresh and are often damaged, stale, or near spoiling. The poor nutritional quality of the charitable food provided by Second Harvest food bank was a continual and well-nigh-universal source of complaint among food security activists in organizations around the Charlotte metropolitan area. Second Harvest allotted food to food pantries and community kitchens not by calories or nutrient content but by the sheer number of pounds of food that could be certified as donated from specific large-scale corporate food retailers, as is true for food banks elsewhere in the country. In contrast to this national norm, Friendship Trays' capacity to prepare meals for citywide distribution that included fresh local produce was indeed an exceptional feature. As of 2019, the network of Friendship Gardens supplying food to Friendship Trays had grown to include an urban farm operated on the grounds of a Charlotte high school and more than one hundred gardens throughout the county (Duong 2019).

Assessing the Turn by SFC toward Supporting Community Gardening

What is particularly interesting and impressive about SFC's effort with Friendship Gardens was not only its ongoing (and as of 2019 still growing) achievements in working to provide high-quality fresh foods to thousands of Charlotte's food-insecure residents but also the approach that its leaders adopted from the start in organizing its effort. Fred Pearson, speaking of the start of SFC's effort with community gardens, noted, "We really didn't want to take on the responsibility. We wanted to, but we didn't have the expertise to take on responsibility of creating urban gardens or community gardens, because currently we are a very white organization, and our experience in doing a little bit of research is [that] the worst thing in the world you can do is to be a bunch of white people to go in and tell a minority community how to operate. So we started as funding folks that were already active, and some of them worked out; others just kind of—did not." Pearson's awareness of the racial and class privileges of SFC's leaders and members and of the imperative for SFC that successful organizing by those with white and class privilege can only be done "with" and not "for" or "on behalf of" working people of color and other marginalized populations was evident—not only in this statement but in the course of the SFC strategy of consultation and support of Friendship Gardens that evolved. Bill Fletcher, when asked by Sarah Johnson whom Slow Food ought to serve, replied,

> That's a big debate, because the whole local food movement, the organic movement especially, can get into an elite [pattern]. That's great. It's awesome for yuppies. . . . But if you're a single mom, you're going to get what's affordable. And that's part of why I love the school [garden] work and the community garden work, because I think . . . it's for everybody, . . . teaching people to grow more organic food. . . . Okay, so some folks can't go to the farmers market. Why isn't every backyard a garden where you have organic veggies? I mean, that's affordable.

When Sarah interviewed SFC's leaders in 2010–2011, she asked how they squared their commitment to small farmers with their extensive work promoting urban gardening through Friendship Gardens over the previous two years. They explained that the mission for SFC was to educate Charlotte's residents to energize them to move beyond mere consumption to become "coproducers" with farmers by providing residents with hands-on experience growing their own food, that is, to become gardeners. Empathy and support for small farmers, that is, would emerge out of the numerous embodied experiences of Charlotte's gardening consumers/coproducers. Thus, SFC leaders believed that various groups needed to learn about the value of locally produced fresh food: students through their school gardens; neighborhood residents through their community gardens; members of church congregations who sought food to provide to poor, food-insecure people, through their churches' gardens; members of various "therapeutic" organizations (e.g., battered spouses' shelters, addiction recovery houses, halfway houses) through the gardens their clients could grow. SFC leaders acted on their convictions: for years, they have provided encouragement and their own labor in managing and coordinating Friendship Gardens harvests, worked to provide financing for the gardens' network, and brokered technical assistance and access to seeds for the gardeners of the city.

Fred Pearson described the arrangements that had made possible the provision of locally grown, fresh food as part of meals for Charlotte's homebound population as a virtuous circle that drew in multiple institutions in need of connection and community: "The whole building of the gardens to grow food is kind of my pet project, and I'm driving it. To me, what's most exciting about it is it does—well, it does many things, but the three—let's see if I can keep it all in my head—one is relationship; one is the original object of the grant, which is grow food for the Meals on Wheels program; and the third is to create cohesion or community—put a face on the community of urban gardeners." Thus, Fred observed, the Mecklenburg County Health Department's nutrition programs; the county's Wipe Out Waste program to promote composting; a local hardware store that donated seeds for gardens; Jail North, whose inmates germinated the seeds; SFC's role in distributing the seedlings to school gardens and community gardens; and the tie-in to the Community Culinary School ("which teaches people who are having barriers to employment"—that is, have been in prison)—all these were connected via the Friendship Trays and Friendship Gardens program. "So we find that as the relationships come together, they are really kind of daisy chaining."

Pleasure with Responsibility?

The mobilization of scores of community gardens cultivated by hundreds of people throughout the city of Charlotte and Mecklenburg County to grow and harvest produce for Friendship Trays is a stellar achievement with regard to provisioning fresh, local, and nutritious meals to the nine hundred housebound daily recipients of Friendship Trays—one of which the food activists and people of Charlotte should be proud. However, we wonder whether ultimately this will compensate for their failing to deal with the increasingly precarious economic situations of small farmers in the Charlotte region, who confront land scarcity, expensive farmland threatened with development, and the lack of attractions and economic support for young adults attracted to the demanding and hard life of small farming. All these characteristics of a threatened small farm economy are present in the Charlotte metropolitan region of the city and its surrounding necklace counties, as described in the literature cited earlier.

Late in 2011, SFC leaders did in fact revert toward more active support of the regional small farm sector, albeit in a modest way. A small fund was raised by SFC fundraisers to pay for workshops and lectures to be offered to local/regional farmers who were thought in need of education. SFC leaders asked the local/regional farmers they knew about how such an education fund might be used, and by the end of Sarah's fieldwork in Charlotte, they had already followed the farmers' advice and scheduled an expert to speak to farmers about maintaining soil fertility. Other events to educate farmers were being planned for the future.

When in 2010 Sarah interviewed members of Slow Food Charlotte, including its founder, they were somewhat uneasy about their new relationship to community gardening through the Friendship Trays project, but not because it diverted effort from assisting small farmers. Had the convivium's activism become instead too much about "saving the world"? Although the founder and other SFC members of the Friendship Trays network showed their enthusiasm for the project ("my pet project," said Fred Pearson), they also missed the pleasures of sociality that came from dining together before the crisis that led to the split in the SFC board. As a result of the split, as Bill Fletcher put it, "we became so serious. I mean, all we did was our food desert stuff and our gardens and our school gardens. And at a certain point, Jim Parise, at the end of one meeting, he was like, 'You know, I miss the fluff. We used to have fun. We used to just hang out and enjoy each other's company.' And so it was kind of a wake-up call for us, to be like, 'We need both of those things. We need to enjoy yourselves.' I mean, enjoyment of table is part of our founding, our underlying concepts." Fred Pearson, when Sarah asked him why "keeping it fun is part of Slow Food's goals," referred to the same board member's complaint about the loss of pleasure ("fluff" and "fun") during the work on Friendship Gardens and commented, "We need to keep doing that. So it's more of not having every project be necessarily something that is morally the correct thing to do—'morally' isn't the right word. But it shouldn't be trying to change the world, but it should be connecting with the community and enjoying ourselves in the process."

This reaffirmation of the pleasures that are at the core of Slow Food's mission in Charlotte, as they are elsewhere in the Slow Food movement, reminds us of the imperative to be simultaneously receptive to yet skeptical of the political possibilities that pleasure provides—even if the pleasures of the tables of Charlotte's diners reinforce the making of new relationships and connections that increase the public's appreciation and knowledge of the local food economy. Was SFC's accomplishment in promoting Friendship Gardens up to the standard that the Slow Food movement has set out? This is that the *pleasures* received from food must always come with a constant sense of *responsibility* toward those who grow and produce it and to the environment and nonhuman species that make farming possible (Petrini 2007), and that responsibility must be continually redefined, given substance, intensified, monitored, and expanded in all sectors of the food system.

In many respects, the experiences of Slow Food Charlotte that are anchored in the pleasures of the table, given the affluence concentrated among the wealthy but not shared among the general population of neoliberal Charlotte, provide the best case to determine whether such pleasures have a place in contemporary US progressive food politics. On the one hand, under the conditions of gross economic inequality and the governing philosophy of neoliberalism, the distribution of pleasures in the current US food system is outrageously skewed in the direction of people with wealth and power, who are predominantly white and

more often than not male. Under neoliberal capitalism, it is remarkably easy for those with wealth to experience the pleasures linked to the
consumption of locally grown, fresh food prepared through great effort
at the apex of its gustatory and aesthetic attractiveness, while ignoring
the hidden costs of such pleasures. After all, all they have to do is use
their market power to buy whatever, within broad limits, they wish to
consume—and disregard everyone else.

If the distribution of these pleasures must necessarily always be se-
lective and partial under the conditions of gross economic inequality,
then the majority of people denied access to these pleasures in food
have every reason to scoff at the degrees of enjoyment of it that Slow
Food seems to discriminate between. Put bluntly, shouldn't nutritious,
fresh food, locally grown, artfully and skillfully prepared in meals that
please the palates of the epicurean eaters of Charlotte and their subset of
Slow Food Charlotte members, be more widely available to the general
population? Isn't access to such food a fundamental human right? And
shouldn't the conditions and relations of production under which this
food is grown be more consciously interrogated in order to support the
livelihoods of local farmers, when these livelihoods are the precondition
for the enjoyment of a meal by an upper-middle-class gourmet?

On the other hand, the international Slow Food movement and Slow
Food Charlotte make a good case on behalf of the pleasures that are intrinsic to human sociality, community, and cooperation, which include
the pleasures of eating well. And the argument is persuasive that the
Slow Food movement has shown that the pleasures of eating food together, like other human pleasures, are crucial to forming the positive
human relationships whose presence makes collective efforts to transform the world possible. But this will happen only as long as pleasure
and responsibility are consistently aligned in movement efforts to transform food systems in more sustainable and socially just directions.

Pleasure, Responsibility, and the Urban Food Commons

We submit on balance that Fred and Bill and the other leaders of Slow
Food Charlotte engaged in reflective discussion and ethical decision-
making around the making and sustaining of a new urban commons of
fresh and nutritious foods. As Gibson-Graham (2006, 88, 95–97) argue,

the question of the commons lies within the scope of ethical decision-making within the diverse community economy. Common property is crucial to human flourishing: "The commons can be seen as a community stock that needs to be maintained and replenished so that it can continue to constitute the community by providing its direct input (subsidy) to survival. . . . Whether to apply, maintain or grow the commons is a major focus of ethical decision-making" (Gibson-Graham 2006, 97). The efforts by Slow Food Charlotte and its leaders first to bring into existence and then to undertake over many years to expand an urban commons of fresh and nutritious foods through its Friendship Gardens project represents the extraordinary value of what can come out of such ethical decision-making. Designed and financially supported by SFC's leaders for the use of Charlotte's food-insecure population via Meals on Wheels, this urban commons in produce also drew on the labor of residents of Charlotte's poorest neighborhoods.

A cynical observer might suggest that the sudden shift by SFC's leaders in 2008 to undertake this project was an opportunistic decision to maintain their reputation in the face of criticism by Sonya and the other resigning board members—criticism consistent with the international Slow Food movement's public political shift toward supporting global Southern farmers and food systems at the time (see Andrews 2008, 148–164). We remember Bill Fletcher's comment about impression management: "This is perfect—the garden, the food goes into Friendship Trays. I mean, that rings like every bell." However, the motives for a decision matter far less than the effort and reflection with which a decision is implemented. Not only were we deeply impressed by SFC's efforts over years to attract funding and provide support for the Friendship Gardens project, but we also found crucial Fred Pearson's awareness of the imperative that when one is a powerful, dominant outsider working with vulnerable racial groups, one must work *with* and not just on behalf of these groups, recalling his observation, "currently we are a very white organization, and . . . the worst thing in the world you can do is to be a bunch of white people to go in and tell a minority community how to operate." This is precisely a sign of ethical reflection in a specific domain of social responsibility.

There are no a priori guarantees or a general "pass" promised by theory that ensure that the politics of pleasure and consumption will be progressive. Each situation should be subjected to rigorous scrutiny about the degree to which ethical debate and decision-making take place. In the situation analyzed here, despite SFC leaders' ethical commitment to meeting the food security needs of Charlotte's poor and vulnerable population of color, the same cannot be said of their failure to confront, or even to speak of, the desperate conditions of small farmers undergoing displacement and marginalization within the metropolitan region that represents Charlotte's foodshed.

We conclude by returning to the theme of food's pleasures. Under the contemporary conditions of neoliberal capitalism, which corrode the relationships leading people to work together in solidarity and instead set each individual at war against every other, fragile human relationships need attending to and nurturing. The pleasures surrounding food are fundamental to human flourishing. Just because some people are denied the right of access to such pleasures does not mean that their intrinsic value for flourishing should be denied. Instead, access to the pleasures surrounding food must be extended to everyone.

* * * *

In this chapter, we examined the relationship between the role of pleasures taken in food by members of the Slow Food Charlotte convivium and their ethical discussions and decision-making around the urban food commons whose produce is provided to the city's food-insecure population.

First, we recounted the establishment and early history of SFC in the early 2000s as an appropriation by upper-middle-class epicures in Charlotte of the discourse of "slow food," taken from the political consumerist Slow Food movement and then expanding to North America. We posed the question, How could Slow Food Charlotte's commitment to the pleasures of dining for the few be reconciled with the broader landscape of racial and class inequalities for the many that characterized Charlotte's finance-driven economy, if it could?

Second, we described how SFC's leaders sought to offset its identification with members' conspicuous pleasures created by dining on expensive, nonindustrial "slow food" by undertaking projects of educating consumers about the value of supporting local farmers growing such food.

Third, we recounted how a schism in 2008–2009 within the SFC board of directors centered on the accusation that SFC was no more than a "supper club" that failed to support local farmers when it came to the economic precarity they experienced. This schism led SFC's remaining leaders to reevaluate their public celebration of food's pleasures and to turn toward working to generate social and financial support for a network of community gardens across Charlotte to provision the city-wide Meals on Wheels program with fresh local produce.

Fourth, our analysis of the context of this schism within the racial and class politics of food provisioning within the city led us to celebrate SFC's turn toward supporting the food needs of Charlotte's poor, housebound population. SFC's leaders' explicit acknowledgment of their white, male, and upper-middle-class privilege and their willingness to cooperate with, rather than seek to lead or represent poor and nonwhite Charlotte residents as Friendship Gardens got under way stood out as an exemplar of ethical discussion and decision-making around the building of an urban food commons that would provision a large number of food-insecure Charlotte residents with locally grown produce.

Fifth, we also noted SFC's leaders' neglect of the precarious economic situations of the local and regional farmers they had previously supported—connecting this to a class- and race-based failure by SFC leaders to challenge the gentrification of Charlotte's agricultural hinterland by the finance and construction industries, which are hegemonic in the city's commercial life as a national financial center.

Finally, SFC's successful project to support the launch and expansion of community gardens within the Friendship Gardens network to provide fresh produce to Charlotte's large-scale Meals on Wheels program, Friendship Trays, achieved the goal of both community food security and sustainable food activisms. Engaging in labor to secure financial and material support for the Friendship Gardens project, SFC's leaders did so with a sensitivity to their own white, upper-middle-class privilege, which allowed them to work with and follow the lead of neighborhood African Americans and other minority gardening activists in Charlotte, thus helping empower the activists with more control over their own neighborhoods' food provisioning, in line with community food security activist objectives. In so doing, SFC leaders also sought to set into place a lasting process of provisioning Friendship Trays with freshly grown local produce, a "scaling up" process of creating an urban food commons consistent with sustainable local food activism and Gibson-Graham's (2006) vision of the importance of a commons in the making of a community economy. However, their failure to recognize the disappearance of small-scale farms within Charlotte's food hinterland or to strategize about how to remediate it points to the limits of their ethical commitments to sustainable "slow food." The record of Slow Food Charlotte's food activism is thus a mixed one.

7

F.A.R.M. Café and Classed Lives at One's Table

Well I think there's a lot of hunger in this county, like in every county, that doesn't necessarily go noticed. I think there's people that they're not on the books of maybe the Hunger Coalition [local food charity], or they're not getting direct assistance, but they're there. And they either have too much pride or feel like they're going to be okay. I know through the school systems that there are a lot of school kids who are going hungry these days. So I see that need. The need for community I see all the time.

—Renée Baughman, cofounder of F.A.R.M. Café

Making People Feel "Significant" in Mountainous Western North Carolina

The epigraph quotes Renée Baughman, cofounder, executive director, and chef at the F.A.R.M. Café, when our ethnographer Jennifer Walker asked her why there was a specific need for F.A.R.M. Café in this mountainous western North Carolina region. Renée was a middle-aged woman who, after having taught history, worked for years as a chef in restaurants and resorts during the late 1990s and early 2000s as the region developed its reputation for scenic tourism and a large influx of affluent retirees. As someone who grew up in the High Country region with years of culinary experience and a lively intellectual curiosity, she could speak knowledgeably about food matters, and what we learned from her in 2010–2011 has proven illuminating to our research. By 2023, we have had even more motivation to listen to her, given that F.A.R.M. Café, despite many obstacles since its opening in early 2012, has become a prominent regional restaurant known not only for its attractive and nutritious dishes and meals largely prepared from local ingredients but also for its distinctive "pay what you can" pricing, in which its patrons decide how much they will pay for the meals they eat.

One question came up from Renée's reply to Jen's question about the purpose of F.A.R.M. Café that is relevant to the theme of this chapter: Why are many poor people who are faced by food insecurity and even real hunger so reluctant to turn to the charitable food sector for food, and what might an adequate response to this reluctance look like? In this chapter, we set out the remarkable experiment of F.A.R.M. Café, located in Boone, a large town in mountainous western North Carolina. F.A.R.M. Café represents an important achievement. It has simultaneously provided strong support for the livelihoods of local small-scale farmers while crossing over into community food security activism to confront a key limitation of the charitable food sector: how to overcome the class divisions and humiliations that characterize the contemporary charitable food sector and drive poor people away from accessing it. As noted earlier, many people whose low incomes should qualify them for federal SNAP or WIC food vouchers or TEFAP food assistance are often ineligible for such benefits, and many of those who do receive these benefits do not receive sufficient amounts to tide them over until the end of the month—so Renée's observation that many who are hungry in the High Country are "not getting direct assistance" is not unexpected.

Yet, despite the desperation that comes with being food insecure or outrightly hungry, many poor people in the US are reluctant to visit the food pantries and community kitchens where charitable food is on offer. "They either have too much pride or feel like they're going to be okay," as Renée put it. While acknowledging that federal food aid is still far from sufficient, it is important for us to come to terms with the attempts that food activists like Renée Baughman have made to overcome this problem. This is not by allowing people to go hungry or by imposing on them petty humiliations in return for food they need but by "de-charitizing" the emergency food that is available and thereby maximizing access to such food for those who need it. In other words, how might alternatives emerge to the current system of "food bank nation" (Riches 2018) that neoliberalism has imposed on US society? Here, F.A.R.M. Café (whose "F.A.R.M" is an acronym for "feed all regardless of means") has much to teach us.

The answer to this question resonates with many other local food systems throughout the US, because F.A.R.M. Café represents a local exemplar of a networked movement, One World Everybody Eats (OWEE), which increasingly is being recognized for its accomplishments at a national scale, as when its founder, Denise Cerreta, received the 2017 James Beard Foundation's prize as Humanitarian of the Year (One World Everybody Eats 2022). More than sixty OWEE cafés can now be found throughout the US (One World Everybody Eats 2022), and their collective work has come very close to repudiating a key premise of neoliberalism and possessive individualism, that "everyone has to look out for number one, because everyone is on their own."

Challenging the Language of "Givers" and "Takers" through a Profusion of Pronouns

We start our inquiry by being curious about Renée's language. As Renée reflected to our ethnographer Jen Walker on her experiences with her friends in founding F.A.R.M. Café, it was no simple matter to offer poor people food to eat without patronizing or offending them, but that is precisely what most food pantries and soup kitchens did, she told us. In her distinctive voice, she claimed,

> This is going to sound really sappy, but I believe it. It's not just a hunger of their stomach; it's a hunger of their soul. It's a hunger of all of our souls. You can't just keep telling people, "You're worthless." You can't. And if you do it in every model and you do it with food . . . So you got to find a way to make that available so people feel significant. Because if you feel significant, then you don't feel the need to be that person that's taking [from others]. That's what we all want. We get frustrated when people are just takers, then we get frustrated people on the lower levels who seem to be just takers. Well what else are they going to do? Because you certainly aren't giving it.

Let us follow what Renée says more closely. At first, her speaking of "takers"—as when "we get frustrated when people are just takers"—appears, when first heard, to resonate with the neoliberal (and conservative) shibboleth that poor people and people of color believe that they are entitled to government benefits and an easy life at the expense of righteous taxpayers and wealthy citizens, who are successful in "the economy" through their "hard work." Many still remember the infamous moralizing in 2012 by Senator Mitt Romney, private equity tycoon and Republican presidential candidate, when his pitch to wealthy funders behind closed doors was secretly recorded: "47% who are with him [Barack Obama], who are dependent upon government, who believe that they are victims, who believe the government has a responsibility to care for them, who believe that they are entitled to healthcare, to food, to housing, to you-name-it. That that's an entitlement. And [that] the government should give it to them" (Cohen 2012).

However, in light of what we have seen regarding the reception and appropriation of neoliberal language by food activists, we believe that far from approving of the opprobrium toward "takers," Renée is instead challenging it. When she speaks, she first appropriates and then deflects and inverts neoliberal discourse, when she says, "we get frustrated when people are just takers," but immediately repudiates it by qualifying "people . . . who *seem* to be takers." In this exchange, Renée immediately made it clear to Jen that "takers" have no choice—that this is the role the class system casts them into: "Then we get frustrated people on the lower levels who seem to be just takers. What else are they going to do?"

As she speaks, moreover, she switches between the pronouns "you," "they," and "we" and whom each word refers to. What is to be made of this profusion of pronouns, and why does it matter? We want to suggest that she uses personal pronouns to signal that the shift in position between those who give food, those who take it, and those who comment on giving and taking it reveals the moral equivalence of each person to every other. She does this by employing the dialogic strategy of "repeated speech" (Bakhtin 1981), in which she allows the meaning of the speech of someone opposed to the speaker to be incorporated into one's own, thus subverting that meaning: "You can't just keep telling people, 'You're worthless.' You can't."

Renée's metaphor for what hunger is, accompanied with her shift from "they" to "we," also points to this moral equivalence when she says, "It's not just a hunger of their stomach; it's a hunger of their soul. It's a hunger of all of our souls." That is, she insists to the listener that it is a moral impossibility to rank the humanity of those who give food over the humanity of those who receive it—or over the humanity of anyone. When Renée told Jen, "You can't just keep telling people you're worthless," she was saying that humans are interdependent as mutual givers and takers, so a moral ranking makes no sense: those who give are not essentially superior to those who take. The denial of this insight, when a giver of food claims superiority over a taker, leads to humiliation of the taker and a loss of their human dignity. The more carefully we listened to what Renée told us, the clearer became this fundamental insight that informed how Renée sought to structure F.A.R.M. Café as an alternative enterprise to a privately owned for-profit business.

F.A.R.M. Café was founded on the premise that food-insecure people would rather go hungry or pretend that they are not than become "takers" humiliated by a society that does not recognize that their need for sufficient, healthy, and culturally appropriate food is a human right. But if people are going to be adequately fed, Renée insists, then they must instead be treated as people with dignity. The key realization of Renée and the other founders of F.A.R.M. Café was that one of the great tragedies of the extraordinary economic inequalities that divide the United States are the deep humiliations that a neoliberalized food charity system imposes on its recipients, humiliations that destroy their dignity—and that, as a negation, no food system should be premised on this violence.

The Making of F.A.R.M. Café: 2009–2012

Jennifer began our ethnographic encounter with Renée and her friends involved in the founding of F.A.R.M. Café in mid-2010, as their vision of what they hoped the café would become became increasingly clear to them. At the time of Jen's interviews with them in early 2010, they were actively forming links to other food-related organizations, raising funds for the café, publicizing the concept of F.A.R.M. Café to fundraisers and the public, and seeking out retail space for it to operate out of. By early 2012, F.A.R.M. Café had opened in a downtown street of Boone, a university town in mountainous western North Carolina, not long after our ethnographic study wound down.

Conceiving F.A.R.M. Café: A Discussion among Friends in the Same Church

Renée recounted to Jen how the group's discussions progressed from mid-2009 onward. In August 2009, a group of friends, all members of the congregation of High Country United Church of Christ, a liberal, LGBT-friendly nonevangelical, Christian fellowship, got together over a meal to talk about hunger in the High Country region of western North Carolina. One of them was Renée Baughman. In addition to Renée, there was Dave Carpenter, who worked as a chef in a nearby town; Sally Willard, then working as a waitress in a local restaurant and a recent local university graduate interested in nonprofit work; and several others, including a clergywoman and a woman employed by a regional poverty-alleviation nonprofit organization.

Jen observed, "This is not a very wealthy crowd. They have some connections to the wealthy liberal community [in the county] but only because they are contractors who worked on their houses, . . . worked in one of their stores, . . . or have served them at a restaurant where they work. This group of people is not coming from the perspective of an entitled or privileged group but more from the perspective of those who cannot afford to always eat great local food." Reflecting the experience of many people in the (media-denominated) US "middle class," at this point most of the people in the group had themselves experienced some degree of food insecurity at some point in their lives. Moreover, Renée and the clergywoman had worked as volunteers in soup kitchens or homeless shelters for many years and thus were acquainted with the social roles of both people "taking" emergency food and those "giving" it.

During this early phase of the group's discussions, they quickly came to agree that they sought to alleviate the problems associated with food insecurity among the poor population of the county, particularly in its outlying, more remote areas. They thus had many opportunities to reflect on the meaning of poverty in the High Country region and more broadly in the United States. These reflections are important because they get to the core of a debate about what class in the US means as an interpersonal relationship between people designated as "poor," "hungry," or "food insecure" and others who are not but with whom the former have to interact if they are to secure food.

Renée, for instance, thought of the F.A.R.M. Café in contrast to her prior experience as a restaurant cook in a high-end restaurant in Boone. She said, "As a chef, I certainly learned a lot about fresh quality food—how it's available, where it's available. And it's like we've said a lot of times in meetings: it's always available if you got the money to have it. You know, it's like probably working at Green Grove [restaurant] because it was the highest-end place I've ever worked at. It was a real eye-opener that, 'Wow, look at all this great produce.' . . . No one else is going to buy those because they can't afford it." Based on their sense of Christian compassion, Renée and her fellow congregants put themselves in the position of someone who had experienced hunger yet had to confront their dependence on the generosity of others who were not as poor or as food insecure as they were. While not singularly an American trait, having to ask others for food that one cannot provide for oneself (or for one's children) in most settings of food charity in the US signals one's dependency on others and subjects oneself to a barrage of petty, class-based humiliations exacted by the unreflective kindness of others, under the conditions of charitable help.[1] As either memory or intuition of the small debasements inflicted on the poor that impel so many to avoid food charity, the imaginary of Renée and her cofounders of F.A.R.M. Café provides the material for a figured world of what poverty is like and how it might be addressed, if not redressed completely, within a setting in which the shared humanity between givers and takers is on display.

Renée intuited what the life of a poor and food-insecure person was like in the contemporary US:

> It did make you start realizing that it's just another way of—you know, on top of poverty, not only are you . . . in a position of poverty, which has all the stuff that goes with it—that you're not good enough, that you're not smart enough, that you're a failure in the society. And now, . . . even in the help agencies, it's pretty much like, because of that worst stuff, because of that, you get the least of the least. And so then, when you have health problems, well, you can't have health care either. Because you have to maintain this. You're the bottom feeder, and everything we do is going to reinforce that.

Dave Carpenter, a young chef who worked in a nearby town and was one of Renée's cofounders, when Jennifer asked him about whom he expected F.A.R.M. Café to serve, replied that one of the ways a reluctance by rural poor people to be subjected to such humiliations occurred was by their avoiding urban-based charities in a mountainous region where rural poverty for many city dwellers was largely invisible: "I think there's plenty of need as far as the food-insecure people and families, especially in the outer reaches of the county and the surrounding counties. I think it's one of those things that could fall through the cracks. . . . That's one of the challenges I think we're gonna have; it's getting those people in the restaurant . . . having it be approachable enough to where they feel comfortable coming in." Later in his interview with Jennifer, he elaborated on his hopes for whom the Café would serve: "I would say individuals and families that are struggling to find healthy food options, healthy meal options at prices they can afford. I think ultimately, that's our definable base of who we want to be helping."

In the same interview, Dave reflected on establishing a café not only in which people of diverse classes and life experiences might eat together but also, reflecting a broader theme among the F.A.R.M. Café founders, in which better-off patrons might learn something important from those less well-off as they came to talk with one another: "You have a professional or a student or a professor or somebody who comes in and pays for their meal and sits down next to a single mother with three kids, you know, who's using the restaurant as it's intended, to give her family healthy choices at low cost. They're directly seeing where their money's going, and I think that's cool." This view of two people of "different means" sitting across from each other in the café was widely held within the group. As Sally, another member of the group, put it, "We also—community, which is where my passion lies—the building of community is the whole model for 'pay what you can' is. We want to attract people of substantial means that can give a little more to eat alongside of people who, you know, are struggling—can't pay for a meal every day. And so sitting down and having conversation is basically, you know, between . . . the economic classes, . . . building community. Because, we've all got stories, and we've all got, you know?"

Learning of "One World Everybody Eats"

In the course of Renée and the other founders' research in planning for a "community café" in summer 2009, Renée heard an NPR radio program about the "One World Everybody Eats" (OWEE) community café in Salt Lake City, Utah, that the chef and activist Denise Cerreta had founded several years earlier and that was already serving as a model for many community cafés being set up elsewhere in the US. In the Salt Lake One World Café, patrons only paid as much for a meal as they wished to donate to cover the cost of food (One World Everybody Eats 2022). This enabled a wide range of people with different incomes to patronize the café on a more or less socially equal basis. Allowing for payment on a donation basis, whose amount patrons could independently determine and not have to reveal to others, seemed to overcome the problem of the patron being "that person that's taking [from others]"—at least in part.

But only in part—because a flexible pricing arrangement by itself is insufficient for many food recipients to overcome the subjective sense of shame that they still experience for not "paying one's way" and therefore being dependent on others (Poppendieck 1998). This affect of shame held by many recipients arises from the dominant neoliberal ideology in the US that identifies the morals of "the market" with a strong sense of individual responsibility. Thus, the One World Café model also allowed patrons to volunteer to work in the café in lieu of having to pay for their meals. The use of volunteers in this way was not a trivial feature of the OWEE model but instead a radical innovation hidden within its broader array of (what appear to be anodyne) variations on the charitable food model. It has become the key design feature from the OWEE network movement, one that F.A.R.M. Café has adopted and taken in a distinctive direction, as we discuss in the following sections.

Making Holistic Connections

In addition to adopting the OWEE model of "pay what you can," the vision of Renée and other founders was also grounded in their sense of the holistic connections that they thought should be made within the local food system. It was particularly important for F.A.R.M. Café to source the food it prepared from local farms, thus contributing to the sustainability of local small-scale agriculture.[2] Thus, Renée spoke of using the meals served by F.A.R.M. Café as a way of "trying to get people interested in local food, local means of not just . . . purchasing food but local appreciation of what we have in the area and the farmers market participating on that level and getting to know people that are growing your food and how that works and what happens to it."

Renée was referring to an issue that had been widely discussed among local sustainable-farming activists about the need for consumers to be willing to pay sufficiently high prices to local farmers selling through farmers markets and CSAs to allow the latter to make a living from the difficult but independent farming they did. By this point in these discussions, it had become a matter of regional consensus among sustainable food activists that local farmers, most of whom themselves lived at the precarious edge of livelihood with low incomes, were the last people who should be called on to sacrifice their incomes to make the food that other poor people (or charitable food organizations) bought be more affordable by lowering their own prices.

Dave, who appears to have been more experienced than Renée with restaurants' sourcing of food in his job as executive chef, foresaw one of the structural difficulties arising from local farmers' insistence on the high prices they needed to subsist from their farming, an issue that has continued to be a challenge to F.A.R.M. Café since its 2012 founding. He observed that local farmers were difficult for the chef of a commercial restaurant to work with: "As far as the food goes and the sourcing, you know, we'd obviously like to keep it as local as possible, as much as possible. There are challenges to that, and my experience here is it just takes a little more work to make that work, to make that happen and work out well. I know that the farmers here, like, need a fair price for their product, and on the other end, we need to pay a fair price for it." Later in the interview with Jen, Dave became more specific: "Someone told me once— somebody from the farmers market told me their job is to make sure the farmers make as much money as possible and get top dollar for their product. And they can get—you know, they can [sell] arugula at nine dollars a pound at the farmers market; I can't buy arugula at nine dollars a pound at a restaurant. So there needs to be some more ground—like, bridging that gap." In contrast to directly dealing with local farmers, Dave pointed out, the wholesale agents he worked with as a restaurant chef in his current job were relatively easy to work with, because not only could they set a lower price than local farmers but they could also quickly provide a wider set of sourcing options (e.g., different kinds of meats or produce at different prices) due to their high volume of aggregated foodstuffs compared to local farmers.[3] Nonetheless—looking forward to the years of operation after F.A.R.M. Café opened—the imperative of paying a fair price to local farmers has been a challenge that F.A.R.M. Café has successfully met.

Ironically, compared to other alternative food organizations, F.A.R.M. Café has tended to downplay its connections to expensively acquired local foods, even as it has continued to offer them to its patrons. This has everything to do with its broader mission of offering food to "everybody regardless of means." As Jen observed, "F.A.R.M. Café seem to prefer to downplay the local and organic nature of the food they wish to serve, out of fear that would 'brand' the restaurant as a space for eaters of means [and] therefore [would] not be as welcoming to low-income eaters."

Reaching Outward to the OWEE Network

Having heard of and then done research from afar on the One World Everybody Eats model, Renée and her friends were particularly excited to attend the first summit conference of One World cafés, which was held in New Orleans in January 2010. Having previously traveled to New Orleans as members of their churches' mission group to help recovery efforts after Hurricane Katrina in 2005, they became doubly enthusiastic by what they discovered once they attended the New Orleans Summit of One World Everybody Eats, when they met and talked with managers and volunteers from other community cafés established elsewhere in the US. As Renée put it, "Either they were like us, they wanted to start a café, they already had a café, or they had something akin to it that they were doing. That was tremendously helpful in helping us move forward and say, 'Okay, this is what we can do.'"

Forming Local Links: Demonstration Dinners, Fundraisers, and "Getting the Word Out"

Under Renée's leadership, this group began to take advantage of ties they already had with the staff and supporters of other food-related organizations within the region. This was consistent with a definition of "building community" that was based on their making connections and creating greater understanding among divergent economic groups in the area. Members of the group also drew on these different organizations and the people they knew working in them because it was clear that F.A.R.M. Café was to be a unique institution with very specific differences from the conventional model of the privately owned, for-profit small restaurant. This had to do with a variety of needs that F.A.R.M. Café would have compared to the usual capitalist model—with regard to surplus (aka "capital"), labor, connections to its source of supply, and connections to its patrons.

With respect to surplus, F.A.R.M. Café depended on a fundamental ambiguity. On the one hand, F.A.R.M. Café was to take the form of a retail, market-oriented restaurant devoted to serving local and fresh food to whoever came in the door, wanted to eat a meal, and could afford to pay for it at the going price, that is, a price that was competitive with other restaurants nearby. On the other hand, it was a nonprofit organization that also allowed any patron to pay what they could afford, even if less than the recommended price, while asking those who could pay more than they usually might to pay more than that price. Thus, unlike the conventional privately owned restaurant, F.A.R.M. Café could not depend on a predictable level of revenues, based on fixed prices, to cover expenses. In this respect, Renée and the other founders of the café did not expect the café to operate along the conventional budget lines of most restaurants, which in turn implied that if the café was going to open, it would have to raise supplementary funds to cover any shortfalls from day-to-day revenues.

This meant that the founders had to tap into the financial contributions of the "wealthy liberal community" of the county, as well as regional or statewide charitable foundations, for grant support. This led them to organize fundraising events that also served to publicize the café's plans and its goals. These took both the conventional form of a party with a few wealthy potential donors offered hors d'oeuvres, seasoned with an appeal for seed money, and more innovative demonstration dinners modeled on the principles of One World Everybody Eats. At one such dinner, in June 2009 and held at Welcome House, a regional homeless shelter with a kitchen serving food to its residents, its director distinguished Renée and her friends from "community members":

> We wanted to find a way to invite the community to the new shelter anyway. And people have such a misconception of who the homeless are. And now that we have this space—so inviting them to dine. And F.A.R.M. Café is trying to develop their name [in] the community. So it's such a good fit. . . . We had three seatings [meal shifts], and we had just lots of people from the community, and it was great. So F.A.R.M. Café oversaw—they developed the menu, they have a couple of shifts. And we did it all through donations. And they did most of the prep and cooking, but we had a bunch of volunteers come in. . . . And it was just really great food. . . . And people would sit down and were sitting with the residents. It was just really cool.

As to F.A.R.M. Café's labor supply, one of the truly defining aspects of its operation was that it was to have a small full-time paid staff, but the majority of the physical labor of food preparation and presentation was to be undertaken by volunteers. On the one hand, the concept of the "volunteer" allowed those who were food insecure to trade their labor for a meal (ideally, a meal for an hour of their work time), that is, to receive their pay "in-kind." On the other, volunteers from the local population were to provide most of the labor that the café required in its daily operation—and the café's cofounders sought to have this provided in relatively large numbers by students from the nearby regional university. Although it is clear that to use volunteer labor allows for a steep reduction in operating costs compared to the paid wage labor of the privately owned restaurant, it also had a clear educational objective. Renée described her vision for the presence of volunteers at the café before its opening: "I would love for this to become . . . a model of educational resource, . . . [of] how they can participate in their communities by volunteering and by doing things."

A connection with students as a potential volunteer labor force proved straightforward since two of the founding group of the F.A.R.M. Café were faculty members within the Sustainable Development department of the regional university. In addition to assisting Renée in finding and writing grant applications, they facilitated her speaking engagements in classes and lectures and also spoke to students and student groups about the need for the café. When the café opened in early 2012, as has proven to be the case subsequently, large numbers of students volunteered to work at it. Moreover, the founders' personal ties to the university's Sustainable Development department reinforced its initial direction to maintain its focus on sustainable agriculture, that is, defined as "loosely organic, bio-diverse, small-scale, and local . . . and [using] appropriate technology in ways that are helpful for low-resource or otherwise marginalized populations."

As to connections to the café's sources of food provision, local farmers, these appear to have evolved over time, but during our period of ethnography, they were only in the initial stages, anticipating the opening of the café in early 2012. One of the cofounders of the café was a board member for a local farm organized as a nonprofit devoted to sustainable agriculture, and he used his connections to set up an agreement for future produce from the farm and an affiliated regional multifarm CSA to be purchased by the café once it began operation.

As to connections to those local food organizations that the F.A.R.M. Café sought to serve, we have just described its arrangements with Welcome House, a regional homeless shelter. Subsequently, from 2012 to 2022, we found that F.A.R.M. Café had developed further connections to local organizations engaged in charitable food provision.[4]

Finding Space for the Café

By the fall of 2011, the cofounders of the café were searching for space to locate the café and preparing for its opening. At first, they considered locating the café on the premises of their own or another church, but they decided against doing so because they did not want the café to appear to be affiliated with a specific church or its charities, which might be off-putting to some food-insecure people seeking food. They also felt that the unique food-insecure status of some patrons and their sometimes-visible "homeless" appearance might lead to blowback from less tolerant church leaders or congregations, who in turn might force the café to close. Renée said,

> I don't want to hide from the church affiliation of the origins that got it
> started, but I also don't want that to be a hindrance to anyone. This is

not a church-sponsored space by any means. It's a community-sponsored place. . . . I wouldn't want anybody to feel an obstacle to coming in the door because they thought it was affiliated with a church in a way that would keep them away. And I also wouldn't want any church to feel . . . they have a right to make their particular demands and feelings expressed, and I wouldn't want it to be a situation where we were . . . under their thumb, so to speak—like, if they got upset because, yeah, because some moral issue that maybe went against the standards of what they believe in.

Although not quite articulated as such, it also seemed important to the founders to publicize the café as a retail enterprise whose meals and setting would be seen by patrons, whether food secure or not, as comparable in quality to privately owned for-profit restaurants and preferably to locate it in a trafficked downtown street in Boone. In any event, the F.A.R.M. Café began leasing the space in a one-hundred-year-old building downtown, where previously a soda fountain had "served the community as a social gathering place, and F.A.R.M. Café will continue in that tradition by providing a space for all" (F.A.R.M. Café 2018a).

F.A.R.M. Café Ten Years On

It has been a bit more than a decade since Jen Walker met and interviewed Renée and the other founders of F.A.R.M. Café, and we have continued to follow its progress, unfortunately from afar, while over these years it has provided exceptional meals at low cost to thousands of food-insecure people. Over the same years, it has maintained the connections that make a "community," thereby demonstrating other outstanding characteristics relevant to the themes of this book.

Since the café's founding in early 2012, Renée, as its executive director and head chef, has increased greatly the relative proportion of locally grown produce and meats that F.A.R.M. Café uses for its meals, while leveraging its market demand to support the prices that local farmers ask for to maintain themselves. As Jen's 2011 interview with Dave Carpenter suggests, the imperative for any restaurant to continuously acquire the specific produce and other foodstuffs it needs from local farmers poses serious logistical challenges. The commercial food wholesale agent still remains an alternative—unless the restaurant's manager makes a commitment to source locally that overrides the convenience and lower prices that the agent provides. Here, for several years now, although the café receives some donated produce, it has relied on purchasing the vast majority from local sources, "including produce from local farmers, items from local distributors and from regional service purveyors" (F.A.R.M. Café 2018b). Renée states as a moving goal, "Each year, as our budget increases and our operation becomes more financially sustainable, we try to incrementally increase our local food purchases. I think it's really important that people have an economy based on a network of local suppliers that we know we can trust" (Kornegay 2022).

Renée goes on to observe that F.A.R.M. Café not only has consolidated its commercial connections to local farmers but has also done so by treating the farmers who source its food with dignity, in their case by paying them a market price for their produce rather than asking them to donate food or to undercut their own market prices. She states,

> One of the misperceptions about F.A.R.M Cafe is that what we are serving is all donated. . . . It's not. We purchase 95% of what we serve in the cafe. We want to make sure that we get dollars into our local economy. We want to participate in our community, and we want the farmers and growers to have access to those dollars. They're working hard to produce what they're producing, and we don't want to ask them to donate. We want them to also have the dignity of knowing that we are going to purchase from them, just like we are going to purchase from anyone else, and not ask them to lower their standards or price for us. (Kornegay 2022)

As another unique feature for a retail restaurant enterprise, F.A.R.M. Café has drawn on hundreds of volunteers for its labor force, the majority of whom have been students at a nearby university, but many others have been food-insecure people unable or unwilling to pay for their food who have worked as volunteers in return for the meals they receive at the café. Volunteers work as servers, dishwashers, morning set-up and afternoon breakdown crews, prep cooks, and cleaners, while a select few are recruited to serve as greeters to greet patrons, share the daily menu with them, "and explain the mission, concept, and workings of the café" to them (F.A.R.M. Café 2022c).

F.A.R.M. Café's reliance on volunteers may be a way to avoid having to employ full-time staff and pay them benefits. But its use of large numbers of volunteers is also connected to its original pedagogical aims. It encourages students at the nearby university to "get involved" as F.A.R.M. Café interns studying nonprofit management, community development, food production and safety, nutrition, and sustainability (F.A.R.M. Café 2022a). Thus, its extraordinary use of volunteers provides a low-cost alternative form of labor that is critical to a unique institution that successfully meets a variety of needs—not only of the food insecure but also of local farmers and university students and, more broadly, of "the community."

Moreover, F.A.R.M. Café since its establishment has moved into new roles within the local food system of the county and beyond. In the late 2010s, Renée recruited Marta Andrews to work full-time for F.A.R.M. Café to serve as coordinator for Full Circle, the nonprofit's new "food recovery" program. Marta noted why Full Circle exists: "There's so much food out there that is not marketable because it is not aesthetically perfect. There are tons of apples with a scratch on them, or tomatoes that cracked on the buying or have a blemish, that are perfectly good to eat, but that for whatever reason don't meet beauty standards for the market shelf" (Kornegay 2022). This allowed F.A.R.M. Café staff and volunteers to transform this "ugly" produce—food that would otherwise be wasted and end up in landfills—in its commercial kitchen during off hours into ready-to-eat meals. These meals then would be provided to poor volunteers in return for their labor or distributed to local food pantries and soup kitchens (Welcome House and several others) to distribute to food-insecure people or sold commercially through a local nonprofit food aggregator to the public (Kornegay 2022). As of early 2022, Full Circle has diverted fifty thousand pounds of food waste from landfill and donated one hundred thousand meals to provide to food-insecure people in the county and nearby (Kornegay 2022).

Class, the Importance of Dignity, and a New Definition of Community

F.A.R.M. Café in western North Carolina represents a key departure from the endemic but often hidden and almost never publicly discussed class conflict that characterizes economic inequality in the United States. This class conflict is manifested in the charitable food sector that we, following others, refer to as "food bank nation" (Riches 2018). This is a strong claim but a sustainable one. At its best, the charitable food sector consists of hardworking, thoughtful, and compassionate people who are staff and volunteers in food banks, food pantries, community kitchens, and other food outlets working for modest and often little or no compensation. Some of them are charity-based food security activists whose work we have featured in this book. The charitable food sector is also funded by donations of money and of food (in "food drives") made by innumerable generous individuals, corporations, and philanthropies. Together this large, undercompensated labor force combines with the surplus ("capital") provided by donors in food banks, food pantries, and community kitchens in systematic and daily efforts to "pick up the slack" to provide food-insecure people with the food they need to get through the month, given the systemic failure of the US government to do so.

However, there is another side to the charitable food sector which has nothing to do with the personal character of people working in it. As we have argued earlier in the book, food bank nation is a neoliberal project, one that positions poor working-class people, and especially people of color, as "losers" in the division between a large number of working people in the United States and a small elite of "winners," those whom one politician has called "the one percent." Bluntly put, at its most pauperized extreme, the working class consists of poor people who are also food-insecure or even hungry people (and have precarious lives in other ways—their housing, health care, etc.), and we have learned that they constitute between one-seventh and one-sixth of the US population, depending on the current employment rate. The dominant political doctrine of neoliberalism constructs these millions of people as losers who are unable to be independent, self-sustaining adults, and this is what makes them failures—people who are "dependent" or "parasitic"—on a small number of hardworking, worthy, and wealth-generating "entrepreneurial" "job creators," or "winners." Neoliberalism thus plays out on the ground in a particularly cruel way in the case of the charitable food system, in what can only be called a kind of class violence perpetrated against poor people. Its message, constructed and delivered by political elites, free-market economists in certain think tanks, conservative news and social media, some educators, and church officials to this large number of poor people is a crude, if implicit, one. We can decode it from our analysis of neoliberalism and the food charity sector (Holland et al. 2007; Poppendieck 1998, 2014; Riches 2018; Long et al. 2020; Lohnes 2021).

This message goes something like this:

> You are individual and collective losers because you cannot function as independent adults who provide for yourselves by finding and earning money in paid work. This is a sign of your fundamental failure as human beings. Nonetheless, we are kind, and will do what we can within our discretion to give you the food you need to survive. In the process, we will make sure that you are treated as our dependents subject to our whims. We do so by offering you food as charity, but only if you give us in return the psychic payment of your humiliation by admitting our superior status and your undeserving position, shown in the way you behave when you come visiting us to beg for food. This is the bargain we offer you, which you are free to reject.

Empirically, it can be amply substantiated that large numbers of poor people reject this bargain. It is also precisely this message that F.A.R.M. Café repudiates explicitly. F.A.R.M. Café offers an alternative vision of the poor, food-insecure, or hungry person, one in which any person in need of food also is in need of dignity and has the right to both. Significantly, those poor people whom F.A.R.M. Café sees in this way include small local farmers themselves. Moreover, a specific figured world of "community" and the place of poor, food-insecure people in it reinforces and underlies this positive vision.

We turn to the two questions we posed at the beginning of this chapter: Why do food-insecure people not seek out charitable food when it is offered to them, and what meanings of community are at stake in how this question is answered?

Confronting the Reality of Class Violence by De-charitizing Food Charity

F.A.R.M. Café, like other community cafés in the One World Everybody Eats networked movement, stands out because of the ingenious way it confronts the great American denial of class-based injustice and the enormous burden of humiliations imposed on the nation's poor people when they seek the food that they and their children need to survive.

In a magisterial, widely cited analysis of the US charitable food sector, *Sweet Charity? Emergency Food and the End of Entitlement*, Poppendieck (1998, 230–255) names a key chapter "Charity and Dignity." In it, she writes, "In the emergency food sector . . . I have been struck by the extraordinary efforts of providers to avoid, overcome, or at least minimize the negative, demeaning aspects of giving help. What is interesting, in light of the proliferation of such establishments in the last fifteen years, are the attempts to create soup kitchens . . . that preserve dignity and offer a warm, welcoming environment. Can this goal be achieved, or are these programs inherently demeaning?" (Poppendieck 1998, 235).

In a sustained argument, Poppendieck (1998) sets out the ways in which most food pantries and community/soup kitchens are organized in ways that damage the dignity of their food-insecure clients by inflicting numerous small and large humiliations on them as they seek food. Here, we review what Poppendieck's progressive criteria are for how charitable food outlets could be restructured to avoid the numerous humiliations they inflict on poor people as they seek out food and to preserve their dignity, and we assess how F.A.R.M. Café fares in meeting these criteria. We do so cautiously and provisionally, drawing on our 2010–2012 ethnography and on news and website accounts (admittedly from afar) of F.A.R.M. Café's operation up through spring 2022. We seek to model this as an open and ongoing inquiry, which continues to pose questions in their ascending order of importance of preserving the dignity of poor people—not just about F.A.R.M. Café but about all charitable food providers.

1. *Avoiding suspicion of clients.* Do managers, staff, and volunteers of the community kitchen/café avoid showing suspicion toward food-insecure clients in their actions and procedures, and can clients freely walk in without experiencing bureaucratic obstacles to service? Do they avoid requiring clients to fill out forms, show personal documents, and so on, before they can be provided with food? Do they avoid requiring clients to be checked for their eligibility (e.g., by a government agency) for food aid either on-site or off? Do they avoid requiring the client to be referred to them through a second-party charity (e.g., a homeless

shelter) (Poppendieck 1998, 235–238)? *F.A.R.M. Café avoids all of these obstacles to food access for its food-insecure patrons.*

2. *Welcoming clients.* Do managers, staff, and volunteers of the community kitchen/café treat clients with friendliness, hospitality, and attempts to be knowledgeable about clients' individual lives and circumstances—as distinct from depersonalizing them in the name of efficiency? Are clients made to feel welcome? For instance, are there places for clients to sit inside, where the heat and lights are on? Is the ambience when they visit a cheerful one? Do clients have access to bathroom facilities? Are they addressed respectfully in a welcoming manner and by name, for example, as "Mr." or "Ms.," instead of "Hey you" (Poppendieck 1998, 239)? *At F.A.R.M. Café, the answer is yes, definitely.* Evidence we have from recent news accounts and videos is that Renée and hired staff treat food-insecure patrons with the same level of respect as would be found in other restaurants—with friendliness and attention to people's names and offering the same facilities for all patrons irrespective of their food security status.

3. *Providing clients with choices.* Are clients of the community kitchen/café treated as if they have the right to choice and can be trusted not to abuse their right to choose? Are clients treated as adults who are given choices? Do community kitchen/café managers exhibit a trusting attitude toward clients (Poppendieck 1998, 240–243)? *For F.A.R.M. Café, the answer is an unqualified yes*: F.A.R.M. Café patrons, regardless of their food security situation, are provided with choices through identical menus with the same meal options, and the "pay what you can" practice is one that shows trust toward patrons, irrespective of their food security status.

4. *Treating food-insecure clients like restaurant patrons.* Do community kitchens/cafés show features indistinguishable from retail cafés with regard to choice for patrons, comparable table service, and attractive setting? Do community kitchens/café staff and volunteers offer clients a range of choices from a menu of different dishes they can have? Can clients come in without reservations, for example, those provided by a service agency (Poppendieck 1998, 247)? *Again, for F.A.R.M. Café, the answer is yes*: F.A.R.M. Café's downtown Boone location, the welcoming behavior of greeters and staff toward patrons as they come in, the furnishings (the open dining space where people of "different means"

can sit together or near one another), the brightness of the setting, the artwork on the walls, and the menus all connote an attractive setting of a retail restaurant within Boone's competitive market for patrons.

5. *Avoiding explicit/obvious class or racial distinctions.* Is there an absence of social class or racial distinctions between clients and those who serve them in community kitchens/cafés? Are clients no different racially or with regard to their visible class status from either staff or volunteers? Is there an absence in roles between "active givers" and "passive receivers" so that it is impossible to determine the dependent status of the recipients? Are clients treated the same with regard to their class status as the staff or volunteers, through "a common meal that blurs . . . the distinctions between givers and receivers, providers and clients" (Poppendieck 1998, 248)? Are there no evident differences in the qualities of the dishes being served (Poppendieck 1998, 248–251)? *For F.A.R.M. Café, the answer is yes in response to all these questions.*

At F.A.R.M. Café, the same dishes and meals are provided to staff, volunteers, and patrons; the question is how much patrons pay for these on the "pay what you can" model. The presence of patrons who are moderately well-off or affluent who will sit near and speak to food-insecure clients (and vice versa) is encouraged by the norms of the café's manager and staff and by the ambience and physical layout of the café (e.g., a counter and large tables where unacquainted patrons can sit together). These indicate the explicit attempts by the café's managers and staff to diversify the class composition of its patrons. There appears to be no racial difference that would distinguish staff or volunteers on the one hand from patrons on the other.[5] Volunteers, whether they are food-insecure patrons working for meals or university students working as interns, appear to have the same choices of volunteer roles (e.g., server, kitchen assistant).

6. *Giving explicit economic value to clients' volunteer labor and valuing their presence.* Do clients have opportunities to contribute through volunteering their labor at the café? Do managers, staff, and nonclient volunteers recognize client volunteers as distinct and valuable contributors? Do managers, staff, or nonclient volunteers realize that they personally benefit from the presence of client volunteers because of their knowledge about class and social difference? (Poppendieck 1998, 251–252)? *For F.A.R.M. Café, the answer is yes.*

Patrons are invited to offer their labor as volunteers in return for a meal. The labor of volunteers is valued as they work together and "share their lives" with staff and other volunteers. As Marta states, "If someone is unable to make a donation based on our suggested cost of $10 per person, they can exchange labor for a meal. We want people to be part of our community, to participate. We want to invite them to be with us because we believe there's dignity in working beside each other and sharing our lives over a meal. One of the easiest ways to start talking to people is over food" (Kornegay 2022).

Staff also realize that they benefit personally from interacting with food-insecure volunteers, as does the café's enterprise as a whole. As an example, Renée notes how surprised she has been to discover how many food-insecure patrons she has met suffer from mental health challenges:

> I wasn't seeing them, which I'm not proud of, because I didn't have to. You can easily go through life with your eyes averted from what's right in front of you. . . . [I] didn't know so that many people don't have a safety net, or don't have a place to be. They're not in the main paragraphs of life, but they still have value. They can't do a 9-to-5 job because it's not working for how they see life, they see it through a different lens. *What I didn't expect was people who would come and stay all day long.* I thought people would come, get a meal, maybe work as little as possible, and leave. But once they discovered a place they felt was safe, a place they could start to build community with others and work and get a meal, *they would stay and work the rest of the day.* I never saw that coming. (Kornegay 2022; emphasis added)

7. Valuing clients' knowledge as information to managers and staff that can be used for ensuring their dignity. Do café managers and staff recognize the knowledge that clients have as volunteers to help the café perform its function with more respect and dignity for clients? Do café managers and staff realize the value of the volunteer labor that their food-insecure patrons (and other volunteers) provide? Do they recognize that the café, through its volunteers, can serve more people and/or increase the time that it is open to serve more clients (Poppendieck 1998, 250–253)? *For F.A.R.M. Café, the answer is yes.*

Renée appears to reflect a broader sentiment among the staff at the café when she says that instead of a food kitchen, she "wanted to be involved in something that was more about getting to know and see people for who they are": "Let's cook good meals together, let's talk to each other about our lives, and maybe that way we can figure out how to create some change" (Kornegay 2022).

Renée and Marta recognize that volunteers play major roles in the food-prep, clean-up, and service labor that makes the café's operation possible. Crucially, the volunteer labor of food-insecure patrons is seen as having an economic value through the indirect equivalence of "paying it forward": "Anyone can work an hour in the café in exchange for a meal. In addition to our volunteers, we have 'pay it forward' patrons who pay more than the suggested donation for their meal, which helps cover the cost of those who cannot pay at all" (F.A.R.M. Café 2022b).

8. Do food-insecure clients occupy roles of authority as managers or on boards of directors, of a community kitchen/café? *At F.A.R.M. Café, no, not that we can find.*

From this review, we can see that the founders, managers, and staff of F.A.R.M. Café have gone to extraordinary lengths to "de-charitize" the café to avoid the many indignities and humiliations imposed by the charitable food sector on poor people in need of food—what constitutes "symbolic violence" (Krais 1993)—but nonetheless real violence—inflicted by that sector on those who are the poorest people of the working class of the United States.[6] The gentle but nonetheless insistent and continuous repudiation of this violence by Renée, the other founders, and past and current staff of F.A.R.M. Café marks it out as exceptional and indeed as a significant and promising departure from the neoliberal underpinnings of food bank nation and the corporate-dominated mainstream food system of the US. We say this despite concerns we have about F.A.R.M. Café's dependence on corporate and foundation sources of surplus, realizing that the shift toward an alternative food system will only occur through the strategy of constantly shifting the target toward greater food justice.

F.A.R.M. Café and the Diverse Community Economy

With regard to the diverse community economy, in the terms of Gibson-Graham's (2006, 71) typology, F.A.R.M. Café is an example of an alternative/capitalist hybrid enterprise; that is, it is a nonprofit enterprise

dependent on several sources of value. Like a commercial for-profit restaurant, it draws on the varying revenues from its retail sales from the
meals it sells but also on the salaried and wage labor of its chef and other
staff members. But unlike the commercial restaurant, it also draws on
the alternative paid labor of its food-insecure volunteers (meals as in-
kind wages), on the unpaid labor of student volunteers, and on outside
funding from wealthy families, corporations, and foundations. What
makes this improbable combination work are the ethical decisions about
need and surplus that Renée, board members, and staff of F.A.R.M. Café
have made over many years.

For Gibson-Graham, the determination of *what is necessary and what
is surplus* is not fixed by biology or culture but arises from ethical decisions that distinguish necessary labor from surplus labor (or the necessary payment for labor versus surplus value). These decisions translate
"into the level of consumption directly accessed by the producer and her
dependents, the volume of social surplus that is privately or socially appropriated, and thus the level of social consumption indirectly accessed
by society at large" (Gibson-Graham 2006, 89). Renée and the other
leaders of F.A.R.M. Café decided that it required necessary labor to operate in ways that appeared to be like a commercial restaurant, so that
the meals it offered could vary in price depending on the "pay what you
can" precept, which allowed food-insecure people to be patrons. This
meant that certain kinds of necessary labor would have to be paid the
going salary or wage rate on the regional labor market, for example, for
the manager/director, chefs, and other full-time staff members. However, F.A.R.M. Café's use of volunteers allowed it to employ far more
socially necessary labor time in the form of volunteer labor, the labor of
either patrons acting as volunteers who received the in-kind wage of a
meal (for an hour's work) or volunteers whose labor was unpaid, as in
the case of university students.

The question may also be asked: What are the sources of value that
allowed volunteers who were alternative paid laborers (i.e., the food
insecure) and unpaid student volunteers to survive? This is a crucial
question. We suspect but cannot demonstrate that food-insecure volunteers who were paid in-kind also received a combination of paid wages
from part or full-time employment, self provisioning, and government
transfer payments like Social Security, while the unpaid volunteers who

were students also earned wages from part- or full-time work, parental support, and (increasingly) college loans. This question is subject to different answers over the long term—partly, politically determined ones—and should also be the object of ethical discussion and decision-making by F.A.R.M. Café.

However, despite the presence of these alternative paid and unpaid volunteers, combined with the café's revenues from its paying patrons (who are selectively urged to "pay it forward"), there remains insufficient surplus to meet F.A.R.M. Café's expenses. Therefore, for it to continue, Renée and F.A.R.M. Café staff engage in considerable unpaid labor to raise funds from wealthy donors, which often entails major outlays of labor for meals prepared for funders and the like. But what is crucial here is that these are conscious decisions that Renée and the F.A.R.M. Café board have made that are considerably different from the capitalist enterprise.

Renée and the F.A.R.M. Café board have also engaged in ethical discussion and decision-making about how F.A.R.M. Café is to use the *surplus* it generates. First, as Renée would say, the whole point is for F.A.R.M. Café to generate sufficient surplus to "Feed All Regardless of Means," so that poor people can be fed with dignity. Second, as noted earlier, Renée states, "Each year, as our budget increases and our operation becomes more financially sustainable, we try to incrementally increase our local food purchases. I think it's really important that people have an economy based on a network of local suppliers that we know we can trust" (Kornegay 2022). Begun in 2009, the long-standing ethical discussions and decisions made by Renée and her colleagues have made it possible for F.A.R.M. Café to square a very recalcitrant circle: How can an enterprise both increase the economic viability and sustainability of local farmers while also providing poor people with the food they need while protecting their dignity?

* * * *

In this chapter, we reconstructed the founding and building of the remarkable F.A.R.M. Café in Boone in the Appalachian High Country region of North Carolina. First, we reconstructed the genesis and successful development of F.A.R.M. Café as due to a combination of factors: the charismatic and energetic leadership of one of its founders, Renée

Boughman, the inspiration that she and the other founders of F.A.R.M. Café drew from the national One World Everybody Eats movement, and her ability to make connections to other local activists and raise funding. The moral logic of F.A.R.M. Café was captured in the question that its founders asked: How do you make people "feel significant" in a situation in which they either sacrifice their bodies to hunger or their identities to loss of dignity?

Second, we analyzed the social context and interactive styles set by F.A.R.M. Café's leaders that allow poor food recipients to avoid the petty indignities and humiliations of food charity by keeping their food-insecure situations confidential and offering them alternative paid labor in the form of a meal paid in-kind.

Third, we found the sensitive efforts by F.A.R.M. Café leaders and staff to de-charitize their enterprise so promising that we examined it as a model for how to avoid humiliating working-class people in need of food by going through Poppendieck's (1998) inventory of steps that any food-distribution system can take to prevent inflicting in-person class-based violence on food-insecure people.

Fourth, it was also important for us that as F.A.R.M. Café's leaders de-charitized the restaurant, they undertook, from its founding onward, to purchase from local farmers and to sustain their economic viability. From the café's early years to the present, its staff have made a consistent effort, to the extent supply and their surplus allowed, to purchase locally grown produce and meats at the prices that local farmers asked for, as quoted earlier, "to make sure [they] get dollars into our local economy . . . [and for] the farmers and growers to have access to those dollars."

Fifth, we demonstrated that F.A.R.M. Café's founders and staff exemplified the standards that must shape ethical discussion and decision-making around two key questions that are crucial to advance any diverse community economy: What is necessary? and What is surplus or wealth? (Gibson-Graham 2006). Both questions generated conscious discussion among F.A.R.M. Café's founders, staff, and volunteers around issues such as the labor necessary to operate an enterprise, what can be used as surplus, how surplus might be acquired from outside the enterprise when needed, and in what ways surplus should be used so that the enterprise furthers economic flourishing within a community.

Finally, F.A.R.M. Café's model shows how a nonprofit food enterprise can succeed in simultaneously attaining the objectives of both sustainable farming and community food security activisms. By developing an enterprise working style that allowed it to "Feed All Regardless of Means" when it provided food-insecure people in the locale with attractive fresh local food while preserving their dignity, F.A.R.M. Café's leaders and staff empowered them in accessing culturally appropriate and sufficient food on their own terms, a crucial community food security activist goal. Over many years, F.A.R.M. Café also promoted sustainable local farming and foods by increasing its purchases of foods sold by local farmers at their asking price and by creating its Full Circle program, which converted farmers' "ugly" unsold produce into prepared high-value meals that reduced large amounts of food waste.

8

Cornucopia Landscaping, Abundance, and Food Justice

One of our core beliefs is that . . . liberation through abundance is possible with these sort of overyielding fruit trees,
berry bushes, things that even if the residential client may
not realize it, he or she is putting in a community garden:
they are in effect because there's so much food that's going
to be there in the next five to ten years that they're going to
have to learn how to share no matter what.
—James Sokolov, founder, Cornucopia Landscaping,
speaking to Anatoth Church gardeners, Orange County,
North Carolina

"We All Live in a Food Desert. . . . However, There Are Entrepreneurial Opportunities around Food"

Because of the shortfalls in government food benefits and the charitable food sector, a large number of food activists have become deeply
concerned about community food security, the idea that large numbers
of poor (and not so poor) members of urban populations should have
access to locally grown, culturally appropriate foods in sufficient quantities by learning to grow such food.[1] While some activists have sought
to draw on the nonprofit sector to develop the urban food infrastructures that improve community food security, other activists have relied
on commercial, surplus-generating (or profit-making) enterprises to
pursue this goal; most such activists have owned or worked in small
farming or food-processing businesses. In this chapter, we examine a
small urban farm cooperative to make sense of the activism that this
latter group has undertaken.

In what follows, we first consider how urban food cooperatives as an
alternative kind of commercial enterprise can further community food

security—an alternative to the large-scale for-profit food corporations that provide the bulk of globally sourced, industrial food to the US population. Second, however, our examination of one such urban farm cooperative, Cornucopia Landscaping (CL), that operated in our field site in Durham County, North Carolina, also led us to confront another key issue that faces food activists in cities and towns across the US: How do alternative food organizations led largely by middle-class white people work—or not work—collaboratively with poor people of color, activists, and residents, who are greatly overrepresented in urban neighborhoods at risk of gentrification and thus subject to displacement?

Both issues are central to the concerns of many urban alternative food activists, as they have reflected on the connections between food, wealth, sustainability, and social justice. One major concern of activists is that enormous monetary wealth flows out of the locales where eaters/consumers/workers live, in the form of food-corporation profits destined for private shareholders and institutional investors. This extraction of wealth by corporations over time can and has contributed to impoverishing the populations in these locales, particularly since low (and government-subsidized) industrial food prices have not compensated for declining real wages of food workers (in manufacturing, distribution, retailing, restaurant, and fast-food industries) in the "race to the bottom" brought on by neoliberal globalization since the 1980s.

This dilemma of the present moment has been recurrently raised by past activists for at least two centuries: How can as much wealth as possible produced by people who have labored to produce food (and other commodities) who live in a locale be retained within that locale and used to invest in its future flourishing? How can the gross economic inequalities of the industrial food system (e.g., the huge salaries and stock options of food-corporation CEOs and managers, now hundreds or even thousands of times greater than the annual wages of a typical food worker) be reduced?

In response to these two questions, many food and other local activists have turned to the long, successful history of food-producer and food-consumer cooperatives, not only in the United States but in Britain and elsewhere (Knupfer 2013). Cooperatives, when democratically governed by their worker-owners, have been shown to be effective and long-lasting institutions that retain "surplus"—conventionally called "profits"—within the locales in which they operate and distribute it on a more or less equal basis among worker-owners, broadly sharing the prosperity generated, even as their cooperatives compete successfully in commodity markets vis-à-vis capitalist enterprises (Gibson-Graham 2006; Webb and Webb 1907, 1921). At the same time, cooperatives are not without their strategic and operational challenges (Gibson-Graham 2006, 106–111). How a small urban farming cooperative in Durham, North Carolina, might manifest these characteristics at a time of generalized economic stagnation in the United States, looming global food shortages, and the mother of all crises—climate change—should thus be of great interest to many people.

A second set of questions arises from what an increasing number of food activists and scholars of food studies are beginning to realize: that local food activism has primarily been a racially white-dominated initiative, creating various kinds of "white space" that threatens the potential for widespread cooperation between food activists, the majority of whom are white, and the large number of poor people of color, including activists, who reside in urban neighborhoods throughout the US (Ramirez 2015; Slocum, 2007, 2010; McKittrick 2011, 2013; McKittrick and Woods 2007). These different kinds of white space—community gardens, farmers markets, community-supported agriculture schemes, small-scale urban farming projects, food-aggregation hubs—threaten such cooperation because of an implicit exercise of white privilege that governs such space (Ramirez 2015, 751–752). These white-defined and white-controlled spaces, the white uses of them, and white-invented discourses about them generate a specific racial geography—and preempt the alternative spatial uses, imaginaries, and representations around food by nonwhites.

These issues have led critical food geographers to ask in particular what a "Black food geography" might look like and how it might be the basis for African American–led food-justice projects (Ramirez 2015, 750; McKittrick and Woods 2007). Investigation of such Black geographies in their historical and cultural contexts suggests strongly that a large number of African Americans do have such a vision: "the site of the plantation . . . embedded in the American black ontology" (Ramirez 2015, 750; see Woods 1998; McKittrick 2011, 2013; McKittrick and Woods 2007). The US southern plantation—a system of rationalized enslavement, industrial-based ecological simplification, and value extraction—not only has been a basic site of production replicated throughout the territorial United States and US imperial sites overseas (e.g., Hawaii, Philippines, Puerto Rico) but was also "the birthplace of an African-American ethic of survival, subsistence, resistance, and affirmation from the antebellum period to the present" (Woods 1998, 27). This historical insight strongly suggests that large numbers of African Americans see a nightmarish figured world of "the plantation" as the model for white power exercised almost ubiquitously over and against them—in agriculture, in industrial employment, in sharecropping, in schools, in prisons, and (even) in universities. These are all hierarchical institutions controlled from the top down by whites, especially white men, and governed not only by explicit rituals of violence but also by implicit norms that define, represent, and govern spaces owned by whites even as, as Woods points out, the plantation has also called forth an African American ethic of survival and resistance. In strong antagonistic reaction, large numbers of African Americans have developed specific fears and anxieties and strong ethical commitments to their own empowerment around land and its meanings—and this applies not only to rural farmland in the southern US but also to urban land throughout the US.

Given the existence of these Black geographies, the challenge for white food activists as they seek to work not just *for* or *on behalf of* but proactively *with* African American activists and urban residents as their equals in power has to be, "Is the form of 'justice' that food and other social justice activists practice simply a politics of inclusion that upholds power asymmetries stemming from the plantation?" (Ramirez 2015, 750). Or can white activists transcend this limitation? This is the dilemma this chapter poses in the case of CL. Durham, North Carolina, was one of our sites where we witnessed this dilemma being worked through in encounters between white and Black activists.

Durham, North Carolina, and a Vision of Abundance in "a Time of Scarcity"

By the time of our 2011–2012 fieldwork, Durham County had a population of 267,587 people, of whom 128,787 (or 48.1 percent) were white, 105,516 (39.4 percent) were African American, 36,240 were Latinos (13.5

percent), and 6.1 percent were all others. Durham County was highly urbanized, with 94.4 percent of its population designated in the 2010 Census as "urban" (US Census Bureau 2010a, 2010b). The city of Durham is located within the Research Triangle region, so designated for its own major research university, Duke University, one of three constituting the "Triangle" in question. Before the 1990s, the city of Durham was designated as "The City of Tobacco," but it subsequently has rebranded itself as "The City of Medicine," known for its research industries around pharmaceuticals, health care, medical technologies, agrochemicals, and digital-platform corporations. As such, Durham County has become known for its affluence, its large number of highly educated professionals, the influx of many new residents from the North and Midwest, and its attractive lifestyle for retirees (Holland et al. 2007, 18–34).

Despite the city's overall affluence, Durham by the end of the first decade of this century was characterized by high levels of economic (class) and racial inequality. Its population designated as living in poverty (i.e., below the federal poverty line) included 47,993 people, or 17.9 percent of Durham's population, of whom, in turn, 51.9 percent were African Americans (US Census Bureau 2010b). If people were poor, they were by and large also food insecure. Among the neighborhoods of the city of Durham where a large number of poor people of color were concentrated was Northeast Central Durham (NECD), where Cornucopia Landscaping was operating. Large numbers of African Americans had experienced neighborhood and school segregation created by Jim Crow laws and Federal Housing Authority "redlining" that made NECD unbankable from the 1930s to the 1960s, followed since then by de facto segregation reinforced by white flight and discrimination by the real estate industry (De Marco and Hunt 2018).

From the early 2000s up to time of our fieldwork in 2009–2011, Northeast Central Durham was on the eastern edge of a wave of gentrification spreading out toward the east and north from Durham's downtown. This period was first marked by rapidly rising housing prices, followed by the 2007–2008 subprime mortgage crisis, leading to traumatic subprime foreclosures of homeowners, business closings, and job losses that rapidly increased poverty in NECD, just as we began our fieldwork in Durham. The federal poverty rate in US Census Tract 10.01, which covered much of NECD, increased from 38.0 percent in 2000 to 61.4 percent by

2008–2012 (calculated for US Census Tract 10.01 from De Marco and Hunt 2018, 17, table 2).[2] By 2010, gentrification had resumed its eastward spread and was then starting to encroach on NECD—so that if we look forward to 2016, the price of housing in East Durham had increased threefold since 2012, from $37 to $100 per square foot (De Marco and Hunt 2018, 22). With a large number of NECD African American working-class residents still un- and underemployed from the subprime mortgage and financial crisis, by the time we began our ethnographic research there in 2009, they not only had very low or no incomes but also had reason to fear being displaced from their housing due to increased house prices and rents. Most of those who purchased and moved into the distressed properties from which African Americans had been displaced were relatively affluent whites. This has been the recurrent pattern of urban gentrification in the postwar history of the United States (Johnson 2019).

Thus, a review of Cornucopia Landscaping, a small cooperative landscape enterprise in Durham, North Carolina, which has centered its food activism in Northeast Central Durham, allows us to address two themes: how food cooperatives can reduce urban community food insecurity and what dangers exist when those who lead these cooperatives (or other alternative food organizations) unconsciously exercise class and racial privilege. During our fieldwork, our ethnographer Patrick Linder brought to our attention something that James Sokolov, the college-educated, white founder of CL, said to a meeting of food activists about the food-supply chains that lead to stores like Whole Foods and other retailers: "We all live in a food desert. . . . The East Coast has only a two-week supply of food, and any sort of major disaster could cut off those sources. . . . However, there are entrepreneurial opportunities around food." James elaborated on how CL could play a role in such a transformation: "Our main goal is to get people prepared, get them into a position where they can easily transition to a time of scarcity—times with fewer resources—and increase self-reliance."

What we find most intriguing within this figured world of the future are two aspects of CL's approach as articulated by James and other worker-owners of the CL cooperative. One was CL's attempts to formulate a strategy for preparing people for a probable future of food scarcity in ways that address the deep economic and racial inequalities that make the lives of working people and poor people of color in the United States increasingly precarious and food insecure. This is a strategy that CL has fully tried to implement in its daily workings and has built into its business model. By attempting to teach working people and people of color how they can grow food in abundance through the methods of permaculture, CL has sought to provide them with the means to earn their own incomes and ensure their food security (on permaculture, see Holmgren 2004; Rhodes 2017; Ferguson and Lovell 2014).[3]

The key term is "abundance": as James put it, "One of our core beliefs is that the liberation through abundance is possible with this sort of overyielding fruit trees [and] berry bushes." Far from the problem being food scarcity, "there's so much food that's going to be there in the next five to ten years that they are going to have to learn how to share it, no matter what." CL also has combined this vision with what its owners-workers see as a viable model for a small-scale, labor-intensive enterprise that provides not only food but also employment and income to residents of poor neighborhoods. CL thus puts forth a figured world of self-sufficient working people and people of color with increased incomes, abundant food, and the means for producing it.

James stated CL's goal this way: "Our primary goal is to get more people to grow more of their own food where they live. . . . Definitely one of our goals is to increase people's resiliency and self-reliance, get to know their neighbors, and also as far as workshops, giving people practical tools to do things together or DIY[do it yourself], DIT [do it together], DIY." Part of this process, and the second original aspect of CL's strategy, was that it incorporated the replication of its own cooperative model into the food-related work it engaged in with poor residents of Durham, North Carolina.

According to the CL vision, the cooperative model it implemented in its own practices was a counter both to the false promises of large-scale industrial food production (e.g., the myths of techno-abundance and of cheap food), on the one hand, and to what James called the "nonprofit industrial complex" of food-related nonprofit organizations devoted to the objective of making the US urban population food secure, on the other (on the nonprofit industrial complex, see Incite! 2007/2017). While, according to James, the industrial food system is dangerously fragile, exploitative, and unsustainable, in contrast, nonprofit food-related organizations depend for their economic survival on private foundation and government grants. But this was not a viable long-term strategy for activists when government support was unreliable, grant-chasing consumed too much of activists' time, worthy causes competed for limited private and public funding, and the whole process demoralized large numbers of food activists (see Incite! 2007/2017).[4]

For working people, including people of color, CL also sought to provide them with an economically viable alternative to capitalist food enterprises—the small-scale food-growing and landscaping cooperative—that could teach them how to grow food in abundance through the methods of permaculture and agroecology. In CL's case, it attempted to create new cooperative enterprises like itself, whose successful operation, as its worker-owners envisioned—were such enterprises to be multiplied and scaled up many times over—could bring about widespread, fundamental social change by ensuring material food abundance. In this model, cooperatives make "community" through the cooperative-building process. Thus, CL put forward a figured world in which cooperatives are self-replicating and become central to community, and their owner-workers and other community members can ensure their own food security, independent of being the citizens of a nation-state.

Finally, in addition to these two remarkable aspects of CL's figured world, it stood out from the majority of other food-related organizations we studied by seeking to bridge the activist domain of local sustainable food and farming with that of community food security. We thus felt that we had much to learn from Cornucopia Landscaping.

Water, Water, Everywhere: Constructing the "Edible Schoolyard" at Andrew Tyson Elementary School

In the summer of 2009, the staff of Cornucopia Landscapes helped the Andrew Tyson Montessori Magnet School parent-teacher association (PTA) in Durham design and install a new outdoor project called an "edible schoolyard," or garden. This garden—eventually to consist of three separate areas, a fruit garden and arbor, a courtyard garden of perennials and annuals, and a series of raised beds that grew vegetables—came out of the collaborative arrangement between Cornucopia Landscapes and the school's PTA. While PTA parents and volunteers provided most of the labor for the construction of the fruit garden and arbor that formed the first elements of the "schoolyard," CL staff as "garden experts" undertook responsibility to design it and oversee its installation (WRAL News 2009).

A key objective in the construction and planting of the fruit tree and shrub area and its arbor of shady trees was to deal with the problem of water flow. As the website of Cornucopia Landscaping observed, "One of our first observations was a tremendous amount of water that moved through the area. Directly below three downspouts, dumping thousands of gallons of water from the school's roof, water right away denuded soil into a stormwater drain 100 feet away" ("Case Studies—Water Management School Garden," CL website, 2013).[5] The objective was to divert this flow of water away from eroding the soil next to the school building, to slow its movement in order to use it in the new garden's fruit tree and berry bush area, and to store the water at the top of the garden so that it would drain gradually into the lower area of the garden. This was more than a technical challenge of adjusting drainage, measuring drainpipes, or designing slopes. It also represented a challenge to social relationships, for, as the CL website put it, "a new garden would have just washed away, without some problem-solving between Cornucopia Landscaping and Andrew Tyson PTA" ("Case Studies—Water Management School Garden," CL website, 2013).

The initiative for an edible schoolyard came not from CL but rather from Joy Stein, the parent of a Tyson School child who had read Alice Waters' book *Edible Schoolyard: The Universal Idea* (WRAL News 2009). As part of Joy's initiative, she and the PTA of the school called on CL as "garden experts" who "helped with the design and crop selections" (WRAL News 2009). This was consistent with one key requirement of CL, according to James, that when it engaged in pro bono design and installation work for a community, the idea for the project had to come from members or residents of the community itself.

This realization by the founder and other worker-owners of CL represents a key insight: growing food is never just a biological or technical process. It is deeply enfolded within social relationships of cooperation and assistance that must be renewed and sustained for this process to be successful over time. Moreover, these relationships should build on initiatives or at least desires already mobilized. With PTA parents and neighborhood volunteers and CL staff all working together, a successful edible schoolyard was built: "We dug a large rain garden at the top of the slightly sloped space. A rain garden is a shallow depression planted to filter and slow rainwater. A beautiful array of moisture-loving plants thrive in rain gardens—native pawpaw, beautyberry, and hibiscus to name just a few of the towering beauties. The berm at the bottom of the rain garden creates a nutrient-rich edge that is a perfect home for high-bush blackberries" ("Case Studies—Water Management School Garden," CL website, 2013).

What resulted was the abundance of permaculture: "The edible landscape at Tyson is loaded with perennial edibles. Planting perennials that live for decades is an investment that pays off through time. Fruit trees and berry bushes yield prolifically—at maturity, the plants at Tyson will collectively produce more than 250 pounds of fruit per year. We designed as much spring and fall ripening fruit as possible for harvests during the school year. Now persimmons, blackberries, edible dogwoods and much more grow with each generation of students who pass through the school" ("Case Studies—Water Management School Garden," CL website, 2013).

Installation of the Tyson School's edible schoolyard was but one example for CL's worker-owners of the broader transformation to which CL was committed—not only helping others build new infrastructure for common use but also catalyzing more extensive processes of social and natural transformation, to get things going and growing, to "jump-start" a process. As one of its owner-workers, Pete, put it to us when we interviewed him in early 2011, much had happened at Tyson School since the successful installation of the edible schoolyard in 2009:

> So they've got . . . a school garden that we helped install and a court-yard, and they've expanded it. . . . We've also done a portrait area, and we, . . . parents and people, did a lot of work [for] the project. They've also been instrumental in the parents getting the teachers to take an interest in the garden and to get the education [about the garden] and even get some of the food into the school. Like once a week or something, [they] have salad greens from the garden . . . for lunch or for snack. . . . I think that's really important work, you know, . . . in our primary schools.

Note that Pete here spoke of a dynamic in which CL has continued to play only a minimal role, because the momentum needed to keep the edible schoolyard going has arisen from the pedagogical and nutritional seductions of an edible schoolyard that teachers, students, and parents at the school themselves have continued to sustain.

Cornucopia Landscaping and the Cooperative Model

By the time of our fieldwork, CL had already established itself as a small commercial enterprise widely known and admired for its innovative permaculture and landscaping designs, highly respected by other local food activists for its pro bono work, and winner of a regional weekly newspaper prize awarded it for the high quality of its services. By 2013, it had installed more than 125 edible landscapes like the one at Tyson School. CL described its edible landscapes as gardens that "simultaneously grow vegetables, fruit trees and berry bushes while helping regenerate the land. They harvest and utilize rainwater, build soil, and provide habitat for beautiful insects and animals. They create abundance both for human communities and the environments they are tied to" ("About Edible Landscaping—What the Heck Is Edible Landscaping?," CL website, 2013).

The design and construction of the Tyson School edible schoolyard illustrates two important characteristics of the Cornucopia Landscaping cooperative as we discovered it during our fieldwork. First, there was the business model that animated this cooperative's work in Durham. Founded in 2006, CL stood out from other small-scale food activist organizations that we encountered. It was organized not as a one-person business or as a capitalist enterprise hiring employees for wages but as a producers cooperative. It thus belonged to a long tradition in the US of rural food-producers cooperatives. Yet we could say that, instead of only producing food, it also simultaneously produced (i.e., designed and worked to install) urban gardens that yielded food and the "growers" of these facilities, like the students, teachers, and parents of Tyson School. As James put it when talking to the gardener of the church garden, "We try to grow growers. . . . Most landscaping companies make 60 to 80 percent of their money on maintenance. We started out with a pretty firm stance about helping people as much as possible, answering questions, . . . encouraging them to participate in the installation, but [then] . . . we wanted to pass it off to them and have them . . . maintain it."

What made this possible depends on what distinguishes a cooperative from a capitalist enterprise, and it is important. Unlike the latter, in a producers cooperative, any surplus generated by income from its market-oriented work does not go toward enriching nonworking shareholders but instead is divided between an income fund for worker-owners for their economic support in lieu of a wage and the cooperative's necessary reserve (sometimes called its "rainy day fund"), which is set aside for use as surplus to be used however its owner-workers decide. Cooperatives, like privately owned capitalist firms with shareholders, own property and compete in markets to realize surplus ("profits" in capitalist accounting) from sales of what they produce. But unlike capitalist enterprises, whose managers trumpet their legal "fiduciary responsibility" to pass all surplus to their private shareholders, worker-owners in a cooperative can use a portion of its surplus for social projects that they decide, for example, would create meaningful work and income streams for residents, enhance urban neighborhood environments, or provide food to marginalized populations.

Alternatively, instead of maximizing and then reallocating cooperative *surplus in its money form*, worker-owners of a cooperative instead can collectively decide to allocate their labor time not only to work for their own financial support but also to work on projects that benefit the locale and its population. Whether in terms of co-op owner-workers' income earned or equivalent amount of labor time they expend, this capacity remains their choice, and this is what the owner-workers of CL have done. As James put it to our ethnographer Patrick Linder, "We've gotten to the point now where, at least when we first started it might be six days a week chasing clients, but now it's like we get about two, sometimes two and a half, three days, to focus on the [pro bono] work where we live." What they earned collectively while competing in the market underwrote the otherwise unpaid labor they did on such projects.

Indeed, the economic democracy within cooperatives and their worker-owners' capacity to allocate part of their surpluses to improve the lives and living conditions of urban residents have been characteristic more broadly of the urban food cooperative movement. However, most urban food cooperatives are consumer-owned rather than worker-owned, and they therefore focus, at least initially, on retailing locally grown produce and foodstuffs to their own members, although some have ventured later into urban and suburban farming as a food source and source of employment for residents (Knupfer 2013, 190–203; Kauffman 2017; Auborg 2020).

A second key characteristic of the CL cooperative was that although most of the paid landscaping work it did was with private landowners for relatively high fees, which CL was able to command due to the exceptional quality and high reputation of its designs and installations, it was unique in this locale in that it employed some of the profits it earned and certainly the energies of its worker-owners to freely provide their labor and design talents to the construction of such projects as the Andrew Tyson School edible schoolyard. In fact, over the five-year period since its inception and up through the period of fieldwork, it had done many such projects.

Doing Pro Bono Work with "Some People [Who] Have More Time than Money"?

It is important to distinguish the work that CL's worker-owners did as consulting designer-landscapers for individual, usually wealthy, paying clients from the work they did for community organizations and groups. When it came to these two complementary kinds of work, when Patrick asked James whom CL ought to serve, James replied that there were people with more money than time and people with more time than money: "We ought to serve everybody. . . . I think that . . . we make a mistake when we start compartmentalizing into, 'Well, they only work with rich people' or . . . 'Oh, they're communist, and they only work [with poor people].' . . . Everybody—some people have more time than money; other people have more money than time. . . . It really is just a factor of how much time we have."

More than from CL's periodic workshops on permaculture and other agroecological topics sponsored by local nonprofit organizations, which people paid fees to attend, it earned most of its income from its commissions from wealthy individuals to design and install an edible landscape on land they owned in return for a substantial fee. In the case of one project it had been commissioned to undertake, James told Patrick that

CL's owner-workers earned $15,000 for approximately four days of their labor. But to install a new edible landscape was one thing, James said, something they did willingly. It was another thing to be asked to maintain it over time by a wealthy individual, something that CL's owner-workers disliked doing. To avoid doing this, it charged its wealthy clients very high fees for maintenance relative to what a "regular landscaper" would charge.

This is where we turn to the other form of work that CL did at no cost in money for community groups and organizations, including those led and participated in by poor people—the people who, according to James, had "more time than money." This was work that CL's well-paid commissioned projects for private clients made possible. This was what James and other worker-owners of the CL cooperative said they much preferred to do with their work time, to labor toward a situation in which "eventually the pie [of surplus] will be big enough to . . . just work down here five days a week [on community projects]": "I'd love that. . . . I'd love to be able to work full-time . . . and . . . in North East Central [Durham] in this little six-block . . . radius. . . . That's why I think we're also stepping up the efforts on the urban agriculture side of things"—as distinct, that is, from working with private but well-paying clients.

In CL's pro bono work for community groups, its distinctive approach was to work with a school PTA, a neighborhood association, a church, and the like, whose leaders or members had previously expressed a desire for a garden/farm or more comprehensive landscaping project. This might begin with a paid-for or free workshop offered by a CL worker-owner to members of the organization. In the next step, CL offered to design the edible landscape while consulting with the organization's leaders and then would oversee the work required to construct the project, which employed that organization's own volunteer labor, as we saw for Andrew Tyson School.

It is crucial to note the implications of this model for relations of power between CL and its pro bono clients. Such an arrangement revalued "upward" the labor that CL's worker-owners put into it as "expert gardening" and not the physical drudgery of landscaping. This was a claim that invoked CL's expertise in pedagogy—whether it was teaching the organization's members how to build a permacultural garden or how to grow shiitake mushrooms. This also, however, could—and, we argue, has at times had—the effect of revaluing "downward" the volunteer labor, the actual drudge labor of digging trenches, hauling rocks, and so on, that CL worker-owners oversaw but largely did not undertake to do themselves. Moreover, in the case of African Americans in Durham, it threatened implicitly to reenact features of a specific older pattern of labor relations—that of the southern plantation, its owner-entrepreneur, its manager and overseers . . . and its slaves.

That said, this pro bono practice on the pedagogy of urban agriculture was by no means merely a public-relations performance—although it generated enormous goodwill widely across Durham—but a consistent practice of its worker-owners. One we interviewed, Pete, put it this way: "So the bread and butter [revenue] of the operation is working on residential projects and individual homeowners. But at the heart of the work is working with community groups and with community members to help strengthen community relationships and community self-reliance, [to] empower and motivate folks to be examples of [the fact that with] their little piece of land that they are contributors to the local food economy."

Reflecting on the purposes of CL more broadly, Pete connected these two distinctive characteristics—having a cooperative structure and contributing as experts to "empower and motivate folks"—to each other:

> That's community education through all kinds of workshops, and it's the support of other community projects that aren't just our home, through different types of collaboration. Definitely there's been a return of any surpluses. . . . The other thing is about being a model of a community-based micro-enterprise or collective model micro-enterprise that's fee-for-service as opposed to supported largely with grant funding. So I think that's an intention that we are aware of. . . . So there's been a return of surpluses . . . to employees throughout its history at the end of the year. We found ways to return some of the surplus. . . . So, you know, the model of being a worker co-op. James's whole paradigm in the outset has been about, How do we create intermediate alternatives to transitioning out of the form of capitalism that we know today?

Thus, through its income sources—projects with private wealthy clients and holding classes and workshops on permaculture and other sustainable agriculture topics (e.g., growing shiitake mushrooms)—CL managed to devote a considerable part of its resources to the reconstruction of the food institutions of poor and working people in Durham. James argues, "It really is just a factor of how much time we have"—and of how they decide to use it.

There is, however, one point about which we disagree with James. People in the US live in a society with gross yet overlapping economic and racial inequalities—where a relative few not only are privileged with financial wealth or at least adequate stable incomes but also disproportionately possess other kinds of unconscious social privilege arising from being white, male, and (recognized as) experts. Some of these few may be impoverished for time, even as they are financially wealthy beyond the everyday dreams of most people. The partial reality of the 1950s–1960s stereotype of the corporate "rat race" has been reinforced since then by competitive consumption striving by Americans who are "overspent" (Schor 1992) and "overworked" (Schor 1998)—and hence starved for time.

Yet, the majority of working people in the US—including those whom CL has sought to teach a new abundance to—confront conditions that not only make their lives precarious and impoverish them but also devour their time as they work and seek work, labor in two shifts or two or more jobs each day, make long commutes on public transport or drive cars in disrepair, spend vexing hours applying for state benefits (SNAP, Medicaid, Social Security disability), while many—especially working-class women—also care for family members who are young, sick, injured, or elderly. Many are women of color whose everyday economic obstacles to their getting by are profoundly exacerbated by impediments of institutionalized racism, classism, sexism, and ableism. They are not "people with time," despite James's claim.

Many among this majority also aspire to and occasionally achieve the satisfactions that come from, say, activism that contributes to food security among the communities they belong to. Here, CL's vision that working people are instructible "people with time" surplus available to use as volunteer labor under CL owner-workers' direction does not fully comprehend the intimate interdependencies that exist for working people between their labor for their own subsistence and for those they care for, their vulnerabilities to everyday violence inflicted on them by others with power and privilege, their remaining capacities to work for the food security of their communities, and the costs of maintaining their own dignity.

The question is, therefore, Can an innovative food cooperative like CL, consisting of white and highly educated owner-workers, come together with working people of color—who actually as individuals do *not* have "time" in abundance—to work as equals with the latter as cooperating participants in a project to successfully construct a new community food security infrastructure, for example, an edible landscape or a hub garden? Here, as in the example that follows, the case of CL has much to instruct white middle-class activists about.

Whose Community? Whose Food Security?

In our discussion of Cornucopia Landscaping, as elsewhere in this book, the word "community" repeatedly appears. As we began discussing CL's cooperative work, we began by distinguishing descriptively between its paid landscaping contracts with private individuals and its "community projects" done with "community organizations and groups." But here it is necessary to unsettle our own use of the term. This is because the term "community" has many meanings and associations because its use by local food activists allows it to be mobilized in many ways. Its use can reveal possibilities for mutually respectful collaborative work among the members of a population toward goals collectively decided on—or its use may obscure important differences and reinforce inequalities within the population and police the boundary between members and outsiders, or simultaneously, it may even do some of both.

In what follows in our discussion of CL and its efforts on behalf of poor people and people of color in Durham, we found that the meanings of community, of how it was to be built, and of how food security for its members was to be attained were subject to contestations between CL owner-workers and the working people/people of color they sought to help. But these contestations were not academic: the capacities of poor people and people of color to ensure their own food security proved to be at stake, as we discovered when we examined CL's attempted collaborations with residents in Northeast Central Durham. Put another way, the definitions of community food security, and of how it was to be achieved, were tied to relations of unequal power and access to material resources.

Disputing the Meaning of Massive Harvest Farm: "Community Garden" or "Urban Farm Cooperative"?

In late 2010, a group of Durham food activists met together several times to plan to establish a county-wide land trust for holding land in Durham County in perpetuity for agricultural use. Among those attending one such meeting were James of CL and Lavondra Jones, both of whom had worked with Jeff Nowell, executive director of Great Tasks, a for-profit consultancy that specialized in assisting the start-up of sustainable enterprises, especially around food. Lavondra was a middle-aged African American woman who lived in Northeast Central Durham and owned a small house and plot of land called T-Nina Manor. Among other things, her residence served as a "halfway house" for African American youth experiencing legal and personal difficulties. Also present at these meetings was Saliya El-Amin, a middle-aged African American woman and cofounder with her partner, Joyce Wilkerson, of the urban gardening business Green City Action. Not long previously, Saliya had stepped down from a well-paid full-time position at a regional consulting and research firm to found Green City Action with Joyce.

At one meeting, these like-minded activists decided that they could not afford any longer to wait during the group's long and detailed discussions to set up the land trust, before acting themselves. Our ethnographer Patrick Linder described how what occurred came about: "James quickly became impatient with the interminable meetings entailed in starting an organization of this sort [the land trust], and when he saw others around the table that showed his sentiment, he drew them off into a group that was dedicated to 'doing something.' This was the birth of the Massive Harvest Farm project," involving James of CL, Saliya of Green City Action, and Lavondra of T-Nina Manor.

The Massive Harvest Farm collaboration began to get under way in the next few months of spring 2011. We heard of it being featured in the Eastern Triangle Farm tour in March, when Patrick talked briefly with Saliya, who described it as based on "cooperation."

We lost contact with the project for several months. Then, in mid August at a meeting of food activists that Patrick, Kevin, and Don attended, we heard that Massive Harvest Farm was being structured as a cooperative farming enterprise. Attending, Maggie Elspeth, one of CL's owner-workers, described CL's role as helping start Massive Harvest Farm, where she and other CL owner-workers worked with neighborhood youths and taught them to grow food; Massive Harvest Farm had set up a small CSA to help pay for materials but also "gave away as much [food] as they could."

In late August, Patrick interviewed Saliya of Green City Action. In the course of the interview, it became clear that there was a full breach in the collaboration between Saliya and James and other CL owner-workers at Massive Harvest Farm. Early in the interview, Saliya commented, "We have done work with Bountiful Backyards. . . . We did back away from that relationship in the way that it looked—because our—we just didn't feel like the way . . . they were organizing was aligned with the way we wanted to do it. The motto was pretty much 'if you build it, they will come,' and—we just wanted to—actually be in community with the people in Northeast Central Durham—and the way that we understand that to be."

Saliya indicated, however, that the problem was not easily remedied because too much ill will had already been created. Speaking about "what it's like for Black folks and white folks to work together," Saliya set out her critique:

> We all came into this system—right, of—privilege, and [some] have majority privilege and power—it is very difficult, in my opinion, for—some white folks, specifically white men—to share power, like, share it. So this is what—so shared power means "I have something that I come with, and you have something that you come with, and together we're in alignment with one another, and then we move forward." *And the way in which most of the white folks who I come into contact with, no matter what job I do— the way in which white folks are used to dealing with Black folks is that we work for them. We are—we work for them.* (emphasis added)

In light of the nightmarish vision of the plantation and its land as a site of white power that figures so prominently in Black food geographies that we report from the literature, we cannot help but think that Saliya, as a relatively well-educated organizer and activist, had precisely the plantation model, or something very like it, in mind.

Saliya continued to mention what this default expectation of white privilege when whites interacted with African Americans looked like, as she saw it:

> I have to believe if we're in the trenches together that our—and our backs are up beside one another, that you have my back, and it makes it—it's very difficult when you know that this person every day is having difficulty with you—because they just don't know how to share power—don't know how to resolve conflict—don't know how to be called out, and say, "What you're doing is effed up—this is not right, and this is the way I think it should be done." And then you can say, . . . "Well, maybe I disagree." . . . And they can say—"Well, maybe I do too." And you can come to some sort of agreement. But if just the mere fact that I call you out on it cascades into—you know, this huge ginormic thing that means we can't plant vegetables and we can't—you know—interact—interface with the community—then I don't want to work with you.

In retrospect, Saliya spoke of her experience with CL as a months-long ordeal of miscommunication, frustration, antagonism, and eventual rupture at the end of summer 2011: "So we partner with another organization, and it was like—we're tired of going to meetings. So it sounded like—again, it was a—in theory, an amazing idea—but walking it out without actually doing that internal work [of creating equality in decision-making] . . . made it difficult. And then by the time you look up, you don't want to do the work anymore. You're like, 'No—these folks have worked my nerves to the last bone.'"

We heard James's side of the falling-out only several months later in November 2011, when Patrick visited James. James told Patrick that CL had reconciled with Lavondra. Her problem had been that she was concerned that Massive Harvest Farm should be more of a community garden, "which was the idea all along, but whatever." James's sarcastic recharacterization of Massive Harvest Farm as a community garden as "the idea all along" instead of an urban farm, combined with his dismissive expression "whatever," suggested that the rift was from James's point of view wide but not irreconcilable.

In a later interview with Patrick in December 2011, James recalled a conversation a week previous that he had with Joyce Wilkerson, whom he considered a friend but who was also the business (and domestic) partner of Saliya of Green City Action. "I was just like, you know, 'I'm really feeling some tension with some of the folks of color, and like . . . I don't know where it's coming from, and can you give me any insight into this?'" He said Joyce told him that "some activists feel like CL takes too much from the community because, like, we're not a nonprofit and, like, we're a business, like—like, that we take energy away." James called this a "really disingenuous argument," since, "well, like so what, like every cooperative in the world is a for-profit business. It's like—it's just working within the framework."

During this interview, James persisted in his hope that Massive Harvest Farm would become an urban farm cooperative: "Well, I've—we more or less kind of—completely backed out of—just being in that space while she's [Lavondra's]—going through what she's going through and sort of lashing out at people and stuff." But he said that much could be achieved—or lost—in the current situation. Although there had been no cultivation in the backyard garden space since September, there was produce still growing and ready to harvest. "Like, somebody could go there tomorrow, and . . . it's all set up. It's ready to go." He still had hopes that the two neighborhood African American youths befriended by CL would try to work with him the next day and that the idea of a small African American youth farm cooperative at Massive Harvest Farm might continue through them: "Hopefully—we can grow to the point where these guys are sort of spinning off—in their own thing. It's one vision I have for Massive Harvest this [coming] summer." Thus, the two youths, along with a couple of others, James explained, "can band together—you know, plant it out, work with me on my planning it out, and then—you have a part-time summer job."

He went on to tell Patrick that through a mutual friend, he persisted in proposing to Lavondra his idea of Massive Harvest Farm as a youth cooperative enterprise.

So I pitched this idea about a youth cooperative to her, . . . I haven't heard back but . . . it's not gonna be a community garden—without—somebody sort of picking up the mantle, you know? She can say that all she wants

[that]—there's not enough Black people there, and she can [say] all that other stuff as much as she wants to, but—everybody who's been there and the kids that have been there know that it's—an awesome place, and—I don't think anybody ever felt marginalized there—except by her.

James then went on to speak of reconciliation, but only if it turned out on his own terms, as if negotiating with Lavondra. CL need not be associated with their cooperative, but she had in return to recognize the importance of the cooperative for the youths who have been working there: "When you remove CL from the equation, . . . she can be up here again, . . . if she can control herself." James went on to visualize the income-earning possibilities that he projected for the several youths who could work at Massive Harvest Farm if Lavondra only allowed them: "Give her 30 percent of the produce and let these kids make—I mean, at really conservative estimate, . . . they could make like seven grand." He simply could not see why the benefits of this compelling business model were not immediately obvious to Lavondra and why she insisted on seeing Massive Harvest Farm as a community garden.

Thus concluded this failed experiment of cooperation between white and African American food activists in a poor neighborhood of Northeast Central Durham. What can we make of this failed collaboration?

We review the two key issues of the conflict we see leading to a breakdown in the collaboration. First, as our chronology indicates, the African American food activists were interested in creating what they called a "community garden," while CL's owner-workers were committed to bringing a commercially successful urban farming cooperative into existence. According to CL, their collaboration would be one based on permacultural planting and tending the long-growing perennial fruit and nut trees that were to be a source of permanent abundance, in addition to cultivating other crops for sale while the trees grew to maturity. Yet one characteristic of permaculture is that these trees required several years of someone's labor tending, watering, and mulching them before such labor literally "bore fruit" (or nuts) that could be sold off or consumed by the youths. But this left the question of where this labor was to come from. Would the poor youths labor for years to attend these trees before they could draw income from them? Would the annual produce they planted every summer provide them with sufficient income in the interim?

In contrast, the community garden to which Lavondra and Saliya aspired, like community gardens in Durham and elsewhere, was based on a vision of an indefinite and renewable process of neighborhood-based cultivators sharing access to the plot and soil of the garden, making sure that each cultivator had access to the resources (tools, compost, seeds, and skills) necessary to grow produce and could use their own free labor to cultivate and, above all, share the food grown with other neighborhood residents, particularly those who were poorest, oldest, and most food insecure. Unlike the cooperative, this was a design that encouraged maximum participation by residents in a project with modest expectations while characterized by flexibility, available voluntary labor, and widely shared benefits among residents.

Second, despite the formal equality between white and African American activists participating in the making of Massive Harvest Farm, in actuality CL's James and its other owner-workers expected they would initiate and plan this cooperative to be a successful commercial venture, one based on the long-term requirements of permaculture, while they taught African American youths (and by extension African American activists) how to successfully cultivate food and market it commercially. This was a "top-down" scheme that CL sought to impose on a highly precarious and historically underprivileged demographic.

This was, after all, a vision based on a kind of expertise brought in to the community and pointed to as an ongoing challenge to the making of community by food activists across class and racial lines. Saliya, Lavondra, and Joyce indicated their deep concerns about the motives and objectives of James and CL's other worker-owners. This was precisely because the cooperative was to be surplus-producing only for a few owner-workers and therefore not devoted fully to a socially just "nonprofit" orientation. The operation of the cooperative, moreover, was being implemented through a structure of power inequalities based on white privilege that threatened to exclude from decision-making precisely those people who had the most at stake in its operation—the poor and food-insecure residents of Northeast Central Durham. CL's vision was insufficiently attentive to the perceived risk of an emergence of an old "plantation logic" based on white control and the subordination of African Americans as semi-unfree laborers—this time, through control by expertise rather than the overseer's whip.

Nonetheless, CL's project with Saliya and Lavondra was grounded in a clear awareness by James and other CL owner-workers of the threats to African Americans of gentrification and displacement, all too frequently brought on by the uninvited appearance of young, white, college-educated gardeners from outside dedicated to improving the diets and lives of poor people of color through starting community gardens in economically blighted urban spaces—a phenomenon common across the urban United States (Slocum 2007; Ramirez 2015) and evident among food activists elsewhere in Durham, as we saw earlier with IFFS's West End Community Garden. As Margaret Marietta Ramirez's case study of "creative," young, white gardeners seeking to create community gardens in Black neighborhoods of Seattle indicates, "the imposition of their privileged presence [as whites] upon a poor community can have detrimental effects. The act of building gardens in low-income communities may inadvertently 'undermine the well-being of people of color' by attracting development that pushes them out of the neighborhood" (2015, 761–762). James, in contrast, had been invited by Saliya and Lavondra to join them in undertaking the Massive Harvest Farm project.

Lincoln Street Neighborhood Farm: Micro-Alternative to the Capitalist Food Economy

Despite the failed attempt to make Massive Harvest Farm a small farm cooperative, it is a tribute to the persistence, good intentions, and political commitments of James and CL's other worker-owners that they continued to work with neighborhood activists in Northeast Central Durham to bring about abundance through edible landscapes. The Lincoln Street Neighborhood Farm (LANF) was started by a chastened CL joining with neighborhood activists in the spring of 2012 in the Lincoln Street neighborhood of Northeast Central Durham. We have followed its history in the local media and on CL's website since then. It was a half-acre vacant lot whose transformation, according to CL's website, sought "to create a replicable mini-farm model that inspires more neighborhood farms and backyard gardens to blossom in the NE Central Durham area, creating jobs and localizing neighborhood wealth" ("Community Projects," CL website, 2018). As James put it, reflecting back in 2013, the objective was "to create an agricultural revolution in

East Durham, so people can provide their own and need less from outside sources."

Neighborhood residents started the Neighborhood Farm with CL by beginning to cultivate a disused plot of land "filled with old bottles and trash" (Isaacs 2015). The half-acre lot of LANF was extensively transformed into a lush green garden through the design and assistance of the worker-owners of CL and by the labor of five residents of the Lincoln Street neighborhood, who worked on it and also received income by cultivating produce and selling it. Other neighborhood community-supported agriculture leaders also invested their labor in keeping the farm going. Neighborhood residents could become members of the farming cooperative by investing thirty hours of labor in planting, watering, weeding, and harvesting the produce and fruit and received a weekly food share in return (Isaacs 2015). It was reported that about twenty visitors each month came in for the Saturday workdays; most were children. Saturday workdays were also opportunities for residents to socialize, listen to music, barbecue, and work together on the farm, while learning from one another the best methods of farming. Neighborhood resident farmers taught the children the process of farming/gardening—from making compost to planting, tending produce, and harvesting their own crops (Exum 2013).

How far did the subjective transformation go this time around? As hoped for, CL's aspirations of engaging community members and transforming them into proud urban farmer/gardeners were at least partly successful. One of the community leaders in the neighborhood told a reporter, "Once the bugs bite you, you'll be finding ways to get back out here. It's relaxing and I learned so much from being out here" (Exum 2013). One observer in 2015 noted that the LANF "focuses on increasing fresh food access and sustainable bargaining practices in the economically depressed neighborhood . . . [but also] strives to provide an outlet for residents to come together" (Isaacs 2015). Although we were unable to follow the LANF ethnographically in these years, it appears that CL and neighborhood activists were able to agree on a hybrid farm that incorporated some of the features of a small farming cooperative with those of a community garden; for example, much of the produce that residents grew was either for their own consumption or sold to neighborhood residents, and residents were invited to use the farm's space for socializing, the education of neighborhood children, and so on.

Despite the challenges of working with community residents that CL encountered, the eventual success of CL's vision of cooperatives set up to work not only for but also with poor people and people of color in order to create community and generate abundance is a tribute to the ingenuity, hard work, and persistence of CL's James and its other worker-owners to help poor people bring about community food security in Northeast Central Durham. CL's antiracist approach shows that food activists are beginning to gain crucial insights about the ways in which race, class, food, and farming intersect in urban spaces. Yet, as the failed venture at Massive Harvest Farm indicates, much work remains to be done on the part of white community food security activists to learn to work with—and to learn from—their African American counterparts under conditions of shared power, expertise, and authority, so that white activists will, in Saliya's words, "know how to share power . . . [and] know how to resolve conflict" with those people who are most vulnerable and in need.

Cornucopia Landscaping and the Diverse Community Economy

Within Gibson-Graham's (2006, 71) typology of labor, transactions, and enterprises in the diverse community economy, the Cornucopia Landscaping cooperative was a noncapitalist enterprise, whose owner-workers received subsistence payments from their cooperative's surplus fund in lieu of capitalist wages.

We have witnessed the trade-offs that James and other CL owner-workers have made between working for the wealthy clients they create edible landscapes for—"to pay the bills"—and the gifts of their time and expertise in permaculture and garden design that they provided freely to community organizations and activists in order to provide them with the more sustainable long-term food abundance that permaculture offers. This illustrates the ways in which they have collectively engaged in ethical debates and decisions around how to use the *surplus* that their cooperative generates. Surplus generated by an enterprise can take a monetary form of wealth or can take the form of surplus labor hours—CL provides an excellent example of how surplus in labor hours can be ethically deployed to further the goal of building sustainable food systems.

By seeking to replicate the very process of making a cooperative enterprise, CL's owner-workers in trying to pass this process on to other groups in the local food economy, brought an invaluable reflexive dimension by learning the model of how to form a noncapitalist enterprise to their own practice, and sought to disseminate it to community groups. Despite the failed cooperative venture of Massive Harvest Farms, James and his fellow worker-owners have shown the value of persisting in attempts to develop an antiracist practice by attempting to work *with* poor African Americans in Durham, not just *on behalf of* them, as their more recent Lincoln Street Neighborhood Farm project demonstrates.

* * * *

In this chapter, we dealt with Cornucopia Landscapes as a cooperative food enterprise whose owner-workers were devoted to transforming the public spaces of Durham's poor and vulnerable people into long-term, sustainable "edible landscapes." First, we described the process through which CL's founder and other worker-owners sought to impart to its partners the values of cooperatives as they worked to create these edible landscapes. This dual aspect of CL's work—transforming the material setting while sharing cooperative practices with its partners—was challenging. We showed how redeployment of its owner-workers' surplus labor time, accumulated by its projects for wealthy clients, toward its pro bono projects on behalf of groups that were relatively disempowered might go smoothly, as it did in the building of the "edible schoolyard" at Tyson School.

Second, we recounted how the CL founder's pro bono attempt to transform the Massive Harvest Farm project in Northeast Central Durham into a commercial farm cooperative did not go well. This was a project set in a neighborhood where a majority of poor, food-insecure African American residents resided. CL's efforts fell afoul of the commitment by two African American activists to create Massive Harvest as a community garden. We showed that this was not merely a misunderstanding of what the operational status of Massive Harvest should be but a far more serious conflict between CL's white founder and these

Black activists, who saw CL's staff as exerting "plantation master"–style white privilege by unilaterally seeking to determine Massive Harvest's direction. In contrast to CL, the two activists sought to mobilize neighborhood residents to relieve residents' food insecurity and give them more control over their own food through their own voluntary labor in the garden and giving produce freely to other residents in need.

Third, we were impressed as we found that CL's response to the Massive Harvest fiasco was to learn from it and, chastened, move on to learn to work more collaboratively with neighborhood residents in CL's next project, the Lincoln Street Neighborhood Farm, a hybrid enterprise that mixed cooperative principles with highly valued forms of local sociality.

Fourth, following Gibson-Graham (2006), we proposed that CL's owner-workers provided an important model for how food cooperatives could engage in rewarding ethical decision-making about how surplus labor time accumulated by a cooperative enterprise might best be used for community benefit.

Finally, we concluded that Cornucopia Landscaping was a successful "crossover" in which James Sokolov and CL's other owner-workers found ways to simultaneously achieve the goals of sustainable farming and community food security activisms. CL's owner-workers implemented permacultural designs and labored to create long-term edible landscapes with fruits and nuts that could be consumed or sold by community members, adding to the sustainability of local food sources. At the same time, CL undertook (ultimately successfully) to work with poor African American residents to form a small-scale urban farm cooperative as a solution to food insecurity and income loss among residents, while CL disseminated the model of urban farming cooperatives based on a more just relationship among cooperative worker-owners than the exploitative employer-worker wage-labor relation found in capitalist enterprises.

9

Clarketon Community Garden and Racial and Class Divisions

The [Clarketon] community garden initiative began in 2010. As of spring of 2012, it included twelve acres. The idea of the garden is to increase food security of the area and promote a communal relationship between residents based on their mutual need for fresh food. Food grown in the garden is free to Clarketon residents. Youth, both young men and young women mostly in their teens, are heavily involved in the garden and activities associated with it. The young people work in the garden, learn beekeeping from the beekeeper associated with the garden, and help put in gardens at the homes of people who need assistance. Especially exciting to them, the young people sell produce at the Tarboro Farmers market.
—Willie Jamaal Wright, ethnographer, field note, 2012

What Willie Jamaal Wright, our ethnographer, did not need to note for himself in the field note quoted in the epigraph was that the "community" he referred to consisted almost entirely of poor, rural African Americans living in a region of coastal North Carolina known since the early 1800s for its past landscapes of Black enslavement, cotton plantations, Civil War battles, Emancipation, Reconstruction, Jim Crow, white terror, tobacco sharecropping, the civil rights movement, and, throughout, profound racial animosities that separated the Black and white populations of one of the poorest areas of the southeastern US.

In our research site of Halifax, Edgecombe, and Nash Counties, as elsewhere in the rural southern US, African Americans have demonstrated their long-standing militance in seeking community autonomy, dignity, and self-respect in the face of hostile and dominant white elites and their followers, and they have provided many exemplars of efforts toward an antiracist politics. In a previous book (Holland et al. 2007,

213–220), we followed the social activism of the Concerned Citizens of Thornton in Halifax County, an organization of rural, poor African Americans who had a sustained history over decades of struggle on behalf of the area's African American farmers to hold onto their farmland and receive equitable treatment from the USDA and on behalf of the area's residents to be protected from the downwind and downstream toxic effects of animal wastes from CAFOs located nearby. To generalize, similar efforts elsewhere by African Americans, American Indians, and Latinx populations have taken place across the rural US in landscapes devalued by racial capitalism (Wright 2020).

It is not surprising, therefore, that no social project among all those we studied interested us more in this respect than the Clarketon Community Garden (CCG) in the small town of Clarketon, population 365 people, located in rural Edgecombe County, North Carolina, some forty miles as the crow flies south of Thornton. The CCG simultaneously shows the ways in which activists who were themselves members of poor and vulnerable rural populations could successfully create community food security, while working toward sustainable food provisioning in the only way open to them in a setting where African Americans, despite hard-won but still tenuous political rights, have been systematically dispossessed of their means of livelihood, particularly their farmland, by decades of race- and class-based violence carried out against them by local powerful white elites and their followers.[1]

In this chapter, we consider the Clarketon community garden project in relation to how Clarketon activists have furthered community food security and sustainable food objectives within their own diverse community economy, given its own material poverty and in the face of chronic racial- and class-based violence. In our other three sites in North Carolina, we saw that our theory about the moral logics of local food activisms went far to capture the widespread trends, social projects, and organizing around local food and farming that we witnessed and listened to during our ethnography. To the extent that we were aware of the racial- and class-based inequalities we found in these sites, we paid attention to them, sought to bring them into our chapters' analyses, and showed how they shaped food activism. Nonetheless, the findings in this chapter are a stark reminder that if this theory has failed to take into account encompassing material racial and class injustices that constrain the development of sustainable local food economies, then our theory would need to be transformed or added to. In this chapter, we have done so.

Moreover, although we have found particularly valuable what J. K. Gibson-Graham (2006) have to say about diverse community economies, which challenges the assumptions of conventional capitalist development theory, Gibson-Graham fail to theorize the power of racial capitalism (Robinson 1983/2000; Wang 2018; Nonini 2021) that has overwhelmed the potential of stigmatized racialized populations and their local food economies to flourish. Capitalist economic growth under neoliberalism has propelled the processes of attempted endless economic growth, ecological devastation, creation of gross economic inequalities, and most crucially, climate change—but it also bears with it the further costs to vulnerable racialized populations of their being dispossessed by violence and experiencing the most extreme exposure to the disasters of climate change. Although Gibson-Graham (2006) remind us that people seeking to bring into existence alternative community economies must challenge the conventional neoliberal theory of capitalist growth, we must also understand that this theory has in turn served as a rationale for the violence of racialized corporate states and private actors that prevent political and economic democracy in the US. Therefore, our theorizations have gone beyond Gibson-Graham's (2006) poststructuralist approach to a more comprehensive account of class and racial power, which in part we began earlier with our analysis of the power of the alliance of industrial food corporations and the compliant corporate state vis-à-vis local food economies.

In this chapter, we continue this analysis as we seek to transcend the limitations of Gibson-Graham (2006) in this respect by conceptualizing the limits that racial capitalism imposes on the possibilities of development of sustainable farming and community food activism among racially vulnerable populations, with the situation of Clarketon Community Garden and its rural African American activists in mind as we do so. As Clarketon activists engaged in collective ethical decision-making around necessity, surplus, consumption, and the commons that directly called the assumptions of capitalist development theory into question (Gibson-Graham 2006, 167), they had to do so under the pressures of racial and class oppression that constrained the forms such decision-making could take. We thus examine the ways in which the Clarketon community garden project sets out a model for such decision-making through the ways in which its activists discussed, debated, and reformulated *shared alternative meanings of surplus, consumption, and the commons* but also as they confronted the broader class and racial injustices perpetrated against the residents of Clarketon.

The Making of a "Landscape of Production" in Decline: Racial Antagonisms and Class Inequalities in an Age of Globalization

Edgecombe County is located in the historic "Plantation Belt" of eastern North Carolina. In Edgecombe and adjoining Nash and Halifax Counties, there are still places named "X Plantation." These places and others nearby were sites, from the early 1800s onward until World War II, where whites forced enslaved, and then semifree, sharecropping Blacks to pick the cotton and tobacco that were the commercial staples of rural eastern North Carolina. By the new millennium, the successes of the civil rights movement meant that the overt racial animosities arising from more than a century of slavery, followed by sharecropping, Jim Crow politics of terror and segregation, and Black outmigration from Edgecombe County and nearby, had in the wake of the successes of the civil rights movement given way to a new, more subdued, in some respects more just, but still tense politics of racial and class relations.[2]

This transition in Edgecombe County and nearby had not been an easy or peaceful one, as the bitter retrospective views of some Black people in the county illustrated, especially as they looked back to the early 1960s during the legally mandated integration of white and Black schools in the region. Jamil, in his fifties, a resident of Little Raleigh in Rocky Mount, whom our ethnographer Willie Jamaal Wright had befriended, recalled his experiences as a Black student entering a local middle school in 1962 as it was being integrated:

> I was one of the first Blacks to go to Edwards. . . . That was junior high. And we used to have to go walk to them poor white neighborhoods. Police was there to escort, so like, they were [calling us racial epithets]. [They'll] fuck with you, curse you out, all kinds of shit. They had that little walk

[with signs saying], "N-word, git out my yard." . . . The white boys used to hang over at 7 Eleven, and we used to hang over at the Zip Mart. But we used to meet at Zip Mart, and the police used to meet us over there, and then—then they could to take us to school by the white boys. . . . They had to escort going through them [white] projects, going that way to school. . . . Aycock Park—there's a park over there, Aycock Park. Say, like, back then, say, like right now, if it was six o'clock and it got dark [and] you got caught over there, man, they do whatever they want to do. That's why a lot of people, like older than me, we didn't play ball [and] all the extracurricular shit because when time was up [school was out], we had to leave. Because when the sun came down, man, man, man, them white boys used to chase us, and we used to chase them.

By 2009, when we began fieldwork in eastern North Carolina, race relations had settled into an uneasy pattern of social distancing, tense interpersonal interactions, and mutual suspicion between most whites and African Americans in Rocky Mount and surrounding counties, despite a few close interracial friendships. Class inequalities were reflected in and exacerbated racial differences in the region. A small white elite of business owners, managers, and large farmers were set over a relatively large working population of people who labored long hours for very low wages when they could find work, although many could not. By the end of the first decade of this century, at the height of the Great Recession, Rocky Mount, North Carolina, had gained the dubious reputation of being one of the "ten poorest cities in the US," based on its per capita income, unemployment, and percentages of its population receiving food stamps (i.e., SNAP benefits) and public health care (i.e., Medicaid), according to an analysis undertaken by *Forbes* magazine (*Forbes* 2009). The Great Recession of 2007–2008 had merely made a long-standing situation of lack of jobs, historically disempowered and undereducated rural residents, and disparities by race in access to jobs even worse than before.

Under neoliberal globalization, US federal trade agreements (e.g., NAFTA/USMCA and WTO) and deregulation of capital flows during the 1980s–2000s promoted deindustrialization and the offshoring of jobs from the US. From being one of the most industrialized states in the US in the mid-1980s, North Carolina became one of the least, when it experienced a net loss of almost 360,000 industrial jobs during the years 1993–2013, due to certified "trade-related" job losses from NAFTA and the entry of China into the WTO (Public Citizen 2016; Bureau of Labor Statistics 2016). Like many other areas in rural North Carolina, Edgecombe and Nash Counties were "landscapes of production," which were previously known for their light industries and commodity farming but had disproportionately experienced serious deindustrialization and job losses due to neoliberal policies during these years (Holland et al. 2007, 18–34). From 1993 to 2013, Edgecombe County lost a net 2,687 trade-related industrial jobs, affecting 5.4 percent of the workforce, while Nash lost a net 3,755 trade-related jobs, or 4.6 percent of its labor force (Public Citizen 2016; Bureau of Labor Statistics 2016). Furthermore, an economic analysis of trade-related job losses in a nearby county (Robeson), were it applied to Edgecombe and Nash Counties, suggests that *at least* as many nonindustrial as industrial jobs were additionally lost in these counties, as local businesses dependent on the spending of laid-off industrial workers had to close down (Hossfeld et al. 2004).

By 2011, when our fieldwork in Edgecombe, Nash, and Halifax Counties began, the combination of class inequality, high levels of poverty among the population, a depressed local economy, and residential racial segregation was graphically evident in the physical appearance of downtown Rocky Mount, a town of slightly more than sixty thousand people. The town was bisected by the north-to-south county line that followed the Atlantic Coastline Railway tracks between majority-white Nash County to the west and majority-Black Edgecombe County to the east. As one walked southward down the transect of these tracks through Rocky Mount's downtown, the relatively prosperous western side of downtown showed new stores and well-maintained city government buildings, schools, and churches, while the eastern side of downtown displayed closed shops and ramshackle or abandoned buildings, particularly in the older impoverished neighborhood of southeast Rocky Mount (Barber et al. 2008). This contrast was a revealing, if imprecise, index of the relative wealth and class status of the populations in the western and eastern areas of the city and the spatial disparities in business activity, property taxes, and county government services that depended on these taxes. People were, if not flourishing, then getting by in

Nash County to the west, while doing far less well in Edgecombe to the east. Clarketon was in Edgecombe County, not far from Rocky Mount.

The Clarketon Baptist Chapel, Its "Youth," and Its Garden

The Clarketon Community Garden was founded in 2007 with one acre donated by an African American farmer for use by members of the Clarketon Family Life Center (CFLC), a project of the Clarketon Baptist Chapel church. Four years later, by the time of our fieldwork in 2011, the CCG had expanded to include a one-acre "youth" garden and an eleven-acre community farm. The "youth" (plural) as Clarketon congregants called older children and teenagers, worked in the garden with the assistance of the church's adult garden manager and other staff and joined adult congregation volunteers in the garden to cultivate collards, other greens, tomatoes, and onions that grow well in the summer weather of this lowland region of eastern North Carolina. By 2012, more than fifty African American youth from the Clarketon area had become active participants in this project during their middle- and high-school years. They were learning not only about agriculture and food firsthand but also about jobs and business opportunities associated with agriculture. This flourishing space of experimentation around local food and its potential for the making of a more just future became, for those of us working collaboratively on local food activism in North Carolina, an astounding site of learning and discovery, as we followed the fieldwork of our ethnographer Willie Jamaal Wright in Clarketon and nearby Rocky Mount.

It is relevant to understanding the significance of the CCG that these youth came from the families of poor and historically marginalized rural African Americans in the county and nearby. Many of the youth had family members who were incarcerated or had been previously. Unemployment among adults was very high not only due to economic decline in the region but also due to the presence of elite white networks of local officials and business owners and managers that "exclude African Americans from all but the most menial jobs" (Holland et al. 2007, 176–177). Discrimination against Blacks in employment was exacerbated by historical failures to provide public services like good public schools and bus transport, which placed additional burdens on poor, rural African

Americans in Edgecombe County and nearby. Moreover, large proportions of children, youth, and adult members of the community suffered from obesity. Among older adults, obesity was prevalent, as were serious diseases brought about by it, and many were unable to work due to diabetes, high blood pressure, and so on.

These factors meant that prior to the appearance of the garden, for local youth there appeared to be few options to earn income, none of them attractive: either join a gang—in this case a real-life choice, not a confirmation of a mainstream media stereotype; leave the area along with support from family members and kin to venture to Raleigh or to cities to the north; or with luck, find wage work nearby at below minimum wage, such as at a local fast-foods outlet.[3]

Origins of the Clarketon Community Garden: "We Gotta Do Something to Really Begin to Look at How We Deal with This Process"

The pastor of the Clarketon Baptist chapel, the Reverend David Hemmings, took up his position nine years before our fieldwork began in 2011. He soon noticed that the residents of Clarketon and members of his congregation appeared to suffer from obesity and a variety of food-related ailments. Five years after he arrived, he was able to have his entire congregation undergo health screening at a local hospital, and it was discovered that 70 percent of congregation members (children and adults) were overweight and that an additional 10 percent were morbidly overweight. This led him to found the church's nutritional, exercise, and gardening programs.

Widespread obesity and endemic illness were not the only problems facing Clarketon residents. When Rev. Hemmings arrived, he discovered that the youth in the area faced many obstacles to prospering. He recalled, "When I began to work with the children of that community [Clarketon] and began to track the history of what happened to children that grow up in this community, most of Black males ended up incarcerated, . . . [with] multiple family things going on. We started doing some home visits and [were] really taken aback by how people looked on Sunday but how they lived. And so, I really just start saying, 'We gotta do something to really begin to look at how we deal with this process.'"

This is what led him to start the garden with his congregation members and to work to attract youth in particular to join adults in the garden work. By the time of our 2011 fieldwork, this particular struggle had been won. As Rev. Hemmings spoke with humor about the youth, "They keep showing up. I don't know if I'd show up if I was doing to them what I was doing to them. And they be doing their little push-ups [from early morning calisthenics in the garden], and they be crying, . . . [mimicking] 'This is hurting me so bad.' 'You a mean preacher!'"

A Garden's Pedagogy: Sustainable Farming Begins with Sustaining Youth

Rev. Hemmings and the staff of the CFLC recalled and demonstrated to our ethnographer Willie Jamaal Wright the many-sided pedagogical experience of the youth working in the garden. The youth were taught the fundamentals of nutrition and the importance of fresh food and the nutritional dangers and economic costs associated with a diet of fast foods—lessons that adult members of the congregation learned as well. Charles Rawls, the garden manager, and volunteers taught the youth how to evaluate, purchase, plant, transplant, and harvest produce and showed them such practical skills as testing the soil for pH, driving a tractor, and learning the best growing conditions for the different vegetables. James Jones, another member of the congregation, a retired military veteran, was a beekeeper and taught the youth beekeeping, and by the time of our fieldwork, the youth had kept three hives and collected 150 pounds of honey during the prior year. Shelters From the Storm, a volunteer house-rebuilding project like Habitat for Humanity, sponsored Clarketon Baptist Chapel and other local churches in their efforts to construct better housing for poor seniors in the community. CFLC volunteers taught the youth the skills of carpentry, and at other times, the youth were asked to take on the task of starting gardens for seniors who were disabled.

Rev. Hemmings described to Willie Jamaal a typical summer morning in the garden and how these connections were made by the youth by daily practice and learning not only about gardening techniques and physical exercise but also about social interdependencies that the industrial food economy obscured and even destroyed:

We meet at six o'clock in the morning because a lot of our youth are obese. We started out—we went from five years old to twelve [years old]. They do hundred sit-ups, a hundred push-ups, [and] jumping jacks. The five- to twelve-[year-olds] do a two-mile run afterwards. The twelve- to seventeen-[year-olds] double that for a four-mile run. Our goal is that they will want to build their core, lose that weight, and will learn what it's like to eat healthy. They do that from six o'clock to seven thirty. At seven thirty, they going to chapel time. We have chapel time. After chapel time, one group goes to the garden, one group works on end-of-grade test, one group starts working on social skills and looking at social dynamics, and another group does breakout sessions. When one group brings the food from the garden, another group prepares it for the meal that day. So, everything they do is contained on the campus 'cause we want to teach . . . that you can't go out and buy food. You gotta maximize your food, as you gotta cook your own food. You not gonna come out of poverty buying [fast food from] spending money. You gotta stop spending and start producing.

Rev. Hemmings's last injunction to the youth to "stop spending and start producing" reminds us, as it did the youth, that there are important questions of ethical decision-making to be answered regarding how surplus should be used and consumption defined within any community economy (Gibson-Graham 2006, 90–95)—issues we return to shortly.

Considered in light of experiences like these, it is clear that were we merely to enumerate the future occupational skills and health and nutritional and agricultural knowledge that Rev. Hemmings and the CFLC staff instilled in the youth and tried to get them to master, we would miss a vital dimension of their experience working in the garden. Central to the garden's meaning for youth was the identification of work in it as a need that people met through cooperation and interdependencies. As Rev. Hemmings put it, "It is a need socially because we come together and put something together to help people to have food—which, when people eat, is a social gathering." What mattered was not only the content of what they learned but also, and more importantly, their awareness of who taught them, who cared for them, and what roles those who taught and cared for them occupied within the church and the community as a whole. The youth's mentors included not only Rev. Hemmings but also

Charles, the garden manager (who ploughed the fields and raised starts of produce for the garden and to give away to community members); Mr. Jones, the beekeeper; Trent, a tractor mechanic; and several retirees who assisted the youth and sought to meet the broader health and nutrition goals of the CFLC. They were all, as were the youth they taught and supported, members of a local moral community of poor African Americans whose solidarity represented a crucial resource as they confronted a milieu of hostile and unsympathetic whites on an everyday basis. Central to the figured world of Rev. Hemmings and the Clarketon congregation, in this respect, were the nurturing qualities of the food grown, shared, and eaten within this community.

Mediated by the physical setting of the garden itself and the cooperative efforts of its cultivation, which brought together the youth, volunteer congregation members, a few paid staff members, and Rev. Hemmings, the youth "learned how to learn," which placed the gardening work within the frame of meeting their obligations as youth to themselves, to each other, to their families, and to the community. They saw an emphasis by congregation members and volunteers like Charles and Mr. Jones on a work ethic of persistence and patience, focused on the skilled crafts associated with farming and construction, and also, as Rev. Hemmings put it, on "respect for self and the earth and show[ing] Black males who sell drugs that there's money to be made in growing food."

Youth at the Clarketon Garden saw that retired members of the congregation served as volunteers after long lives of labor in factories, the military, and government employment in the North and had returned to Clarketon in part to teach them what they knew. Youth found that they were expected to look after each other if one suffered the frequent setback of a death or sickness in the family, the loss of a job, or the like. They discovered that the CFLC's acceptance of Trent, the tractor mechanic, who was a white man, demonstrated the church's commitment to racial equality. They saw that they were valued by the church's pastor, staff, and congregation members as having potential for dignified livelihoods based on further education beyond the life of gangs and probable imprisonment and beyond the stultifying, dead-end, and poorly paid wage labor that most whites saw them as only fit for. Within this experience, they found themselves confronted with choices about the meaning of surplus, money, and personal worth.

Ethical Decision-Making, Part 1: What Is Surplus, and What Should Be Done with It?

A key part of the youth garden project was the youth selling part of the produce they harvested from the garden at the nearby Tarboro Farmers Market and the local public-access television station, allowing the youth to earn money from their efforts. However, what they did with the money they earned was not up to them, for they had to turn over part of the money they earned to a special fund, where it was saved for their future college expenses. Local philanthropists and the Clarketon congregation pledged several thousand dollars in matching money to the fund as incentive to the youth. The rest of the income they earned from produce sales allowed them to contribute to their families' income. Moreover, not all the produce they cultivated was sold for their own benefit. Another portion of the garden produce they grew they gave to seniors and other food-insecure members of the congregation.

Two important questions were being asked through the choices the youth were faced with in determining their own connections to food, money, and social relationships, exemplified by considering the value of fresh food for nourishment of family, church, and community members, on the one hand, and the monetary income received for the produce that the youth sold, on the other. They were asked to reflect on the question, "What is the value of the produce you grow and of the work you put into growing it, and what is the meaning of the money you receive for doing so—and when should you receive money for growing food?" One could say, following the feminist geographers Gibson-Graham (2006, 90–95), that the youth were being confronted with the need to engage in "ethical decision-making" around surplus. Such an interrogation of what surplus is and does is an increasing imperative for any group that seeks to build a "diverse community economy" challenging the neoliberal economic doctrine that economic growth and the creation of more monetary wealth are always good in their own right. Gibson-Graham (2006, 193) argue that communities can pursue alternative routes to development when they use "surplus as a force for constituting and strengthening communities—defining the boundary between necessary and surplus labor, monitoring the production of surplus, tracking the ways in which it is appropriated and distributed, and discussing how it can be mar-shaled to sustain and build community economies."

This is precisely what Rev. Hemmings and the CFLC asked youth working in the garden to consider in evaluating the money they made from it. When is it just to take fresh food—something you produce through your own labor but also something only made possible through your interdependence with many other people who care for you and to whom you are responsible—and to convert that into money only for your own personal use? When you sell something you produce, what should the money be used for—for today's pleasures or next year's opportunities? When does justice demand that you forgo turning the food you grow into cash because it is needed by your family or by sick, aged, or disabled members of your community? And is "forgoing" or "sacrificing" an ac-curate description of such gifts to the people on whom you depend and who depend on you for life? How these questions were being asked in the Clarketon Community Garden by Rev. Hemmings, staff, and volunteers such as Mr. Rawls and Mr. Jones and, above all, were answered by the youth working in it was a critical feature of the garden as an ethical and political project—and as a community food security initiative.

Ethical Decision-Making, Part 2: How Should We Think about Consumption?

In Gibson-Graham's proposals for the design features of a future com-munity economy that will be an alternative to a capitalist economy, they ask us to reflect on the meaning of consumption in challenging con-ventional capitalist development theory. They contend that we must see consumption "as a potentially viable route to development rather than simply its end result" (Gibson-Graham 2006, 193) and continually inter-rogate what it means to see consumption at home not as waste but as a kind of investment, rather than seeing investment of capital for profit outside a locale—and its extraction and "return" home—as the be-all and end-all driving development. For example, when a tenants' orga-nization in western Massachusetts established a cooperative to build an affordable housing project within its locale, could this be viewed best as "consumption" for residents or as "investment" of capital—or was this the wrong question to be asking (Gibson-Graham 2006, 184–186)?

Could "investing locally" be superior to "investing globally" when the conventional development story emphasizes "comparative advantage," in which each locale best trades with others, exporting and selling what it specializes in, in order to buy what is most efficiently produced by outsiders—even though this subjects vulnerable local economies to a loss of wealth through outflows of capital in return for services they might in fact provide for themselves? When is consumption wasteful, and when is it productive? We might ask "how well-being can be created directly, without resort to the circuitous pathway of export-led trickle-down development" that characterizes the mainstream capitalist modus operandi (Gibson-Graham 2006, 184).

In the case of the Clarketon Community Garden, we reflect on these questions by describing what superficially appears to be a trivial incident that occurred one day in summer 2011, when sixteen youth driven by Rev. Hemmings and our ethnographer Willie Jamaal Wright went on a field trip to a nearby city to tour a model successful small-scale sustainable farm and, on their way back to Clarketon, stopped for lunch at a Hardee's, an outlet of a fast-food chain located in North Carolina. Most of the youth had not brought any lunch money or their own lunches, and Rev. Hemmings ended up paying for their lunches. Willie Jamaal's fieldnotes commented on what happened next:

> Hemmings shared with me that it felt it was a "shame" that the children did not see the difference between "producers" and "consumers." In his eyes, most of them and their families are "consumers." Not only are they consumers, but they spend most, if not all, of their discretionary income outside of their communities. This he views as a moral imperative within Clarketon because he wants these children to be of the mind-set of a producer, making goods to sell instead of constantly buying from others. Today's meal at McDonald's and Hardee's provided perfect examples for his talk. Rev. Hemmings shared with them that on this day, they spent an average of $100 on meals and incidentals. Most of this cost could have been avoided had they brought their own lunches. For those who paid for their lunches with money given to them by their parents, Rev. Hemmings asked, "How do you plan on paying your parents back for this money? . . . I want this to be the last time that we go on a trip where we spend more money than we make."

Here Rev. Hemmings questioned the nature of consumption by the youth. What was consumed and where it came from was manifest in the approach of the youth as "consumers" who either failed to take their own lunch from home or, for a few, depended on their parents for lunch money. Just as Gibson-Graham (2006, 184–186) argue that the development of a community economy requires thoughtful decisions about the conditions under which consumption within the local economy should occur—which would lead residents of a community to form a cooperative to work together to construct their own housing instead of paying outside developers through the repayment of debt to outside finance—so too did Rev. Hemmings enjoin the youth to reflect on the costs of passively expecting lunch to be provided by someone else, which required the wealth to purchase it (either the church's, from Rev. Hemmings, or from their parents) to leave the local economy. This warning was close to home: the youth were well aware of the exploitation and abuse of young African American adults working in corporate fast-food outlets in Edgecombe County, because their older family members had experienced such mistreatment. Instead, Rev. Hemmings asked the youths to be "producers" who made their own lunches and to realize that their individual decisions had collective consequences that they needed to be aware of and accountable for, so that no longer would they spend more money than they make.

While presenting the dilemma of how the consumption of lunch was connected to its production—what kind of lunch was made, who produced it, and at what cost—appears a trivial situation set side by side with Gibson-Graham's case study, in both cases the same issue of how to create collective economic self-reliance vis-à-vis powerful outsiders came to the fore for discussion. In both cases, decisions people could make to move beyond being merely passive consumers dependent on large-scale capitalist institutions whose operations they do not understand (outside developers and financiers in the first case, fast-food corporations in the second) to take more control over their economic situations (by forming a building cooperative and producing their own food, respectively) pose ethical (and political) questions that subvert the conventional development theory of neoliberal capitalism. But, going beyond Gibson-Graham's (2006) theory, Rev. Hemmings was also reminding the youth in effect that for them, as poor rural African

American youth, to be a "consumer" meant to surrender one's wealth to an antagonistic institution of racial food capitalism—the corporate-owned fast-food outlets that exploited so many local African American youth and left them with dead-end jobs. Rev. Hemmings's practice thus provides a key clue to theorists about how to proceed.

An Afrocentric Moral Logic: "Relationships" That "Help the Plant Stay Connected" and the Commons

The questions posed by Rev. Hemmings to the youth in the course of their everyday garden cultivation, calisthenics, and nutritional education indexed a broader vision of society that he sought to bring to fruition, what he called an "Afrocentric way of living." To Rev. Hemmings, an Afrocentric way of living emphasized "barter"—the direct exchange or sharing of goods and services among people who live in conditions where money is scarce or absent as a means for acquiring the necessities of daily life—not monetary exchange in which goods are sold between people in return for cash. When Willie Jamaal asked Rev. Hemmings what the most important goal of his garden project was, he replied, "The most important goal is . . . bringing back that money is not the only way. Money is not the purchasing power. But learning the bartering is the purchasing power. Because, if you learn to barter, you don't—you can do that in Africa."

For Rev. Hemmings, an Afrocentric way of living contrasted with the "Eurocentric mind-set" characteristic of the dominant white society, with its inequalities—a mind-set that promoted a "one up, one down," zero-sum economy that required someone to suffer in order for others to prosper. This, according to Rev. Hemmings, is the mind-set of calculation that determines a person's worth based on the cash nexus or, put another way, on a person's class position and their access to money. The Eurocentric mind-set was one of calculation and the creation of domination through hierarchy, as Rev. Hemmings put it, "of less than, greater than, . . . the struggle [between] the wealthy and the middle class and the poor": "I want to live where there are no classes. There are human beings." In contrast, he stated, "When money is the driving factor, then there's a lot of people excluded from the table. The job [that is, wage labor and exploitation]. That's how we get poor people. When you barter, there ain't nobody poor."

Rev. Hemmings's model of the Clarketon Garden as a node of community life grounded in barter was thus one of wealth, one in which not only was produce shared between people, but so too were knowledges, not only horticultural knowledge (of how to prepare soil, when to plant, how to control insects) but also socio-moral knowledge—whom one could rely on under what conditions and whom one must be accountable to under what conditions—accompanied by the labor that validated these knowledges. For Rev. Hemmings, as the sharing of produce and reciprocity that tied youth to their families and older people demonstrated, the networks of mutual assistance and nurturance that existed in an Afrocentric community outside the capitalist cash nexus, with its calculations of monetary gain, made it no surprise that "there ain't nobody poor."

Here the key concept of an Afrocentric vision was "relationship," which was to be set over and against "resources" or, put another way, commodities that were bought and sold. "Bartering is about relationship." According to Rev. Hemmings, the Afrocentric vision valued human relationships over resources, to the point that any relationship built solely on the instrumental connections that people had to commodities and money destroyed that relationship; whereas, in contrast, when relationships based on sharing were put first, only then did resources and genuine abundance follow. "So I don't kill the relationship for the resources. Because they know the resources is a resolve [result] of the relationship. Right now we live in a country that—we believe that the money is the purpose of the relationship. So we believe we gonna kill the relationship, and if we got money, [then] we'll get relationships. Those are not healthy relationships." The Afrocentric moral logic thus favored the value of the collective as a community over individual self-interest.

This vision, in which the growing and giving and exchange of food through barter as exemplified in the Clarketon Community Garden and others like it, provided the model for a better world defined by collective welfare based on relationships and not on individuals striving to accumulate resources and money over and against one another, which destroyed relationships. This was also a vision at the heart of community food security. At the same time, this better world still had to come to terms with the world of money and commodities in which it was situated. The Clarketon garden project, which taught youth how money they acquired through their labor could be put to work for their education so that they could later serve their community as well as benefit personally, allowed for this. As our ethnographer Willie Jamaal observed, "This scholarship [money] will encourage students to go to college to 'serve,' not make the most money. Rev. Hemmings believes that individuals who take on certain professions to make money and not to serve 'are what you call building a thief' because they only work for money."

According to Rev. Hemmings, relationships between humans had intrinsic value in their own right and should not be reduced to instrumental connections between humans and the commodities they desired. Accordingly, how such relationships could be sustained under the corrosive conditions of racial capitalism and social division in this rural area of eastern North Carolina were what Rev. Hemmings sought the Clarketon garden to simultaneously demonstrate on its material side and to project as an ideal into the future.

Rev. Hemmings was emphatic that relationships in the Afrocentric vision were based on the sustenance that people provided each other as members of a community. The raising and giving of food to people in need was his central metaphor for such sustenance. When the responsibility for growing food was widely shared and such food was given to others, relationships flourished. However, when food was taken or commodified for individual use at the expense of others, sustenance ended because the relationship was no longer fruitful. "People who do not appreciate the plant" are those who took for themselves from the garden grown by and for the collective, but without thought for others, and were like thieves. As Rev. Hemmings expressed it, "People go out and cut the plant gonna lose productivity. They're only going to get to eat from that plant one time. Anybody to help that plant stay connected, and [if] they only take off that plant what they need, the plant will reproduce itself. We are relationships that do not reproduce because we have cut away from this creator. So there is no creativity in there anymore. The only thing they can do now is probably steal somebody else's creations." Put thus, Rev. Hemmings moved beyond metaphor (where taking from the garden "stood for" human disconnection) to say that humans were part of the natural (and spiritual) world of "this creator."

The trope of stealing the whole plants from the garden, thus rendering it sterile and useless for others, informed the Afrocentric vision of Rev.

Hemmings. It was based on his early experience as Clarketon's pastor, as he sought to draw not only youth into the garden project but also the adult congregation and community members, who fought among themselves in poverty and lacked "social skills" and "rarely talked with one another." This came to a climax in an event that he recounted to Willie. One night, a man backed up his truck to the garden plot and collected not only the produce from the plants but the entire growing plants as well. Rev. Hemmings recounted this pivotal experience to one visitor from a funder to the garden program:

> [A] guy back his truck in there, cut down everything, and carried it away. The kids were like—they called me from work and was like, "Everything is gone." And the guys said, "Well, you said it was a community garden." We really had to deal with the mind-set that when you live in poverty, you really do not believe there is enough to go around. So when you get a chance to get something, you take it all. And so, when to redefine and rework [the] community. And now we see a lot more sharing going on. There's a lot a partnershiping going on, but we're still dealing with a lot of that fragment stuff that comes out of when you live in a community that have generations of dysfunctional stuff going on.

In this story of the theft from the garden and the redemption of those who took from it through the effort to redefine and rework the community, the restorative vision of African American society recovering from "generations of dysfunctional stuff" due to white domination was manifest. This was also the basis of a sustainable local food system—one whose population was deeply impoverished and had to look to themselves for food and other kinds of mutual aid in confronting the hostile dominant world of whites in power beyond the Clarketon community.

Ethical Decision-Making, Part 3: Sustaining the Commons of a Community Economy

Rev. Hemmings's claim that "the resources is a resolve [result] of the relationship" led directly to the question of the commons, the collectively generated and shared material and social wealth fundamental to the economic survival and flourishing of the members of a community over time (Nonini 2007, 1). The wealth of a commons may be composed of diverse kinds of use-values—natural resources, built infrastructure, social skills and labor, intellectual and cultural assets, economically valuable species genotypes, and more (Nonini 2007, 1–8). The key point is, as a noted anthropologist put it, "A community economy makes and shares a commons. . . . Without a commons, there is no community, without a community, there is no commons" (Gudeman 2001, 27). Common property resources—such as freshwater, forests, and arable land, either built up or replenished over time through shared labor by community members and drawn on for use by them—are what outside commercial corporations, financiers, and speculators empowered by neoliberal politics have sought to appropriate for their own profit by dispossessing the prior users of the commons of these resources.

There are similarities between the generative (indeed lyrical) thinking of Rev. Hemmings about the relationship between the garden and the human relationships that could sustain or destroy it and the ethical decision-making about the commons that Gibson-Graham (2006) point to. They write that alternatives to conventional capitalist development mean "creating, enlarging, reclaiming, replenishing, and sharing a commons, acknowledging the interdependence of individuals, groups, nature, things, traditions, and knowledges, and tending the commons as a way of tending the community" (Gibson-Graham 2006, 193). Gibson-Graham (2006, 188) go on to refer to, among other things, a case study of community gardens as a form of commons:

> Nuestras Raices (Our Roots) is a network of community gardens located in the industrialized city of Holyoke in the Pioneer Valley of Western Massachusetts, where one third of the city population is Puerto Rican. Since 1993, this nonprofit community organization has been converting vacant lots scattered throughout downtown Holyoke into community gardens, reclaiming urban space for the people's use. With active community involvement, Nuestras Raices' mission is to build on what people already know (agriculture, for instance . . .) And to deliver what they specifically need, like food security and a place for young people to learn skills and stay off the streets. The gardens currently involve over 100 families and each garden plot produces up to $1000 worth of produce each year, adding a substantial amount to the food budget of a family below the poverty line.

Summarizing how Nuestras Raices created a commons, Gibson-Graham (2006) note that it developed its network of community gardens as a collectively governed resource, whose use-values members drew on for food, even as they sustained it through their labor over time. Maintaining the garden commons allowed Nuestras Raices members to engage in productive activity that met local needs directly, had multiplier effects that spun off small-scale community-based enterprises, and provided greater well-being and social connections within the social and economic space associated with Puerto Ricans in Holyoke and the Pioneer Valley (Gibson-Graham 2006, 188–189).

Similarly, Rev. Hemmings spoke to Willie Jamaal of a very similar commons—albeit smaller in scale than the urban gardens network of Holyoke and perhaps more spiritually marked—when he noted the key reproductive logic behind the garden commons in Clarketon: "Anybody to help that plant stay connected, and [if] they only take off that plant [only] what they need, the plant will reproduce itself. We are relationships that do not reproduce because we have cut away from this creator." As important as what he said, we have seen, were the ways in which he embodied his metaphor for the commons in his everyday practice as teacher and leader of the youth and the CFLC staff in their quotidian labor together in Clarketon's community garden.

Encountering White Racism and Going beyond the Diverse Community Economy: The "Blues Epistemology" and Plantation Bloc Residues in Edgecombe

This story of the theft of the garden can, as a working hypothesis, be placed within the historical context of a "blues epistemology," or what Rev. Hemmings referred to as an Afrocentric way of living, that has arisen historically in opposition to a "plantation bloc" of concentrated rural white power grounded in racial supremacist logics directed against the livelihood of African Americans since the years of slavery (Woods 1998). It is certainly the case that up to the present in Edgecombe County, African Americans have faced the hostility of a quite potent white power bloc, considerably transformed since the antebellum years but still operating in the region (see Holland et al. 2007, 60–65, 73, 173–178). This plantation bloc has taken the form of a networked elite of corporate managers, local officials, and large landowners who have made decisions and set policy, as well as local police, backstopped by the "prison industrial complex" (e.g., nearby prisons, the Caledonia Prison farm) that has disciplined and threatened to engulf large numbers of poor rural African Americans in violence in the county. Large-scale, white-owned commercial agriculture remains a major feature of the county's economy.[4]

Clyde Woods (1998, 29) writes, "The blues bloc consists of working-class African-American communities in the rural south and their diaspora. The ontology, or worldview, embedded in these communities has provided a sense of collective self and a tectonic footing from which to oppose and dismantle the American intellectual, cultural, and socioeconomic traditions constructed from the raw material of African-American exploitation and denigration." Woods notes, following the insight of the novelist Richard Wright, that both the Black church and the blues emerged in rural areas of the South "where black political and economic institutions were subjected to constant surveillance and often destroyed" (Woods 1998, 32).

We propose that the Afrocentric moral logic of Rev. Hemmings exemplifies the blues epistemology to which Woods refers. Particularly salient and relevant were two aspects of the blues epistemology that Rev. Hemmings articulated in his practice: first, the knowledge of a world where the African American community was assailed on all sides by a hostile white majority, and second, the imperative in the face of such outside hostility to sustain the integrity of the "social dynamics" of the Black community that suffered racial abuse (Woods 1998, 29–39).

For Edgecombe County, the field notes of our ethnographer Willie Jamaal Wright confirmed the long, fraught racial and class history of the region, by making it clear that rural African Americans in Clarketon were still aware in the post-civil-rights era of the presence of white hostility toward them, for which a defensive response was necessary, based on an insistence on collective Black dignity and integrity. For instance, Rev. Hemmings told Willie Jamaal of Pete Logan, one of the county's Agricultural Extension agents, a white man, who thought Rev. Hemmings's effort to build the CCG to be amateurish and poorly thought out. As Willie Jamaal recorded it, Rev. Hemmings believed that this man "does not know what to do with him [Hemmings] and his initiative [the garden] because it is led by Black people for Black people. He believes that Pete, like other whites in the area, and many whites in general, do not know how to work with a Black man who is self-determined and is not seeking their benevolent blessings." To make matters worse, Rev. Hemmings felt that Logan patronized him when "Pete tried to explain to him, 'a farm boy,' how to manage a field, grow crops, and deal with pests. He recalls asking Pete about how best to deal with a pest infestation in the smaller garden, and . . . Pete pulled out a book to identify the pest and the best way to deal with them. For Rev. Hemmings, the simple fact that Pete had to go to a book for a reference was a sign of his ignorance and incompetence."

Moreover, Rev. Hemmings revealed to Willie Jamaal that he had found it difficult to develop cooperative relations with the white clergy of other Protestant churches in the area, despite their shared Christian faith. On another occasion, a reporter for one of the white-owned county newspapers came to visit the garden and interview the youth about their experiences. After interviewing an extremely articulate twelve-year-old about his work in the garden and his hopes arising from it, she asked Rev. Hemmings if she could interview one of the other children, "who had not been coached on what to say." Rev. Hemmings "took offense to this, stating that whites in that area are not used to Black people being intelligent and having agency over their lives, especially Black children. He furthered that if white children were interviewed and responded the way [the youth] had, there would not be any question whether the child was instructed on what to say."

The blues epistemology, moreover, has rested on the foundation of "the constant reestablishment of collective sensibility in the face of constant attacks by the white plantation bloc and its allies, and in the face also of a daily community life that is often chaotic and deadly" (Woods 1998, 29–30). African American communities in the rural South, after all, have been under extreme stress, could falter, and were thus always a work in progress. Repeatedly, the interactions between Rev. Hemmings not only with the youth but also with the congregation volunteers, the few paid staff, and residents of Clarketon illustrated his concern that if "relationships" defined by sharing and noninstrumental connections were not reconstructed within the church and community, then the "dysfunctional stuff" faced by generations of African Americans in the area would lead to what Woods calls a "chaotic and deadly" life.

As we noted earlier regarding the episode in which theft from the garden occurred, Rev. Hemmings saw this as not just a problem with the specific person in question but a broader issue, in that residents "rarely talk with one another" and lacked "social skills." After the theft of the produce, Charles, the garden manager, became upset and was about to ban community residents from using it. Rev. Hemmings noted that Charles's anger brought on his own intervention, in which he spoke with Charles about the need to keep the garden open to all Clarkston residents and to educate them instead about how to maintain it rather than destroy it by picking only the fruit of the produce and not the entire plant.

Such internal conflict invariably led to criticisms by volunteers, congregation members, and community residents of the leadership style of Rev. Hemmings himself. To Willie Jamaal Wright, Rev. Hemmings stated, "Black people are known to kill their leaders," in reference to backbiting and criticisms made of his actions as the church's leader behind his back. Rev. Hemmings referred to Blacks being passive and even complacent—not taking personal initiative, while attacking their leaders for doing so—as a "sickness" in the Black church and the wider Black community, a sickness that it was necessary to cure. Willie, citing another local Edgecombe Black leader, referred to this as a "slave mentality." Thus, community building and fostering self-discipline and initiative among the members of the community applied not just to the youth but also to the adults and was exemplified not only in the gardens program for the youth but also in the response by Rev. Hemmings to the theft of garden produce.

It is here that another feature of the blues bloc set against the plantation bloc that Woods describes comes out. Woods points out that arising from the experience of being surrounded and attacked by hostile forces for long periods of time, there evolved a distinctive orally transmitted culture in which "orature"—the performative skills that articulately express a resilience and toughness of spirit through the spoken word, song, and dance of blues bloc leaders, whether "bluesmen" or "preachers"—has been key (Woods 1998, 34). What Woods (1998, 16–21) calls the "blues epistemology" has been the result: an embodied and expressive performance of orature by these leaders who spoke of hope and compassion for people who were suffering and combined these with sarcasm, wit, and irony aimed at those who caused suffering.

Repeatedly, Rev. Hemmings's deft and persistent rhetoric of reconstructing community through an "Afrocentric" vision, on the one hand, and his trenchant responses to ambient racially based attacks on his congregation and himself, on the other, expressed the tensions and achievements of this epistemology-in-practice. Such orature as a rhetorical form for molding community extended to the most mundane of tasks, such as reaching collective decisions about how to undertake gardening tasks. As Willie Jamaal observed,

> The governance structure of the Clarketon garden initiative is in line with its anti-Eurocentric operating model. When I asked Rev. Hemmings about the meetings that [are] held in regards to the garden, he informed me that meetings are held as they work. Ideas and discussions are processed as the parishioners work. The overall purpose of these on-site meetings is to develop strategies to improve the garden, be they growing strategies, season planning, insect management, and marketing strategies. . . .
>
> There is no set method of decision-making for the garden. As the community works together, they talk and also decide on the next and best course of action for the garden. [Rev. Hemmings]: "We work on our relationship. We work the produce. We don't have any problems. We know the work we're putting in. We know our outcomes, and we trust each other." When asked if there were any formal methods of arriving at decisions (i.e., Robert's Rules of Order), Rev. Hemmings responded, "The day we come to that place, we're at the place of failure because our relationship, we have fallen at an all-time low. When I have to bring somebody in the room that ain't in the room to keep us in a row, . . . 'cause, I don't know Robert. He don't know us. It's like the church needing the police. If you get to that level, you just need to fold."

Thus, Rev. Hemmings, through these mundane, everyday practices around gardening, brought together youth with senior adults within the community. In so doing, he sought to redress the "sickness" of complacent attacks by African Americans on their leaders (and the "chaotic and deadly" life of which it was a symptom) by undertaking an orature of spiritual leadership aimed at exalting the embodied practices of community that the Clarketon Community Garden, its youth, and its adults stood for. The cost, however, of his charismatic leadership was to lead some members of his congregation to passively engage in divisive backbiting—precisely what he sought to oppose. Nonetheless, his achievements masterfully illustrated crucial aspects of blues bloc culture that variants of the blues epistemology have sought to address—not only in Edgecombe County in the 2000s but throughout the rural South over the past two centuries.

How Racial Oppression and Class Violence Shape Food Activisms

We return briefly to the question of the adequacy of Gibson-Graham's (2006) theoretical account of the diverse community economy in situations where class inequalities, racial hatred, and race-based expropriations (e.g., the criminalization of large numbers of Edgecombe County African American youth) divide members of the diverse community economy from one another. Whereas Gibson-Graham's (2006) poststructuralist approach suggests that a common vision of the diverse community economy can emerge among community members around their shared critique of a dominant neoliberal or capitalist discourse—of what Gibson-Graham (2006, 55) called "capitalocentrism"—we find this view naïve, to put it politely. In US society, there are clearly long-standing racial and class divisions that make any such shared imaginative vision of a diverse community economy very difficult and, in some settings, even impossible for large numbers of people who live near each other to share.

Unfortunately, these divisions always intrude into local community life, and the question is whether these divisions and the oppressions that go with them are acknowledged, are counted up, and become the basis for restorative justice among community members. *Then, once restoration is under way,* shared ethical discussion and decision-making along the lines Gibson-Graham propose become possible across recognized class and racial differences. This is a rare situation in contemporary US locales and not one we studied in North Carolina.

More frequent are local situations where these racial and class divisions exist but where people of different racial and class statuses (perhaps, e.g., as self-described "middle class") have shared objectives that bridge these divisions and allow for ethical discussion and decision-making to take place, even though some people are excluded from, or exclude themselves from, these ethical rapprochements, while some class- and racially based abuses continue unabated. Our three research sites of Charlotte, Watauga and Ashe Counties, and Durham represent this situation, in various ways and degrees. In each of these sites, sustainability farming activists, charity-based food security activists, and a few community food security activists worked separately. Yet occasionally, a very few activists "crossed over" to become both sustainable farming and community food security activists, as we showed earlier. We hope that food activism can become, even more than it has already, one basis for class and racial rapprochements and successful restorative justice efforts in these locales.

And yet there are locales like Edgecombe, Halifax, and Nash Counties, where racial oppression and class antagonisms not only exist but are overt, with the result that vulnerable poor people of color must erect defensive institutions and practices to protect themselves materially and culturally and, if possible, over the long run, within a political arena dominated by a "plantation bloc" of wealthy and powerful whites—one that had led to the "blues epistemology" we found to characterize Clarketon's Rev. Hemmings and his staff (for other examples, see Woods 1998).

In such situations fraught with deep social divisions, community food security with a focus on ensuring that members of the assailed population have adequate amounts of culturally appropriate food available, by making sure that they able to grow it, will be the paramount goal of food activists. This depends, as Rev. Hemming's pedagogy demonstrated, on members of the community turning inward to rely on one another to develop their own farming or gardening capabilities while connecting to the distribution of surplus among themselves and beyond their community through market exchange—hence the centrality for Rev. Hemmings of focusing his efforts on Clarketon's youth. Sustainable food activism in its usual land-extensive form has to take a backseat when members of a vulnerable community under long-term siege are cut off by their poverty and by violence from access to means of livelihood, like farmland, that are essential to "performing" sustainable farming. Instead, they are forced to intensify their efforts at community food sufficiency, for example, on the eleven acres of the CCG in Clarketon—land that, collectively as a congregation, they *did* control. This, we contend, is the very real manifestation of sustainable food activism in such settings.

* * * *

In this chapter, we examined the extraordinary achievement of the Clarketon Community Garden and Rev. Hemmings and the CFLC staff in working with Black youth in rural Edgecombe County. We first described the troubled racial history of this region in coastal eastern North Carolina. Rev. Hemmings, who organized the garden as an enterprise project for the town's African American youth, did so in the face of extraordinary racial tensions and divisions that characterized this very poor area in one of the lowest-income counties in North Carolina.

Second, we learned that despite attempts to reconcile with local white community leaders, Rev. Hemmings found this to be impossible in the face of their distrust or even animosity. He instead organized Clarketon's youth and his church's staff around an amazing garden enterprise that operated on two levels. On one level, it yielded the material bounty of produce for the youth, their family members, and food-insecure elders and provided youth with income to save for their college expenses. On another level, the garden was a pedagogical undertaking that taught its youth the moral vision of an "Afrocentric" economy of co-responsibility and mutual support—one consistent with both sustainable food and community food security moral logics.

Third, we were struck by the adroit and farsighted strategy of Rev. Hemmings around teaching the youth and CFLC staff the meanings of *consumption, surplus*, and the *commons* through ethical discussion and decision-making led by Rev. Hemmings. These were processes that took not only a discursive form but also one of their embodied practice of everyday labor "in the ground" of the garden. Through this process focused on the garden, Rev. Hemmings was able to increase the solidarity of CFLC's staff and congregation with its youth and with one another.

Finally, we proposed that Gibson-Graham's (2006) theory of a "diverse community economy" can only take us so far in trying to make sense of the possibilities for food activism within a social setting characterized by deep racial animosities and class divisions, such as that which we found in Edgecombe County affecting Clarketon's residents. To move beyond the limits of Gibson-Graham's (2006) poststructuralist approach in the case of food activism, it is essential to turn toward a political economy perspective that allows us to make sense of interconnected class power and racial domination and to show how these work to affect local food economies and food activisms, as we have begun to do in this book.

10

Assessing the Transformative Significance of Food Activism

In the groundbreaking book *A Postcapitalist Politics*, the feminist economic geographers writing as the collective Gibson-Graham (2006) build on second-wave feminist thinking and ten years of community-engaged research/action to present an expansive politics for change. The authors, in a move remarkable for theorists of political economy and the political, pay close attention to the micropolitics of social and economic change, to the processes of embodiment and subjectivation that go on in local spaces of practice, and, though less examined in their book, to the self-authoring and cultural production that shape history-in-person and contribute to the formation of intimate and collective identities (Holland et al. 1998). We see in their work a general theory of the necessary elements of significant, ground-up social and economic change that resonates with our findings, and we draw on their book to infer a set of criteria that can be used by activists and academics alike to assess and reflect on the significance of a given social movement. We enter into a conversation with Gibson-Graham using the inferred criteria to assess the significance and lessons of local food activists and the emergent local food movement in North Carolina.

Gibson-Graham not only conducted community-engaged research; they brought the lessons from their participatory research back into their theorizing to model a transformative activist stance toward the production and circulation of sociocritical knowledge. We, too, take such a stance both in conceptualizing human development and change and in understanding the transformative potential of social science: we see our efforts as resonant with those of other researchers/activists who use their activism and disciplinary training—anthropology in our case—to produce sociocritical knowledge that contributes to a "collaborative historical becoming" (Stetsenko 2008, 471) through ongoing efforts to build a just and equal society (Stetsenko 2008, 471; Langemeyer 2011, 156).

Our first goal is to develop an analysis that can be used for reflection by participants in contemporary local food movements. As argued in María Isabel Casas-Cortés et al. (2008), we recognize social movements and activism as producing sociocritical knowledge alongside, with, and sometimes against researchers. In tandem with the first, our second goal is to establish and develop the criteria for transformative significance derived from Gibson-Graham's work and ours to provoke discussion about the social and economic significance of any change effort, especially those undertaken by social movements.

A "Politics of Possibility": Gibson-Graham's Theory of Change

Gibson-Graham's 2006 book *A Postcapitalist Politics* is a response to what some intellectuals have named "a crisis of the model of civilization" (Sousa Santos 2006). Working in that context, looking for options, Gibson-Graham proposed a politics that aims not only to challenge the hegemony of capitalism as an economic system but also to think about politics in a different way—as a politics of possibility. Both of these objectives entail a vision of social transformation that challenges the perspectives and actors heard from most often—those from the Left and the academy—in discussions about social movements. A message of the book is that counter to the usual view from the Left, the transformation will not be only or mainly the result of the "context" or the "structural conditions." Drawing instead on ideas emerging from the World Social Forums and from movements such as the feminist and Zapatista movements, Gibson-Graham focuses more on possibility than on probability, on acting ourselves into alternative worlds. They believe in the paths that hope, dream, and utopia open for those that have been situated *abajo y a la izquierda* (down and to the left, as the Zapatistas positioned themselves).

Bypassing older ideas of how the social transformation of capitalism might occur, Gibson-Graham recognize in the Zapatista and other movements the emergence of a new political imaginary. This new political imaginary entails a progressive politics, a reconfiguration of the subject's position and role, and a shift in the grounds for assessing the efficacy of political movements and initiatives (Gibson-Graham 2006, xix). They emphasize, in contrast to waiting until control of the state is secured, practicing a politics of the here and now. In order to construct alternatives, noncapitalist economies must be enacted in the present (Gibson-Graham 2006, xxi) so as to enable processes of becoming in place.

Gibson-Graham's tripartite political vision includes a politics of language, a politics of the subject, and a politics of collective action. These politics are aimed at destabilizing capitalism in the first place by crediting the poststructuralist recognition of the importance of discourse. They insist on denaturalizing the "economy" and the dominance of capitalist discourse. This process entails retheorizing capitalism and seeing economic difference in a double move that (1) recognizes the existence of economic alternatives left invisible by a hegemonic ideology that frames capitalism as the sole, current economic possibility and (2) supports the emergence and expansion of alternatives through the development of a language of economic difference. Liberating and cultivating "noncapitalist" economic practices involves "widening the field of intelligibility to enlarge the scope of possibility," while at the same time dislocating the (discursive) dominance of capitalism (Gibson-Graham 2006, xxxiv).

For Gibson-Graham, besides the need for a new language of economy, there is need for a new self and new practices. Destabilizing capitalism requires the self-cultivation of subjects who have greater openness to change and who can desire and enact other economies. Determined not to substitute an analysis of the power of capitalism for a theory of a strategy to overcome or go beyond it and always vigilant for likely human experimentation and diversity, Gibson-Graham nonetheless recognize the attachments and ongoing seductions of capitalist practices and discourses. They devote considerable attention to the "reluctant subject" as both theorist and community member.[1] They theorize the need and possibility for practices of ethical commitment and self-transformation to help the subject overcome attachments to the old economy.

Finally, with regard to collective action, there must be a collaborative pursuit of economic experimentation both within the community and more broadly with extracommunity supporters such as academic researchers (Gibson-Graham 2006, xxiii)—and, we would add, some political party and social movement organizers.

To reiterate, Gibson-Graham are unusual—and consequently resonate with cultural-historical activity theory and related social practice theories of identity—in that they bring into dialogue theorists of the political and those interested in embodiment and the micropolitics of everyday life, enabling both to better understand and support conditions for positive social and economic transformation.

The Criteria: Key Elements of Transformation

Nowhere in *A Postcapitalist Politics* do Gibson-Graham distill what they consider to be key elements of social and economic transformation. In the interest of moving forward in our project of promoting and debating the assessment of the significance of change efforts, we have inferred such a list from their theoretical discussions, briefly sketched earlier, and from their discussion and rationales for their community-engaged research and interventions (see table 10.1).

Although the Gibson-Graham book refers to a post*capitalist* politics, we have read the elements of transformation needed as a pathway to significant change no matter what the domain. We are thinking, then, of a "post-X politics," where "X" is "agroindustry" or "fossil fuels," so "post-agroindustry politics" or "post-fossil-fuels politics."[2]

Using the Gibson-Graham list of critical elements for transformative social change, in the case of contemporary food activists in North Carolina, we ask to what extent they have, as assessed by the criteria, undertaken a *politics of possibility* and significantly transformed the social and economic arrangements of the food system. Such a politics assesses progressive changes, however small, that have already occurred and

---

TABLE 10.1. Key Elements of Social and Economic Transformation, as Derived from Gibson-Graham's *A Postcapitalist Politics*

For a critical mass to undergo politicization, the following must occur:

1. Recognition of a structure of domination, some of its elements, or at least critical reflection on a crisis of the status quo

2. Identification and enactment of a politics of possibility

3. Creation of alternative discourses/visions

4. Orientation to a collective and a building of community—an "us"

5. Changes made in daily life and everyday practices

6. Cultivation of subjects with the desires and capacities for sociability, happiness, and action offered by alternative social and economic arrangements

7. Ethical commitment and self-cultivation

8. Shifts in subjectivities and identities

seeks openings for further changes in the food system by investigating possibilities in the current conjuncture, which as dire as it might appear, could engender major transformations in the future.

In the following, we bring our findings on food activists in North Carolina into dialogue with the criteria to ask how well the criteria help us assess the significance of the change efforts. We conclude by asking to what degree the Gibson-Graham criteria themselves fall short. The most common criticism of the Gibson-Graham book is that it pays too much attention to agency and insufficiently accounts for the dialectical relationship between it and structure. In the conclusion, in the face of a common difficulty—one that has to do with structural constraints that we encountered in the cases of food activism we witnessed—we suggest two additional criteria.

In the case of the contemporary local food movement in North Carolina, we see a movement that is in a formative, experimental stage and more activity focused than analysis/reflection centered. In danger of engulfment, the movement provokes questions about its significance: Is it simply a consumer trend generating yet another niche market, with entrepreneurially inclined farmers seeking to satisfy newly emerging tastes primarily for profit? Or are more community-economy, less capitalist, values being built into the emerging local food systems? Is the movement fertile ground for a new systemic design for the production and distribution of food or mere "projectism" (Alperovitz 2013) that will leave untouched the power of the corporate food sector to maximize profit regardless of its hidden costs for human health and the environment? Will current food activism reduce the chronic food insecurity that a growing segment of the US population now faces or simply leave aside such issues? And what insights are generated by a dialogue between the case and the key elements of transformation from *A Postcapitalist Politics*? Drawing on the data from the project resulting in this book, we offer some preliminary answers to these questions.

The Local Food Movement through the Lens of the Gibson-Graham Criteria

Structures of domination, important in criterion #1 of the Gibson-Graham list, received little collective attention in the local food movement in our four sites. Concomitantly, activities related to escaping the subjective effects of hegemonic structures through *self-cultivation* (criterion #7) were not in evidence. We return to both later. We begin the discussion of the remaining criteria by focusing on the soft-pedaling of talk about the global agroindustry and the practice, nonetheless, of a flourishing *politics of possibility* (criterion #2).

Enacting a Politics of Possibility (Criterion #2)

Collectively, activists were working hard to increase the availability of local food. They were enthusiastically organizing a variety of vibrant local food-related activities, many of which were relatively new or at least modified in character, including most prominently farmers markets; community-supported agriculture arrangements (CSAs); campaigns to expand markets for local food to area restaurants, schools, and hospitals; educational events such as farm tours; and the development of community and school gardens. And yet, contrary to what might be expected from the Gibson-Graham criteria, we heard little extended discussion in the activist meetings or community events about the agroindustry-dominated food regime.[3] No one seemed to require a dismantling of the self-congratulatory discourses of the corporate food industry before they could embrace the idea of local food as superior. Multiple alternative discourses circulated about local food's positive goods, including better health, denser community relations, better environmental stewardship, and more humane treatment of animals. A *politics of possibility* was palpable. There was every sense—expressed through widely circulating discourses—that superior food could be produced in superior ways and that local effort could make that production happen. The movement offered the *alternative vision* (criterion #3) and related activities to the effect that local food systems could be created to lessen dependency on the globally sourced agrofood industry and decrease vulnerability to its harms.

Moreover, the media and a significant portion of the public seemed to be on board. Locally produced food received good press. The food, the activities, and the imaginaries of the local food movement all had broad appeal. Farmers markets drew seemingly happy crowds; the luscious German Johnson and other heritage tomatoes sold for premium prices, and the importance of getting to "know your farmer" was frequently re-peated in conversations at the events. Eating properly grown local food had become for many consumers a marker of good taste (in both the social and gustatory senses) and positive moral value.

It was true that allusions to the negatives of the globalized, industrial, corporatized food regime were commonplace. And it is fair to say that those who are engaged in any of the local food activities tended to be distrusting of, if not antagonistic to, agroindustry as a source of nega-tively valued food. Outside of meetings, when giving personal interviews and commentaries, many activists conveyed a strong sense of discom-fort or disharmony with the contemporary food system. But such com-mentaries rarely became a focus of meetings or the community events we witnessed. There was little or no in-depth discussion in the meetings and deliberations of the local organizations, for example, about agro-industry's domination of expectations when it came to the seasonality of food or the threat to agricultural crops by genetically modified food. Sustained critiques of structures of domination were by and large left, if spoken of at all, to speakers and media brought in from the outside. They were not put into a collective *local* voice. In other words, critiques and critical examination of agroindustry, its hold on the possibilities of the local food system, its lobbying for the farm subsidies that crowd out small farmers, and its environmental and other "externalities" re-ceived relatively little elaboration in the alternative vision of local food. "Local" was discussed as though it were capable of creating pockets of good farming and good eating despite the machinations of the larger food regime.[4]

Moreover, as already alluded to, there were few movement-generated activities that entailed the difficult processes of self-transformation to a new food economy that might be expected from Gibson-Graham's sixth and seventh criteria. At best, alternative values were implicit in *perfor-mances* of the subject positions offered by new agricultural activities, mar-keting practices, and projects and in the honoring and cultivation of tastes for local food and community events featuring food and agriculture.

As a result, efforts to solve problems seemed, with few exceptions, to occur within the paradigm of commonplace, as opposed to alter-native, economic arrangements. There was strong concern for low-resource farmers, for example, but not much thinking about how to better support farmers other than to improve their business skills and help them expand the markets for their products. There was no thinking about how a community, for example, might somehow help small farmers obtain health insurance.[5]

Shifts in Subjectivities and Identities (Criterion #7)

Per criterion #7, there were *shifts in subjectivities and identities* but not of the sort that might result in the "reluctant subjects" whom Gibson-Graham encountered. Granted, for farmers, the transition to new growing methods and new styles of relating to customers was not without its challenges, but for the nonfarmer activists and the eaters, how difficult is it to experience the superior taste of grass-fed beef or learn the exquisite differences among heritage tomatoes? As explored more thoroughly shortly, the relatively nonoppositional collective stance appears to be connected to the sociohistorical cultural terrain of neoliberalism on which the movement has formed. Among other features, this terrain enables small-scale enterprise and experimentation (Holland et al. 2007). During our study, there was a pervasive sense—except among a relatively small number of people focused on food injustice— that the benefits of local food could be accomplished simply by creating relevant activities, convincing others of the superiority of such food, and supporting the farmers who produced it. Especially with regard to the latter, despite the frequent repudiation of neoliberal visions of efficiency, entrepreneurship, profitability, and productivity that we described earlier, they saw no alternative to neoliberalism's reliance on the market and its zero-sum logics.

Orientation to a Collective and a Building of Community (Criterion #4)

Two of the criteria remain to be discussed. The first is criterion #4: Who is the "us" of the local food movement? This criterion links to the capacity of the movement to undertake collective action and highlights potential factions and contradictions within a movement—the contentious differences that, when not resolved, form into antagonistic identities that animate one another.

Not surprisingly, voice, at the time of our study, was rarely, if ever, given to an "us" composed of all those who were dependent on and dominated by the current food regime. Instead, the "us" was set by the theme of localness. An "us" was construed around the local "community" or "area." The names of cities such as Durham or the recognized rural areas such as the "High Country" were used to refer to a local food system.

Against these local collectivities were arrayed a set of confounding divisions. The most pronounced fault lines of the movement were not between and among different localities, as might be imagined, but along culturally coded lines of class and race. Local food's principal object—locally produced food—was by definition dependent on local farmers, and the market was accepted, as described earlier, as the primary means for organizing the production and distribution of local food. Thus, the (re)building of local food systems was happening amid taken-for-granted market privileges for the wealthy, on the one hand, and their lack by those who were subject to structurally induced poverty and associated hunger, on the other. And the systems were understood to rest perforce on the economic viability of small farmers.

These wealth divides lay at the heart of several disconnections among activists. The people excluded from local food by low wealth are also those most likely to be food insecure. As we noted earlier, for a subset of activists engaged in food activism, serving those who were food insecure was the central goal; for others, in contrast, the food insecure were out of sight and out of mind. The division shows up culturally around the moral priorities to be assigned to potential benefactors of the local food system. Who was the most morally deserving? Were they the small farmer dealing with economic insecurities? The low-income, single parent struggling to find good food for their kids? Or was it the consumer of the area with means to buy good food but little choice beyond the tasteless and often unhealthy products of corporate agriculture and the fast-food industry?

For those who were trying to expand the availability of local food for the morally deserving consumer, low-income members of the community other than limited-resource farmers were often out of sight and out of mind. For other alternative food organizations whose members were concerned about poverty, hunger, and the consequences of low-cost diets, low-income people were visible but challenging to involve in the market and in the organizations. Some activists did attempt and sometimes succeeded in ensuring that local food vendors would accept SNAP and WIC vouchers.[6]

Despite the refusal of the widespread neoliberal cultural framework to recognize structural inequality and despite the lack of familiarity among the general US public with Marxist or other conceptualizations of structural forces—just then shifting during our fieldwork with the Occupy movement's attack on "the 1 percent"—a number of activists in the food movement did concern themselves with people affected by high rates of structurally produced food insecurity and by high incidences of obesity, diabetes, and other diet-related illnesses associated with cheap, mass-produced diets heavy in high-fructose corn syrup. For those activists and for activists with liberal religious values who felt called on to help people in need, the community member of lesser means was considered morally deserving of attention and inclusion in the rebuilding of the local food system. These "social justice activists" or "food justice advocates" made the low-income population an issue in mixed meetings and conferences. They made it difficult for others to forget those who were excluded from the benefits of local food because of limited means. These activists also tended to be the source of experiments with alternative economic arrangements.

Another difficulty spawned by class issues concerned the social image of the movement. The social identity of the local food movement is historically entangled with class. Enough high-end shoppers, who tended to be wealthy, white, and middle or upper class, were sufficiently disaffected from the globally sourced agroindustry, at least the retail end of it that they were in contact with, that they valued alternative agriculture and desired organic and locally grown produce and meats. The social image of these high-end shoppers had become attached to the social identity of the "foodie"—someone who is (or aspires to be) an enthusiastic connoisseur of good food, which now includes local food. This social image, apart and beside expectations of high prices, tended to repel those who were uncomfortable with predominantly white, middle- or upper-class demographics or who did not want to be mistaken for a (want-to-be) foodie (Guthman 2008a).[7]

Race was another potent source of division of the collective "us" in movement locales. In North Carolina, and in the US in general, class, race, and ethnicity are entangled such that relatively higher proportions of African Americans, Latinx, and American Indians are low-income. The distinctive histories of race relationships in the four sites showed up in food activism, but there was a general trend. For those groups that were represented, food activism tended to run on different tracks—one Black, one white—especially in Halifax, Edgecombe, and Nash Counties, our site in eastern North Carolina.

In sum, the social identity with the clearest claim on the space of the imaginary participant in the local food movement was that of the "foodie," a sociohistorically constructed figure who was raced and classed as white, middle or upper class, and vulnerable to charges of elitism. The figure of the "food justice worker" was less familiar and less clearly raced and classed.[8] The cultural worlds of the "foodie" and the "food justice advocate" were distinct. There were tensions between the motivations and concerns of the two as well as differences over the moral priority of supporting the small farmer versus that of serving those who were food insecure relative to the priority of increasing the availability of locally produced food for the consumer. These points of difference revealed contradictions of the movement confounded by class and race within the hegemony of neoliberalism's lust for market solutions. Despite these tensions and as might be imagined given the neoliberal cultural terrain, with its blurring of power, celebration of the magic of the market, fetishization of "choice," and deflection of structural questions, there was relatively little enunciated conflict. Participants put their energy into organizations and activities according to their interests, attractions, and personal relationships and animosities. Instead of working through these contradictions, members tended to avoid conflict, simply drifting away from disagreement.

Cultivating Subjects with Desires and Capacities for Sociability, Happiness, and Action Offered by Alternative Economic Arrangements (Criterion #6)

Perhaps the most intriguing and potentially transformative edge of the local food movement we saw was its experimentation with food-related activities and practices. Some of these experiments were homegrown; others, such as community-supported agriculture (CSA) and community food assessments (Pothukuchi et al. 2002), had been pioneered elsewhere.

Among the considerable variety, some experiments encouraged alternative relationships to industrial food production (e.g., community gardens, container gardens); others, alternative relationships to food producers (e.g., farmers markets, CSAs); others, alternatives to food-related labor (e.g., Crop Mob or Cornucopia Landscaping Cooperative); others, to people who were food insecure (e.g., community restaurants, "Double Bucks" programs of farmers market coupons for SNAP recipients); others, to food security itself (e.g., the Slow Food movement). Arguably, some of these practices, when sustained over time, may become embodied and emancipatory, transforming participants' sensitivities and sensibilities. Indicative of the pursuit of many possibilities, they are too numerous to describe here. Nor does frequency dictate which to analyze in depth; none have been widely adopted in a standard form. A brief look at community gardens will provide a sense of the range and a feel for how the activities are faring at the moment.

Community gardens, in which a group shares a space for gardening, were potentially the most transformative. People are brought into fellowship by engaging in common labor and sharing interests in the challenges and joys of growing food. Significantly, the costs of using the garden space are relatively low, reducing the problem of differential wealth. We learned about some amazing projects organized around community gardens. Even in Rocky Mount, located in the site with the most rigid race relations, Lawrence Farmer, a retired white firefighter, managed to create a community garden in Happy Hills, a low-income African American neighborhood, that brought together Black and white gardeners.

However, as we saw earlier, generalizing the results of particular gardens is unwise, as the community gardens were organized in diverse ways and, even more importantly, associated with many different imaginaries. Some were conceived as spaces of rehabilitation for troubled young men. Others were thought of as a source of good food for low-income people. Still others were expected to create a collective affect with the power to build bridges across lines of class, race, and ethnicity. Another idea, more common among African American activists, concerned re-creating a place for intergenerational knowledge transfer.

Other food activist groups whose members engaged in activities that cultivated sociability, pleasure (if not always happiness), and action include the Crop Mob and Slow Food Charlotte, discussed earlier. Viewed from the perspective of this chapter, the Crop Mob celebrated the solidarity and identity of young farmers working together on the demanding common task of cultivating land for each other, culminating in a dinner for attendees and much socializing. Members of Slow Food Charlotte eventually combined the international Slow Food movement's early focus on the pleasures of the gourmet's table with its new project Friendship Gardens to ensure better local food for the city's poor population, thus subverting (to some extent) SFC leaders' own class and race privilege.

Changes Made in Daily Life and Everyday Practices (Criteria #5)

In 2008, sales of locally produced food in the US had grown to $4.8 billion. While only a fraction of the $1.229 trillion in food sales in the US (Hauter 2012), the figure does signal a change in food purchasing patterns. Activists in our sites had, at least in part, shifted their purchasing habits along these lines, and activist farmers had shifted their growing practices. Beyond those shifts, the people participating in the experiments just described were plausibly changing some of their everyday practices as well. (These trends have continued up through the 2020s, as we indicated earlier.)

Lessons Learned

The significance of the alternative food movement in North Carolina was equivocal. It had delivered many projects of interest but little that might count as significant *systemic* social and economic transformation. It is true that at each of our sites, we found that an enormous amount of energy had gone into (re)building the economic, material, social, and cultural infrastructure necessary to supply the local market. Activist efforts had increased the availability of fresh locally grown produce and locally raised meat and raised awareness of the benefits of local food through events such as farm tours and community gardens. These were significant advances toward economically, environmentally, and socially sustainable systems of agriculture and food production that rely on local resources and serve local markets and consumers. Moreover, some activists had made valiant efforts to incorporate local food provided by the food pantries that partially deflect the hunger resulting from structural inequalities.

But, by and large, the alternative food organizations in our research sites have yet to address collectively issues of "systemic design" and begin the political work necessary to maintain local agriculture and protect it from the machinations of the corporate giants of food production, distribution, and marketing. Few activists had become engaged in mobilizing for government policies to better support the small- and middle-sized farms that serve local and regional markets, to break up large-scale corporations that limit competition in food production and distribution, to ensure that genetically modified plants do not contaminate non-GMO fields, or to regulate the dangerous overuse of antibiotics on factory farms, for example. Nor have they made substantial headway in addressing structural poverty or legacies of racial tensions that exclude people from local food. These are translocal issues that nonetheless intrude on local food systems (Nonini 2013). Food insecurity and poverty increased from 2000 to 2013 (Berner et al. 2016, 6–7). What happens, for example, when even more pronounced problems with hunger result, as some people predict they will?

When viewed through the lens of the Gibson-Graham criteria, the movement is puzzlingly uneven. It embraces a politics of possibility (#2), is replete with alternative discourses and visions (#3), and shifts tastes and identification (#6) to alternative practices of food, farming, and community food events. Yet, while repugnance to the dominant food regime (#1) is voiced, there is little sense of encroachment or danger from the power of that regime to negatively interfere with local efforts. The movement could be described as not having yet coalesced around a robust counterhegemonic collective identity or developed a core set of critical discourses that guide its actions. At a recent book event in one of our sites, a leader of one of the national organizations working on food issues expressed frustration with the local food movement's lack of effort to bring about structural change. The leader referred to the local food movement as the "Happy Food Movement."

Admittedly, it is tempting to judge the movement by the Gibson-Graham criteria and find it wanting. Yet, one of the central messages of *A Postcapitalist Politics* is to always look beyond any particular model of change. We must do so, especially for movements still in early emergence, and use the criteria as a tool for dialogue and reflexivity, not judgment. We must leave open the possibility that the emerging local food systems and the emerging knowledges of food and farming in the twenty-first century that they are producing will eventually prove significant in relation to the global agroindustry. How might this be so? As a first step for probing our assumptions, we ask, How are the conditions for activism and avenues for change during the present era different from those in the time and locales of Gibson-Graham's work? The remainder of this chapter argues that less familiar styles of activism are becoming more frequent in response to the "perverse confluence" (Dagnino 2005) of neoliberalism that creates hardship and discontent while simultaneously providing openings for experimentation and new developments. The challenges and opportunities of these new forms of activism are still unfolding.

Efforts to "Make History" May Now Be of a More Individualized Character

In *Disclosing New Worlds*, Charles Spinosa et al. (1997) theorize that in societies such as the US with capitalist markets and democratic governments, the drive for change leads some entrepreneurs (by way of commercial ventures) and some activists (by means of civic action) to transformative social practice or what the authors refer to as "making history." The impetus to make history comes from individuals actively engaging in the world, experiencing "disharmonies," and working to "disclose" or clarify and create new products and services or new policies and programs that resolve the disharmony. Categorized analytically as a social movement, the activity around local food is supposedly a collective effort; however, based on accounts of personal involvement provided in interviews from our research, the movement could be made to appear more as the myriad efforts of many loosely connected individual activists to make history. Given the distinctive aspects of the food system that attracted alternative food organizations and the looseness of ties between and among them, the movement could in fact be made to read more like a novel with many different, occasionally intersecting characters, plots, and subplots.

Many of the activists and social entrepreneurs in our research were first-time activists. They entered into creating a community restaurant, an organic farm, a farm labor exchange, a community garden, a women's group supporting local agriculture, or a variety of other endeavors as a result of various personally experienced concerns about the safety of the food they were eating, about the people in their community who lacked sufficient food, about anticipated upcoming challenges to community food security resulting from climate change, about the disappearance of small farms, and so on. Most were middle class, white, and often searching for something worthwhile to do or some way to "give back" to their community. For those who took up farming, they described the decision as one based on their social values and not because they sought to become rich.

Notice that these emerging activists tended not to create or join groups with long-standing, well-formed politics and analyses. Instead, if they brought elaborated moral commitments and passions to the local food movement, they brought them from elsewhere—from religious teachings about helping people in need, from civil rights and social justice backgrounds, or from free-market moralities encountered in the business world. They talked excitedly about ideas from experiments they had heard about or witnessed. For example, chef Renée Boughman and her friends were animated to come together by their religious values and their visit to a culinary school in Charlotte that trained people out of prison as skilled restaurant staff to create F.A.R.M. Café, which invited low-income people to dine by donating only what they could afford.

As with Renée Boughman, most of the other activists interviewed were focused on local issues and on building a local project for the town or rural area where they lived. They were motivated by the "disharmonies" they sensed personally—in her case, feelings that low-income people should have access to good food regardless of their means, that the community should be connected through ties of friendship that crossed class lines, that she should be able to use her various skills and background for morally worthwhile purposes. In order to address these disharmonies, the activists, except in a few cases (e.g., Slow Food chapters), did not form chapters of national organizations or seek informal "tutoring" in discourses from regional organizations, as did North Carolina environmental and other voluntary associations in the late 1990s (Holland et al. 2007, 199–231). Instead, the local food organizations and projects were more homegrown with idiosyncratic connections to projects in other areas and varying familiarity with the websites of other food organizations. As already noted, although activists were familiar with circulating books and films critiquing the global agroindustry, those critiques did not become the focus of meetings or acquire a collective local expression. Efforts to "make history" remained more a conglomeration of mostly individualized projects—often led by activists with strong individualist orientations.

Forms of Activism Are Changing in Response to the Hegemony of Neoliberal Cultural Terrain

As we noted earlier, the "perverse confluence" associated with the neoliberal turn (Dagnino 2005) has meant that nonprofits have been able to step into lacunae created by national, state, and local government retreating from social services and public protection. This has opened possibilities for nonprofits seeking to provide services based on dissenting ideologies and practices. In previous work, we described several voluntary organizations of note and pointed out the ways in which they strengthened place-based, local connections to local history and cultural heritage while offering beneficial services and experience in participatory democracy (Holland et al. 2007).

Ten years later, we found the local food movement flowering profusely on an even more entrenched neoliberal social and cultural terrain. Some activists and especially funders were increasingly attracted to what the literature refers to as "social entrepreneurism" (Dees et al. 2002; Dees 2007). Such organizations are less oriented to cultivating and representing the interests and wishes of a group of like-minded members than they might have been in past and more focused on realizing a defined social outcome using concepts from the business world. A successful activist, along these lines, was one who realized a social mission through the entrepreneurial skills of recognizing (commercial) opportunities and searching out resources in both expected and unexpected places. For many commentators, social entrepreneurism and, for some of those in our research, today's activism ideally entails moneymaking ventures that help an organization support its social mission activities and avoid dependency on foundations and government grants. Given this marked shift in forms of activism away from advocacy and protest, it may be less surprising that some of the energy of the movement, especially the part devoted to expanding local food availability, was directed to local enterprises, mindful of potential revenue streams from sales of food, and not well connected to regional and national organizations.

Nor are organizations necessarily interested in forming a common set of goals or ideology within their locality. As with the US Occupy movement that emerged in the fall of 2011, the local food movement at the time or our study honored the diversity of efforts that were taking place. Alliances and interlinked activities were certainly common. Intermediary organizations such as farmers' cooperatives, multifarmer CSAs, and, in a few instances, food policy councils were forming. Despite activists' suspicions and skepticism about the corporate-based global food economy, their aim was not so much to create or even debate a set of shared analyses, values, and strategies alternative to those of the global agroindustry but to cooperate in the expansion of positive activities.[9]

One of the corrosive effects of market rule has been to play up the virtues and knowledges associated with market-rewarded expertise—the expertise of corporate and contracted academic technical specialists and technicians, as revealed by Gwen Ottinger's (2013) ethnography of an asymmetrical and ultimately failed campaign initiated by Louisiana environmental activists against a petrochemical company and its experts regarding evidence of air pollution. Moreover, in response to cutbacks and loss of prestige, the hiring by local, state, and federal governments of experts has shrunk—often leaving corporation scientists and technicians holding the high ground that associates knowledge, progress, and human welfare with profit making and private enterprise. The nonintellectualism (if not outright anti-intellectualism) of the food activists we studied is no match for corporate claims to authoritative knowledge in this uneven contest. This book is explicitly addressed to partly remedy this unevenness and to advocate for a people's movement for sustainable agriculture and a just food system.

In the case of the local food movement, activists are not grappling directly with food "experts" from Monsanto, Arthur Daniels Midland, or other giant agrofood corporations but rather with a vague sense of remote, untrustworthy, profit-hungry global corporations out there somewhere, alongside a sense of hollow communities, broken connections to the sources of their food, and questionable moralities that glibly exclude segments of the community from good food. The activists in our study were making great personal efforts through their projects to bring local food systems into better harmony with values they missed in the world shaped by global agroindustry. Yet they were having to (re)build these local food systems on a neoliberal terrain constituted by uneven but widespread government and corporate discourses and programs espousing market rule. On the one hand, the terrain embeds windows of opportunity by providing resources for public-private partnerships, local ventures, and social entrepreneurial ventures. On the other, it privileges commercial entrepreneurial values; devolves risk and responsibility to the individual in the locale; promises safety, health, and happiness through "responsible choice" from a cornucopia of available products made available by capitalism; and blurs, through public-private partnerships, the distinctions between government and civil society and between government and corporations. On such cultural terrain, targets of blame for a dissatisfying food system appear elusive and paths to affecting them unclear. For many activists, it seems more productive to build local projects that enact better values and hope that the broader public will be inspired by the obviously superior goodness of local food and support its expansion.

Conclusions

Used as a tool for judgment of local food activism in our study sites, the inferred Gibson-Graham criteria see a movement hampered by the virtual absence of collective, critical analyses of the global food regime. Used as a tool for reflection and dialogue, the criteria promoted a step back from the research to consider the context in which the local food movement has formed in North Carolina. The larger picture is one in which a surprising number of people are acting to resolve the disharmonies they sense between the food and agriculture of the dominant food regime and an envisioned alternative local system that produces healthful, tasty food, protects the environment, nurtures community relations, and strengthens community economies. The criteria imply an initial and ongoing need for purposeful opposition and difficult disentanglement from the subjectivities formed in the dominant food regime.

Yet, growing amid the "perverse confluence" that characterizes neoliberalism, the local food movement reveals a possibly different order of events. It has been spurred by a sense of opportunity and possibility that has so far overshadowed apprehension of the need for opposition. Also striking, the movement is extremely decentralized (Nonini 2013). Activists draw information and stimulation from the internet, social media, circulating speakers, and media rather than from ongoing relationships with regional and national movement organizations. Although the work of the movement has downplayed a critical structural analysis of the dominant regime and its hold on local possibilities, movement efforts have managed to shift tastes and sensibilities. Activists have put a great deal of energy and labor into (re)building the cultural, social, and economic infrastructures needed for local food systems. Many nonactivists and activists alike now have an appetite for local food and the alternative discourses and visions of food and farming circulated by the movement. Given a regional, national, or even international group with a critical analysis sufficient to intersect with local forms of activism and make sense to local actors, this new cultural base would seem a key resource for mobilization to change government policies and even the systemic design of the food system.

Gibson-Graham's *A Postcapitalist Politics* is remarkable in that the authors bring the results of their participatory action research back into conversation with scholarly theory. Perhaps with their book and Arturo Escobar's 2008 *Territories of Difference*, which similarly returns the results of participatory action research to academic circles, we are witnessing a transformative development within geography and anthropology. In the spirit of such a transformation, we have derived from Gibson-Graham's book a list of key criteria for assessing the transformative achievements of social movements. We intend the list to be useful to researchers and movement participants for dialogue and reflection as well as an object of debate.

We have employed the derived list of criteria to discuss the local food movement in North Carolina. The contemporary local food movement has been enormously important in expanding the amount of organic and locally grown food available to areas of North Carolina. It has substantially increased the frequency and range of food activities and events carried out in communities where it is present. It has underscored concerns about the agrofood industry and its shortcomings with respect to food safety, diet-related diseases, and hidden environmental costs.

Still, the creative, activity-generating effort of the movement has not yet been accompanied, especially for people oriented to expanding the availability of local food, by a deeper collective structural analysis. The activists of the movement, in effect, appear satisfied with creating pockets of "good food" at least for those who can afford it. Recalling the spirit of Gibson-Graham's politics of possibility and using the criteria as a tool for reflection and dialogue, we suggest, however, after considering the conditions for activism in the contemporary US, that the potential exists for mobilizing a more oppositional translocal food movement that seeks fundamental change in the current food regime. But this will occur only if two additional criteria are added to the list of criteria for transformation drawn from Gibson-Graham (2006).

Active Analysis of Structures of Race and Class within the Local Food Movement

First, there must be specific analytical attention among activists to the structures of class, ethnicity, and race that not only prevail across US society but have also become deeply internalized within the domain of local food activism itself. Recurrently in our analyses of both sustainable food and farming activists and food security activists, we found an inattention or inability to address contradictions that shaped the reach and significance of the movement. Major challenges to forging collective action resulted from the structural divisions of the broader society, especially as exacerbated by the historical contexts in which local food activists and an incipient local food movement are embedded. The movement internalized these structural features, and the dynamics that developed around them inside the movement created weaknesses and difficulties that have not been recognized as such. Critics of *A Postcapitalist Politics* claim that Gibson-Graham pay too much attention to agency and too little to analyzing how structures constrain agency (e.g.,

Grossberg 2010) and how structures create possibilities for new productive power, and we are inclined to agree.

Concerning constraints, our analyses of the movements point to a key avenue by which structures intrude into activism and limit agency. Structures of class, ethnicity, and race—supposedly outside the emphases of the movements—have been culturally recoded in the practices and discourse of activists and within the change organizations themselves. Hierarchies of power and status were re-created with only minimal reflection. We brought up earlier the ways in which class divisions and racial animosities structure food activisms and shape their potential. We suggest amending the Gibson-Graham list to incorporate this important point and other reflections we have presented throughout the text. Change efforts need to reflect on the ways in which structures of power and privilege are internalized in the dynamics of activist organizations, creating contradictions and obstacles that restrict the organization's capacity for effective change. A ninth criterion—*awareness and analysis of the dynamics by which broader structures of privilege are recoded/reestablished within change efforts of the movement and constrain success*—has been added (see table 10.2).

TABLE 10.2. Amended and Modified List of Key Elements of Social and Economic Transformation, as Derived from Gibson-Graham's *A Postcapitalist Politics*

For a critical mass to undergo politicization, the following must occur:

1. Recognition of a structure of domination, some of its elements or at least critical reflection on a crisis of the status quo and its interrelation with other structures of domination

2. Identification and enactment of a politics of possibility

3. Creation of alternative discourses/vision

4. Orientation to a collective and a building of community—an "us" that includes reflection about power

5. Changes made in daily life and everyday practices

6. Cultivation of subjects with the desires and capacities for sociability, happiness, and action offered by alternative social and economic arrangements

7. Ethical commitment and self-cultivation

8. Purposive shifts in subjectivities and identities

9. Analysis and awareness of the dynamics by which broader structures of privilege are being recoded/reestablished within the movement and constraining success

10. Ongoing reflexive dialogues within the movements to revise and rethink their objectives, analyze reality, and reinvent their action

Reflexive Dialogues within the Movement on Criteria for Fundamental Transformation

We also propose to include a new tenth criterion: *reflexive dialogues within the movements that permanently look at and rethink their objectives, analyze reality, and reinvent their action*. We consider that this reflexive attitude allows movements to be more successful in the achievement of their goals. This attitude should include the subjective dimension. Thus, we argue that movement organizers and activists need to consider together the key elements of social and economic transformation that we have identified together, in order to contribute to emancipatory change.

Conclusion

The Present and Future of Food Activism

Where We Are Now: Findings from the (Ethnographic) Present

It is time to pull together what we learned from our ethnographic findings of this book in the first half of this conclusion. Then in the second half, we turn to the future of local food systems.

We began the book by introducing the popularity of locally grown, sustainable foods for an increasing number of people in North Carolina and beyond it in the US. We proposed that an essential part of our local provisioning systems for such foods were activists, most of whom were not themselves food producers and who labored to make local food provisioning both more sustainable and more just. A key contention in the book is that the labors of food activists participating in local food systems represent a major unrecognized gift to the US food system and the US population. We found that there were three kinds of local food activists with fundamentally different moral logics: sustainable farming and food activists, community food security activists, and charity-oriented food activists. All three challenged the neoliberal moral logic of the corporate industrial food economy, although the latter group did so less than the other two because of its co-opted position within the corporate-dominated charitable food sector. The labors of sustainable farming and food activists were devoted to making local food systems more robust and reliable over time. Charity-oriented food security activists sought to make local food systems more responsive to the needs of charitable food recipients. Community food security activists worked to transform local food systems so that community members would be able not only to eat but also to produce culturally appropriate, sufficient food and determine the conditions under which it is produced and distributed.

We argue that the food activists whom we and our ethnographers met, interviewed, and worked next to have undertaken numerous projects that have improved the local food economies of the four sites we studied in North Carolina. Activists have expended much of their unpaid labor and alternative paid labor (i.e., paid below a living wage) in projects that help create and sustain small urban farms, gardens, CSAs, farmers markets, food banks, food pantries, community kitchens, homeless shelters, and much else that have nourished and enhanced the lives and livelihoods of farmers, gardeners, food producers, and poor and food-insecure populations in these four sites.

The remainder of the book has provided support for these contentions by not only demonstrating what contributions food activists have attempted to make to local food economies but also ascertaining whether and in what ways their efforts have succeeded. We have done so by evaluating their contributions through assessing the role of their projects within what the feminist geographers Gibson-Graham (2006) refer to as the "diverse community economy"—not the economy that capitalist or neoliberal ideologies aspire to or conjure into existence but the everyday economy that Americans actually participate in, which has diverse kinds of labor, of exchange, and of enterprise beyond the wage labor, market exchange, and capitalist enterprises that conventionally define "*the* economy," that is, the capitalist one. Gibson-Graham's (2006) analysis of the diverse community economy illustrates the manifest importance of noncapitalist and alternate-capitalist (hybrid) kinds of labor, transactions, and enterprises in projects by food activists to develop local food economies, as distinct from the dominant model of a totally capitalist "free-market economy" of food (see table 3.1). For example, farmers markets have far more vendors who are self-employed farmers, self-provisioners, and cooperative owner-workers than do capitalist employers.[1]

Up against the Alliance: The Global Squeeze of the Small Farmer Comes to a Local Food System near You

As we examined local food activists and the local food systems they participated in, we found that they came up against the constraints placed on the development and expansion of local food systems by the global industrial food corporate alliance and its systematic efforts to limit democratic control over the US food economy. We pointed to the ways in which industrial food corporations have become increasingly integrated into the US corporate state and have used their powers of state capture to further their accumulation of wealth. These food corporations have also sought successfully to limit the capacities of local food economies in several ways (legal, political, discursive) but most effectively by the tyranny of the "world prices" the alliance has deployed in effect to set each farmer on Earth in competition with every other, forcing farmers to offer the lowest possible food prices, even when they are too low for farmers to sustain their livelihoods. This global food market is historically unprecedented. It is centrally relevant to local food systems and food activism in the United States, but with implications that few scholars or members of the public realize.

We can see this when we revisit our earlier discussion of the contemporary "corporate food regime" associated with neoliberal globalization (McMichael 2013). We learned that the WTO "liberalized" food prices internationally by stripping all tariff and other protections from domestic food prices in global Southern countries at the same time that it gave foreign corporations the "right to export" food to these countries, that is, required these countries to allow in their imports. The outcome was "the decoupling of subsidies from prices, removing the price floor, and establishing an artificial 'world price' (*substantially below production costs*) for northern grain surpluses dumped on the world market at the expense of non-corporate farmers everywhere" (McMichael 2013, 54; emphasis added), while corporate food brokers forcibly imported lowest-priced foodstuffs to global Southern states, thus driving tens of millions of domestic farmers in these states out of the market and into pauperism.

We are considering here the WTO, a trade agreement *at the global scale* in which the alliance of industrial food corporations and compliant states plays a major role, as we showed earlier. A crucial shift occurred in the 1990s, when these WTO provisions were directed not just to the global South but also toward the *domestic* US food market, which has "allowed" corporate food producers and brokers to import huge volumes of vegetables, fruits, nuts, spices, and other foodstuffs into the US. The result has been a market-induced "race to the bottom" in food prices among American small- and medium-sized farmers.

It does not take much of an act of imagination to understand the consequences of this "global race" for local farmers—and thus for food activists. Jay Thomas, our articulate Crop Mob consultant, stated earlier one central predicament of this book, but it bears repeating: "It's a foundational problem because from a farmer's perspective, working in sustainable agriculture, they need to be selling their vegetables for more than they are [able to] in order to make ends meet. And from a justice perspective, underprivileged folks need to be able to buy the food from those farmers for less than they're selling it for now." This is no less than the global-scale "predicament" of small farmers everywhere under neoliberalism. The "world price" (McMichael 2013, 54), say, set for tomatoes from Mexican plantations telescopes down to the price being paid in US supermarkets, but this is not the price that the local farmer at, say, a farmers market in Durham or Watauga County can afford to sell their tomatoes at, although North Carolina consumers soon become habituated to that world price. What could go wrong with that?

Consider a local farmer selling her tomatoes at the farmers market. Although she may develop a niche clientele among local "foodies" who will buy her tomatoes, which are far fresher, tastier, and more nutritious than Mexican industrial box tomatoes, the probability is high that she will not be able to sustain her livelihood over time because her living and farming costs are too high and the market demand for her produce too limited (e.g., seasonal) and unreliable. One does not go into direct farm sales to attain a "living wage" or income under these conditions (Paul 2017). The same logic would apply to the other produce she grows, all of which would be subject to downward price pressure from its "world price." Her specific, individual economic precarity, writ larger at the local scale, suggests why being a "local farmer"—like being any small- or medium-sized farmer—is not a long-term viable occupation in the US.

Now, let us consider sustainable food activists. They may work to start up a county's food-aggregation hub for local farmers, or they may collaborate with others in trying to set up a community land trust to keep farmland as such in perpetuity. This is all to the good, but they are never in a position to contest the competing "world prices" that continuously set the floor for the prices of produce or meat and threaten the livelihoods of local farmers.

Perhaps that local farmer selling her tomatoes benefits somewhat from the efforts of food security activist when a farmers market sets up the Electronic Benefits Transfer (EBT) system, which allows her to accept SNAP benefits from poor adults fortunate enough to qualify to receive them. Perhaps the farmers market even has a "Double Bucks" program, like Durham's, which provides a limited matching subsidy to SNAP recipients when they shop at the farmers market. These facilities are the creation of food security activists concerned that poor people be given the chance to purchase higher-quality, fresh local food. All power to them, but SNAP benefits that are consistently stingy do not allow poor folks to buy many such tomatoes.

Local food activists, whether engaged in sustainable food or food security activism, thus expend huge amounts of largely free or alternative wage labor and provide immensely valuable services and create critical facilities that sustain local food systems and make them more equitable, but there is no question about the limitations to what they can currently achieve: "the local" market resonates with the global market created by the corporate alliance. In contrast, large-scale corporate "commodity farmers" receive billions of dollars in USDA subsidies for producing vast quantities of corn and soybeans to feed cattle and (in the form of biofuels) automobiles, because their subsidies have been designated by a pliable WTO as "non-trade-distorting," excepted from the WTO regulations that have prohibited, say, Ghanaian state import tariffs that previously protected the prices of small corn farmers in Ghana.

It is in this political-economic context that we consider the outstanding "crossover" activists and their enterprises. We found that these activists tried and partly succeeded in achieving both community food security and sustainable farming/ food activists' objectives.

Four Experiments in Noncapitalist Practices by Food Activists in North Carolina

While we described the contributions of activists to local food systems that we inventoried earlier, we set them apart from those exceptional activists and the four enterprises they built. We now turn to those activists. These four cross-overs successfully combined the two moral logics of sustainable farming/food and community food security activisms and thus have made major contributions to their local food economies. In the case of all four, their leaders confronted the central predicament that the corporate alliance's global market and world prices for food poses in its manifestation in their local food economies, and they found innovative solutions to it. The leaders of F.A.R.M. Café of Watauga County, Cornucopia Landscaping of Durham, and the Slow Food Convivium in Charlotte have all done so by way of cross-class and cross-racial alliances that improved the capacities of their local food systems to provide more locally grown foods to poor and racially vulnerable residents of their local economies, and they sought to do so in ways that supported sustainable food growing.

F.A.R.M. Café and Cornucopia Landscaping did so very successfully and provided new paradigms, along with similar cross-over enterprises, as indeed they already have with F.A.R.M. Café's participation in the One World Everyone Eats movement and CL's emergence at the same time as other urban permacultural food projects, as in the UK Transition Towns movement (Rhodes 2017, 111). These two enterprises have reconciled sustainable food and community food security activism. In the case of Slow Food Charlotte, its leaders worked successfully to bring about a sustainable garden network and food commons in Charlotte but failed to work to protect the livelihoods of small farmers in the city's surrounding metropolitan counties forming its foodshed.

In contrast to the other cross-overs, the leader of the Clarketon Baptist Chapel's community garden, encountering antagonism from nearby white farmers, experts, politicians, and clergy, had recourse to the blues epistemology and blues bloc solidarity associated with African American resistance to racial oppression in the US South. Given the strong racial and class animosities that were endemic in Edgecombe County in rural eastern North Carolina, Clarketon Family Life Center's leader decided that the only way toward community food security and sustainable food growing was through tying the garden to the self-reorganization of the Clarketon African American community so as to allow its youth and their families to have more economically and nutritionally secure futures around food.[2]

In the course of the ethical discussions among the leading activists in each of the four cross-over enterprises, they created new "figured worlds" (Holland et al. 1998) that imagined sustainable and widely shared responsibilities and interdependencies: "feeding all regardless of need"; creating "abundance" through "edible landscapes"; working for "Afrocentric relations" grounded in reciprocity and barter, not money; and starting "Friendship Gardens" to bring together white, African American, and other nonwhite activists in order to create and maintain an urban food commons.

In each of these three cross-overs, we found that activists undertook ethical debate, research, and decision-making that resulted in viable noncapitalist alternatives of great importance to local food economies. In each of the cross-overs, leaders did so in ways that provided food-insecure people with more equitable access to food resources in ways consistent with their dignity. In two cross-overs, F.A.R.M. Café and Clarketon Community Gardens, leaders were inspired to ensure people's dignity by their religious values; in SFC, leaders were drawn to do so opportunistically by the racial politics of Charlotte; while CL's leader was pressed to reevaluate white, middle-class privilege after a chastening experience with African American community activists. In each of these cross-overs, activists made attempts to work with, and not just on behalf of, working-class poor people and people of color.

It is time to assess the ways in which these four cross-over enterprises and their leaders exemplified noncapitalist alternatives that redefined surplus, necessity, commons, and consumption as means for developing diverse local food economies.

The questions we adopted from Gibson-Graham's (2006, 88) analysis were the following:

- What was *needed* by farmers, consumers, farmworkers, or community residents, as distinct from what was surplus to these needs?
- How should social *surplus* within the food system be appropriated and distributed by local food actors, and which ones to further sustainability or social justice?
- When should surplus be *consumed*—and by whom—to meet people's food needs or be added to the commons within the local food system?
- How a should a community food *commons* be created and sustained, through what measures, and by whom?

As to decisions about *what was materially needed or required as socially necessary labor time* versus what could be surplus (or surplus labor time), Cornucopia Landscaping and F.A.R.M. Café both stood out. CL's leader, James Sokolov, and other worker-owners decided that they could reduce the necessary labor time needed for their costs of living, which they earned by contracting with affluent residents to construct edible landscapes for the latter, and redeploy their surplus labor time (equivalent to what was left over from such projects) toward pro bono projects on behalf of community food security.

F.A.R.M. Café's Renée Boughman and her cofounders decided that for the café to appear to food-insecure patrons to be an actual restaurant with an attractive menu and inviting premises, the socially necessary labor time of a few food professionals (head and assistant cooks, food recovery coordinator) should be combined with that of a relatively large number of volunteers for food prep and restaurant setup and closing tasks, including food-insecure patrons. The labor time for the former, they decided, needed to be paid at the going wage-labor rate for their occupations, while the labor time of volunteers was to receive nonmonetary small but real compensation (i.e., a unique learning environment for university students or a café meal for poor patrons); most of the time of volunteers became surplus labor time. Insofar as F.A.R.M. Café's actual proceeds from selling meals to its economically secure patrons fell short, its staff decided that they also needed to use their own labor time to prepare and present food at fundraiser events patronized by local wealthy liberals—a shortfall in surplus (we point out) that in another economy might be provided by the state.

Regarding activist projects around *surplus and how it should be used*, CL's worker-owners decided that they would direct their surplus social labor time toward specific pro bono permaculture projects, depending on the invitations they received from community leaders and the buy-in and willingness shown by community members to work on a specific project. F.A.R.M. Café and its leaders stood out with regard to their ethically grounded decisions about how to productively use surplus in a variety of forms—monetary surplus received from foundation and corporate grants, surplus appropriated from volunteer labor and the labor of its founder and staff, and the surplus food acquired through F.A.R.M. Café's Full Circle food recovery program that it could sell—all to finance the café's operation. Slow Food Charlotte's leaders sought out surplus for their Friendship Gardens from a local philanthropy, while investing some of their ample surplus labor time in the planning and execution of the Friendship Gardens network.

Clarketon Community Garden's leader showed the most ingenious and inspired actions around surplus. On the one hand, his Afrocentric vision dealt with redistributing surplus through "barter" to support interdependent social "relationships," while his gardening concern for "anybody to help that plant stay connected, and [if] they only take off that plant what they need, the plant will reproduce itself" anchored his sense that surplus always had to be clearly distinguished from what was needed. At the same time, his many discussions with Clarketon youth about the disposition of their own surplus from the garden produce they grew were exemplary: Should such produce be given to family member or needy elders? Should it be sold and the proceeds saved and matched in the youths' budding college funds? Should it be sold and the proceeds used for their own consumption?

As for decisions about *consumption and its uses*, the question had to do with the ways in which consumption *or* avoidance of consumption could lead to production that could in turn generate more surplus to spend on socially and economically valuable assets for the community economy and its members. Among our cross-overs, we saw this most clearly expressed in Clarketon, in what could be called Rev. Hemmings's parable of the Hardees fast-food meal to youth during their road trip, which taught them that their consuming fast food led to the outflow of their (and their families') scarce income to white-controlled fast-food corporations that were inimical to them and that they needed to be "producers," not "consumers," if they were to be in control of their own lives. F.A.R.M. Café's staff decided that providing a restaurant meal to food-insecure patrons in return for the volunteer labor they did supported patrons' consumption as a means of providing them nutrition, of rewarding their volunteer labor time and protecting their dignity against being seen as "taking a handout." Even the learning experience of CL's founder in moving forward from the Massive Harvest fiasco to the successful Lincoln Street Neighborhood farm implied his realization that poor neighborhood residents had overwhelming needs to consume the foods they grew instead of growing it only for sale.

Finally, with regard to *creating and sustaining the commons* of a community, Slow Food Charlotte leaders' efforts to bring about an urban commons in freshly grown produce through the Friendship Gardens have to be noted. Recognizing the need for poor, housebound residents to have sustained access to such produce in the Meals on Wheels meals they received, SFC leaders, aware of their own race and class privilege, were able to work with African American and other minority gardeners throughout Charlotte. This allowed them to successfully scale up the Friendship Gardens program over several years to include almost one hundred gardens. This represented a valuable building of Charlotte's food commons—while bearing in mind that most of the labor that made it possible was expended by the gardeners themselves.

Given that F.A.R.M. Café's raison d'être was to "feed all regardless of means," the café's provision of meals to the public acted as Boone's urban food commons by providing a steady supply of meals to everyone who wished, with the proviso that the commons would be maintained through a formula that made the cash payment of some (affluent) patrons for a meal equivalent to the hour's labor by other (poor) patrons.[3]

The leader of the Clarketon Community Garden project also reinforced the importance of the garden as a food commons for impoverished members of the Clarketon community—both in his injunction that "anybody to help that plant stay connected, and [if] they only take off that plant what they need, the plant will reproduce itself" and in his teaching of youth that poor seniors in the community had claims over the produce that youth might grow.

* * * *

In each of these cross-overs, and also in many other instances of food activism described earlier, there is evidence that food activists have sought to push against the confines imposed on them by capitalist regimes of labor, market exchange, and privately owned enterprises devoted to capital accumulation toward noncapitalist alternatives. In the situations of three of the cross-overs, Cornucopia Landscaping, F.A.R.M. Café, and the Clarketon Community Garden, we contend that they have become small-scale but still very real anticapitalist models for food activists in the present. We recognize that the activists who animated these three cross-over exemplars, and other rebellious food activists (e.g., Crop Mob) whom we described earlier, operated under the demands and constraints imposed by capitalism and made what they saw as necessary compromises with it, but nonetheless they stand out as paradigms for an anticapitalist and arguably socialist practice that seeks to transcend its own conditions of existence. Above all, they stand for small collective examples of solidarity with local small farmers and food-insecure poor populations.

Local food activists like those we have described earlier but, above all, the cross-over activists we have featured in this conclusion would flourish if they belonged to a broader, translocal national movement of food activists—one energized by a critique of the global corporate industrial food sector and committed to forming the transnational solidarity needed to ally US local farmers and activists with movements of small-scale farmers in the global South. Such a movement emerged from 1994 to 2012 in the form of the Community Food Security Coalition (CFSC). The CFSC, as noted earlier, was a national alliance of more than three hundred food related organizations focused on community food security, on food security as such (e.g., as organizations combating hunger), or on sustainable agriculture.

The activities of the CFSC were many. It lobbied Congress successfully to include a Community Food Projects (CFP) program in successive Farm Bills from 1995 to 2012—a program that provided scarce funding for many community food security projects. The CFSC worked to connect community organizations of poor people to other sectors in the local food economy—to local farmers, school meal programs, food banks, food pantries, farmers markets, and CSAs—to increase poor people's community food security. The CFSC came up with innovative participatory research methods like community food assessments, farm-to-cafeteria programs, community garden projects, CSAs to support low-income members, and youth groups focused on urban agriculture.

At the CFSC's annual meetings, we witnessed intellectual and personal connections between activists in sustainable agriculture, the antihunger movement, and community food security movement being formed, common interests discovered and debated, and experiences shared. The CFSC's building of national and transnational solidarities culminated in its 2010 annual meeting in New Orleans, where we saw its first annual Food Sovereignty Award presented to international representatives of the international federation of peasant organizations La Via Campesina (which we discuss later) for its work done on behalf of small farmers around the world, including North America. Unfortunately, CFSC closed its operation in 2012 due to a loss of outside funding.

Indeed, we are in a moment of transition in which the paradigms and the anticapitalist energies of our cross-over exemplars are needed—and need amplification by a national-scale local food movement, which would not be an oxymoron. To see why, we need to briefly set out why the global corporate industrial food sector that farmers and local food advocates outside the US are rebelling against represents existential risks to the human population, and we need to project what this might mean for the future of food.

Existential Risks: The Alliance Is Powerful, the Food System Is Fragile

As we saw earlier, the alliance of transnational food corporations and compliant corporate states has distilled its multiple forms of power into a corporate food regime. This regime has created a global market based on its corporate members' use of their financial and political leverage to restructure domestic markets to dictate world prices to farmers around the world—prices so low on offer to small- and medium-sized farmers that tens of millions are unable to sustain their livelihoods (McMichael 2013, 2023). We have seen that such monopoly power does not stop at the US border but directly threatens the viability of US farms as well and that this is evident in our local food economies. Thus, despite the popularity of local, fresh foods produced via agroecological and organic methods in the US, the US food-provisioning system has become almost completely reliant on the global industrial food sector.

This is a dangerous moment for the US population, as it is for populations in the rest of the world. The alliance of the global corporate industrial food regime may be powerful, but it is on a path-dependent trajectory that generates conjunctural political, economic, and ecological instabilities threatening the viability of the US food-provisioning system and the health and survival of the US population that it is supposed to feed. This is even more the case internationally for populations outside the US and most gravely in the global Southern countries that are increasingly the source of US food imports.

The global industrial food alliance, in tandem with other large-scale capitalist sectors with which it has commercial, political, and investment ties, such as the fossil-fuel industry, large-scale investment banks, and Wall Street capital managers, is subjecting the Earth's human population to five major existential risks that not only have their own distinguishable effects but are interacting in complex and increasingly unpredictable ways. Some of these risks involve processes that have already led to recognized disasters, while others are incremental but are reaching upper limits; some are global from the get-go, others regional initially; some are obvious, others far less evident to most people.

The five existential risks posed by the global industrial food economy are:

1. *Agroindustrial-induced climate change*: The global industrial food economy is dependent on fossil fuels for its food production, transport, distribution, and marketing to the point that it currently generates between one-fourth and one-third of all greenhouse-gas emissions. (For evidence, see the appendix.) Its enormous consumption of fossil fuels has already contributed to global climate change, and the continuation of fossil-fuel consumption at anything like the current rate cannot occur beyond the next one to two decades without major climate-change catastrophes occurring, as predicted by the IPCC report on the grave consequences of global warming at or above the two-degree Celsius increase since the preindustrial era (IPCC 2018). Short of systemic change away from fossil fuels, major threats to the viability of human beings and other species on which they depend (including for food) will certainly emerge on an increasingly hot planet whose climate system is going into overdrive.

2. *Depletion of topsoils and subterranean aquifers*: Industrial agriculture has in its short history already consumed a large percentage of the US's best topsoil and fresh groundwater, and industrial methods will continue to endanger these topsoil and water sources during the next several decades.[4] Nor can US consumers continue to

look elsewhere beyond the US for the topsoils that grow their food: globally, the Food and Agriculture Organization has predicted the depletion of topsoils usable for farming within the present century if current rates of soil degradation and erosion continue (Cosier 2019). Freshwater from underground aquifers has been pumped up for irrigation for decades, and there are signs of complete aquifer depletion in the next few decades.[5] The consumption of these fundamental resources of farming under current rates of use by industrial agriculture cannot be sustained indefinitely.[6] There are interaction effects, as climate change lowers the availability of surface water for farming and consumption due to increased heat and droughts, as in the case of the Colorado River watershed, which is essential to California's agriculture, while torrential rainstorms and hurricanes erode increasingly large volumes of topsoils in prime farming regions.

3. *Pesticide overuse leading to pollinator species extinctions*: Industrial agriculture and its associated use of pesticides have been responsible for the disappearance of large numbers of species of animals and plants, some of which are clearly key species in the reproductive processes of farming, for example, bees, other insects, and birds as pollinators (Center for Food Safety 2022). A failure to pollinate is a failure to reproduce many of the plant species on which humans depend for food.

4. *Industrial animals and zoonotic viral pandemics*: Insofar as the industrial food economy continues to include meat-animal production, the processes that produce these animals not only contribute to greenhouse-gas emissions and climate change but also rely on concentrated populations of meat animals increasingly subject to zoonotic pathogen infections that can be and have been disseminated to and can be fatal to human beings. Examples include swine and avian flu viruses spreading from CAFOs to nearby dense urban human populations in North America, Europe, and China, and the COVID-19 virus, SARS-Cov-2, whose "spillover" into the human population is now traced to the market for "wild" animal foods in southwestern China, which has led to tens of millions of fatalities and devastating losses of economic output (Wallace 2020).[7] Taste for animal meats continues to spread among the middle classes of

the newly industrialized economies (e.g., China, India), while still being firmly entrenched in the US food economy. The probability of future pandemics emerging due to the intersection of concentrated meat-animal production with the reproductive ecologies of new (and already existing) zoonotic pathogens is high and, over the next few decades, almost certain (Wallace et al. 2020).

5. *Overuse of fertilizer phosphates on industrial farms*: Intensified use of synthetic fertilizer over the past four decades has led to widespread runoff not only of nitrogen but also of phosphates into waterways, estuaries, and oceans. The process of continually applying these fertilizers is coming up against the "planetary boundary" for phosphates flows into the oceans, to the point that much further use threatens a "large-scale ocean anoxic event"; and paleontological evidence of such past events "potentially explains past mass extinctions of marine life" (Rockström et al. 2009, 474). Current inflows of phosphates into oceans are approaching such a "critical threshold," with many estuaries and freshwater bodies already exceeding it (Rockström et al. 2009, 474).[8]

Two other factors exacerbate these five existential risks to humans from the current operation of the global industrial food economy. First, the financialization and digitalization of the global industrial food economy associated with the neoliberal era facilitate the expansion of industrial food production into new areas of enclosure against farmers, exacerbating the vulnerabilities noted earlier (McMichael 2023, 211–214). As Jennifer Clapp and S. Ryan Isakson (2018, 159) observe, the "financial transformation of the food system" has had the consequence of "heightened food system vulnerability to economic and environmental stresses." New kinds of financialization of food include (1) speculation on food prices and costs (e.g., food commodity index funds, index-based agricultural insurance) that increases the volatility of food prices (Clapp and Isakson 2018, 29–55); (2) credit schemes pushed by agro-input corporations and compliant states that increase unmanageable debt among farmers to purchase farm equipment, seeds, fertilizers, and pesticides (Clapp and Isakson 2018, 116); (3) "land grabs" in the global South by large-scale financial corporations (e.g., pension funds, hedge funds, and wealthy investors) that lead to the displacement of small farmers from their land, followed by rack-rent abuse from their new large-scale corporate landlords (Clapp and Isakson 2018, 93–97); and (4) consolidation across sectors within the corporate alliance, facilitated by the collection of vast amounts of data and its analysis by digital systems to surveil and control food production (McMichael 2023, 214). The consequences of using these financial mechanisms are increased production and price volatilities of essential farm commodities on a global basis, the continued dispossession of small farmers by agribusiness and finance corporations, and adverse effects for food availability and prices in the US.

Second, the political order of postwar neoliberal globalization under US hegemony is itself unravelling and becoming unstable (Gerstle 2022), with largely unpredictable consequences for US food provisioning. Instead of states' and political elites' unquestioned commitments to governance by international trade agreements (e.g., WTO, NAFTA/USMCA) and international financial institutions (World Bank, IMF) within the global order of Pax Americana, we are entering a period when neoliberal principles are being challenged. Trade disputes and pandemics disrupt supply lines and depress industries; political disorder and authoritarian populisms threaten states' stability; the supply of fossil fuels has become more unpredictable due to the high costs and risks of extreme extraction projects. The economic risks from wars increase, from those already occurring (e.g., the Russian invasion and occupation of Ukraine) and probable in the future (e.g., PRC military occupation of Taiwan) (Gerstle 2022).

Finally, the political responses by states to climate change itself remains highly uncertain, as demonstrated by repeated failed Conference of the Parties (COP) meetings, marked by disputes between nation-states over how to allocate responsibility for climate change. These political conditions create even greater economic unpredictability and volatility, adding uncertainty to the capacity of the current global industrial food economy to reliably provide for the food needs of the US population in the coming decades.

These five existential risks of the corporate industrial food economy to human flourishing and even survival are too great and their probability too high to neglect them. A transformation needs to be made in the current US food economy away from its complete reliance on fossil fuels coupled to industrial technologies, and from food capitalism's logic of infinite capital accumulation ("economic growth") on a finite planet toward more sustainable and equitable alternatives. Our argument is that the US future of food will have to include our local food systems as one, and perhaps the, major such alternative. If this is the case, then we had best think more critically about what the futures of local food systems might be like. Here our study of local food activism is relevant.

Food Activism as a "Real Utopia" of the Future

Our evaluation of local food economies and the role of sustainable farming and community food security activists in them not only provides a record of an important cultural phenomenon in contemporary North Carolina and more broadly in the United States. Beyond these findings in the present and looking to the future, I (Nonini) find it necessary to shift to the first-person singular for the reason addressed in the preface. I propose that cross-over food activism provides candidates for what the late sociologist Erik Olin Wright (2010) called "real utopias" that could be integral elements in fundamental political transformations.

By a "real utopia," Wright (2010, 6) meant a practical exemplification of "utopian ideals that are grounded in the real potentials of humanity, utopian destinations that have accessible way stations, utopian designs of institutions that can inform our practical tasks of navigating a world of imperfect conditions for social change." Real utopias are actually existing institutions that exemplify a more socially just, socialist design for living. Real utopias that Wright (2010, 5) proposed included universal basic income (UBI) systems, which have been tried with some success in many countries; the worker-owned Mondragon cooperatives in Basque country, Spain; city participatory budgeting, which has been successfully adopted in Porto Alegre, Brazil, and spread widely beyond it to other cities in Brazil and North America; and Wikipedia, the online informational commons (Wright 2010, 2–5).[9]

According to Wright (2010), each of these exemplifications of a real utopia is a demonstration of how the people of a locale or country can regain democratic control over their own economic institutions from oppressive capitalist systems. UBI frees people from having to find capitalist employment. The Mondragon cooperative demonstrates how workers can own their scaled-up means of production. Participatory budgeting provides urban residents with democratic access to debate in decision-making over city government budgets, allowing them to bring about vitally needed services and employment to their neighborhoods. Wikipedia's self-correcting capacities to provide accurate information freely show that knowledge can take the form of an intellectual commons and need not be monopolized by profit-making publishing conglomerates or universities.

Wright (2010) considers real utopias in his discussion of "interstitial transformation" as a strategy for transforming capitalism into a more economically democratic political order, that is, socialism.[10] Interstitial strategies, as can be imagined, have to do with establishing a practice or institution in the "interstices" of a larger structure that tolerates its existence: "The underlying assumption is that the social unit in question can be understood as a system within which there is some kind of dominant power structure or dominant logic which organizes the system, but that the system is not so coherent and integrated that those dominant power relations govern all of the activities occurring within it" (Wright 2010, 322–323). Hence, the system tolerates the dissident or discordant institution or practice; for example, Brazilian capitalism tolerates municipal participatory budgeting in certain regions. Thus, real utopias are such institutions or practices that can emerge and be sustained in these interstitial locations, and much of Wright's discussion of real utopias centers on their being situated in this way.

I contend that local food activists have within their practices the makings of real utopias. This book has identified three cross-over groups of activists (and discussed a fourth) who support sustainable food cultivation while also seeking to make local food systems more just in distributional terms and thus to address food insecurity, a major achievement for such small groups. These three cross-overs bear within them the characteristics of real utopias. Their activists embody and promote solidarities based on win-win solutions that bridge divides between farmers or other food producers and poor communities of eaters. They provide support for those who are most subject to capitalist depredations. Even when they withdraw behind their own community lines, as did the Clarketon Family Life Center leader and staff, they do so only when they have no choice. They tend to have leaders with flexible tactics to meet their fundamental goals of equity and sustainability in food systems, while they keep these goals clearly in view. They carefully explore non-capitalist alternatives like the commons, define surplus (and surplus labor time) and consumption creatively, and validate necessary labor in the enterprises they create. They are committed to racial equity and the welfare and dignity of working-class people. They tend to last as groups that make themselves into long-term viable enterprises. They "punch above their weight."

However, unlike the other real utopias that Wright (2010, 2–5) mentions, cross-over food activist practices do not originate from a pre-existing state system (UBI, participatory budgeting) or arise from a once-in-an-epoch capture of a platform niche (Wikipedia), although CL at a much smaller scale celebrates the same cooperative principles as Mondragon. Unlike the other real utopias that Wright valorizes, they are small scale but, like them, committed to generating social solidarities and critical of and opposed to capitalist institutions of domination. If they have a particular weakness, it is perhaps being led by charismatic leaders, and therefore their enterprises face definite issues when it comes to these leaders' succession.

I argue that these activists and their practices represent de facto socialist experiments in alternative subjectivities within preexisting civil society, and this is one of their great strengths. Their subjectivities do not come from being made the "new socialist persons" derived from social programs of a new socialist state that does not (yet) exist. I have no idea how or when a new socialist state will come about. So it is better to posit that new socialist people come into existence as they seek to transform the material conditions they live within, which are currently those of neoliberal capitalism on the wane, as in the case of these cross-over activists.

The Twilight of Neoliberalism and Countermovements against Global Commodification

The decline of neoliberalism and the increasing dysfunctions associated with neoliberal globalization (e.g., growing economic inequality, precarious livelihoods for increasing numbers of people, breakdowns in global supply chains) are becoming increasingly evident. The dilemmas for a waning neoliberalism after more than forty years of its dominance in the United States as elsewhere are acute ones. Its free-market fundamentalisms have destabilized and injured the lives of hundreds of millions of people by demanding that postwar nation-states inflict austerity on people by cutting government programs of social welfare benefits they received. In the US, neoliberal prescriptions have been implemented as state policies that have reduced the power of labor unions, impoverished millions of workers, led to capital flight and disinvestment from locales, and threatened environmental standards. The financial and trade institutions empowered by neoliberal globalization (World Bank, IMF, WTO, NAFTA/USMCA, et al.) have pushed the frontiers of food commodification toward a *global* market of world prices in food, leading to the cheapening not only of food but also of the labor and livelihoods of those who grow it, thus leading to the collapse of small farmer sectors globally—including in the US. At the same time, the winners of globalization—the extraordinarily wealthy "1 percent" (or even the mega-wealthy "1 percent of 1 percent") of the population have done conspicuously well—and are flaunting it.

In Karl Polanyi's classic history *The Great Transformation* (1957), he reconstructed the "disembedding" of "the market" from society from the early nineteenth century to the 1930s that led to massive suffering and discontent and created the conditions that brought about two world wars and the Great Depression in between. Polanyi pointed to the defining features of a disembedding of the market that arose from the commercial invention of what he called three "fictitious commodities": labor, land, and capital (Polanyi 1957, 68–76). Their commodification transformed them from living systems with intrinsic use-values in everyday life to goods that could be bought and sold, leading to the exploitation of labor, the despoliation of land and expropriation of farmers on it, and the misuse of money for credit and speculation—all three causing the dispossession of their original owners (Polanyi 1957, 68–110). Millions of traumatized people reacted to these "freedoms of the market" by trying to protect "society" by turning against the European and North American nation-states and questioned their legitimacy, leading to the rise of fascism and Stalinism (Polanyi 1957, 249–258) and, in the United States, to the socially progressive New Deal (Polanyi 1957, 223–248).

There is every reason to believe that the loss of legitimacy of neoliberalism and the global delegitimization of nation-states today, since the global financial recession of 2007–2008, are signs of the "terminal crisis" in which the US regime of accumulation now finds itself (Arrighi 1994/2010, 381–384) and which most recently has led to the formation of a new state form: the corporate state. This twilight of neoliberalism and the recurring deep crisis of capitalism have also set off similar counter-movements to the crises of the nineteenth and early twentieth centuries (Polanyi 1957), but this time around, they are regional in scale and global in extent, as is the neoliberal globalization that is bringing them about, one generating massive reactions by large numbers of alienated, anxious, and angry people, with many casting about for scapegoats.

Thus, large numbers of people in the US, provoked by authoritarian populist leaders, social media rumors, and media counterrealities (e.g., from Fox News), have increasingly turned against what they see as liberalism and the liberal state, given its failure to align with their moral values around family and patriarchy, markets, God, and patriotism, and have exhibited racial, antisemitic, and xenophobic antagonisms, while some, still a relative few, have gone so far as to manifest their hatred in violent acts against vulnerable people and the neoliberal US state. They have been attracted to the authoritarian populism of the US Right. Moreover, as predicted, right-wing authoritarian populisms are emerging in many other states around the world as regressive countermovements.

The past two decades, however, have also witnessed the emergence of a global countermovement from leftist organizations and populations, although it has not (yet) consolidated—and may not ever. Outside of the US, large numbers of poor rural and Indigenous peoples have come together collectively at an international scale to protect "society" by resisting the most damaging kinds of dispossession that they have experienced at the hands of the corporate food regime, resisting it in the names of the "peasantry" and "Indigenous people." This is La Via Campesina (LVC), mentioned earlier, a federation of peasant and Indigenous peoples associations that number 149 organizations in fifty-six countries, to which hundreds of millions of small farmers belong in Asia, Latin America, Europe, and Africa (Desmarais 2007).

In 1996, at the World Food Summit in Rome, LVC challenged the globalization of food by the corporate food alliance by announcing its principle of food sovereignty: "the right of each nation to maintain and develop its own capacity to produce its basic foods, respecting cultural and productive, diversity," and "the right to produce our own food in our own territory," adding later the "right of peoples to define their agricultural and food policy" (quoted in Desmarais 2007, 34). This provoked an extended conflict between LVC and the corporate alliance and its state and plutocratic supporters in the World Economic Forum (WEF) and occurred just as the public was becoming aware of the threat of climate change. As of this writing, the conflict between LVC and WEF—a transnational class conflict—is ongoing. According to Philip McMichael (2023, 216–217), agroecology is "gaining ground and/or legitimacy in international organizations . . . and among producers—either by political-ecological motivation or necessity, where farmers of varying scales are unable to afford the inflated costs of monopolized agro-inputs . . . [and] counter with seed sharing and rural/urban solidarity alliances."

In the US, many people, especially young adults of color, are consciously aware that they have suffered at the hands of neoliberal capitalism and the corporate state. Some of the disaffected belong to the working class, while others are downwardly mobile, indeed proletarianized, children of professionals and petty property owners; a few are children of elites. They have come together to participate in the labor movement, the Black Lives Matter movement, the women's movement, the immigrants' rights movement, the movement against the Israeli occupation of Palestine, and increasingly the climate justice movement. At present, people from these various movements are finding common cause in coalitions within the climate justice movement, based on ecosocialist principles (Baer 2018). The question is whether the climate justice movement will become the "movement of movements" that it purports to be, whether local food activists will come together within it, and whether it will join in international solidarity with La Via Campesina.

As in the 1930s, these political mobilizations are taking the form of countermovements against the very social crises that market fundamentalism in its current neoliberal manifestation has brought on by disembedding the market from society—but this time through the global commodification of new kinds of fictitious commodities around the natural world and human bodies in it, especially by the fossil-fuel industries, provoking profound revulsion among millions of people.[11]

The combination of climate change, socially unjust suffering, and political instability coming out of the human and natural disasters wrought by neoliberal capitalism has a highly destabilizing dynamic that is already under way. It remains to be seen which if either of these countermovements will eventually triumph as a new post-fossil-fuel economic era begins or whether they will take the regressive forms of capitalist industrial plantation systems spiffed up with "green" imagery and "innovative" "smart" digital surveillance, or of barbarous hybrids such as plantation mass production of food crops by convicts or slaves to provision fossil-fuel-powered enclaves for the privileged wealthy few, or of an ecosocialism grounded in agroecology and social justice, and supported globally by subaltern classes.

Although these processes are becoming increasingly clear to many observers, no one knows how they will play out. Why are they relevant to local food activists as creators of real utopias? I return to the two opposed kinds of countermovements. If the population of the United States were to shift its support decisively toward the right-wing countermovements, there would be little to debate about whether its food activist enterprises could provide real utopias, because their defining moral logics of social justice and sustainability would disappear under the onslaught of allied racist and plutocratic right-wing forces. Even worse, the embrace by right-wing forces of the US fossil-fuel economy, and the fossil-fuel industry as central to "the American way of life," could well lead the US political and economic elites to refuse to slow, much less wind down, fossil-fuel consumption and thus commit the rest of the population living outside the elites' privileged enclaves to large-scale misery and premature death. This would entail an abject surrender to the forces that perpetuate the use of fossil fuels, leading to even worse cataclysms of climate change, driving the Earth system beyond its resource and waste sink limits (Rockström et al. 2009).

For any opposed countermovement grounded in a global coalition comprising the world's nonindustrial farmers and the climate justice movement, in what sense could local food activism serve as a real utopia and play a pivotal role in a countermovement from the left as it challenges neoliberal capitalism? Here the keen observations by the eminent sociologist Michael Burawoy (2020) about real utopias in the context of the new countermovements are relevant. He writes about the new wave of commodification of elements of nature (and the human body, for example, gene sequencing as intellectual property) and why its treatment of them as fictitious commodities offends so many people. He argues that there are three reasons for this:

> The commodification of these entities disturbs our moral compass, for it violates the essence of their existence. . . . Second, commodification is economically dysfunctional: when fictitious commodities are subjected to unregulated market exchange, they lose their use value, even to the point of being unusable, becoming waste. . . . This, too, leads to collective protest. But "fictitiousness" has a third significance. . . . [It is] the very production of a fictitious commodity, that is, the process of disembedding nature, labor, money and knowledge from their social integument—a process that others have called dispossession. (Burawoy 2020, 92–93)

Thus, the left countermovement could be a collective response "inspired by some combination of moral opprobrium, the production of waste and dispossession" (Burawoy 2020, 93).

Then I would ask, Are not such values precisely evident in the moral logics of sustainable farming and community food security activists? Aren't the moral sensibilities of sustainable farming activists offended by the commodification of forests as timber to be clear-cut, farmland sold off for real estate, or the cruelty and wastefulness of industrial animal production? Aren't community food activists deeply offended by the enormous food waste of the US food economy when it allows people to go hungry or by the indignities inflicted by authorities that depersonalize food-insecure poor people?

Burawoy proposes that the significance of real utopias is that they are an index to the decommodified realities that ultimately make life meaningful and with which the members of a left countermovement would identify. They serve as material metonyms for a reality that stands against the commodification of things that by their intrinsic nature should not be commodified—care, money, land, housing, knowledge, and so on (Burawoy 2020, 93). Burawoy goes on to summarize an important claim that is deeply insightful: "Real utopias cannot stop the expansion and deepening of the market, but they can provide the basis of a counter-movement to the commodification of everything—a commodification that is neither conjunctural nor contingent but systematically generated by capitalism" (Burawoy 2020, 93).

Burawoy's insight is that real utopias, like the cross-over activist practices and enterprises featured in this book, are needed by any future left countermovement because real utopias incorporate processes of ethical and democratic decision-making that have been demonstrated to create and sustain practices that are emancipatory, given the oppressive conditions of neoliberal capitalism currently in effect. Under such conditions, local food cross-over activist groups could play important roles by manifesting real utopias that offer experiential prototypes for alternative local food futures, coming out of a progressive countermovement against agrocapitalism and neoliberalism within what may become the future *convergence of movements* in response to the ongoing *converging crises* of neoliberal globalization and climate change—one that dominant capitalist institutions can neither ignore nor overcome.

The outcome of which global countermovement triumphs over the other matters in this present and coming period of unprecedented convergent crises. Agroecological and non-fossil-fuel-based methods of agriculture are the only ones that will allow human life on Earth to be sustained. Like previous struggles over Polanyian countermovements, the outcome will depend on the relative power of one or the other to prevail, which may in turn determine the collective human response to climate change.

It is premature to write about the death of capitalism. However, the delinkage of fossil fuels from the internal logic of capitalist accumulation in the case of the corporate industrial food economy or any other capitalist sector is highly unlikely because the intense concentration of fossil energy combined with human labor in capitalist material commodity production represents a singular boost in "productivity," "growth," and above all "profits" compared to the alternative on offer: renewable energy. Fossil fuels constitutes capitalism's "fix" over the past two centuries to satisfy its addiction to continued growth even as it drives its wage laborers to the point of exhaustion (Malm 2016), and this is a hard addiction to "kick."

While the presence of renewable energy might make possible a long-term graduated process of strategic substitution of fossil fuels by renewables in some sectors of the economy, thus allowing capitalism to stay linked to industrial technologies, the industrial food sector is not one of them. The current industrial food system is precisely that: a system consisting of components that are complexly interlinked, with consequences for climate change irrespective of renewable energy. The industrial production of beef cattle emits massive amounts of methane into the air; deforestation for animal pasturage or to raise feedstocks generates huge volumes of CO_2 on its own; monocropped large fields require huge volumes of synthetic fertilizer that breaks down in the soil into nitrous oxide, a potent greenhouse gas; phosphates and natural nitrates are finite and, as noted in the appendix, already being maximally applied; and one could go on.

It therefore is worth predicting that the politics of the world economy will have to choose one of two paths to follow soon, probably at a fuzzy fork but not much further into this century. If one path is taken, irrespective of renewable energy use, the capitalist industrial food system's voracious and massive resource extraction (e.g., of topsoils and freshwater), overuse, toxification, and wastage will continue, in which case the five existential risks to humans from the global industrial food system mentioned in this conclusion will remain, with the probability that one or more will increase to the point that one or more will take catastrophic forms—not good for humans or other multicellular life forms.

Or, if a global left countermovement prevails, then the agroecological options associated with local and regional food systems could stabilize and then expand to reclaim the ruined territories of the old and by then-defunct plantation systems, while a new semi-industrialized socialist food economy based on local food systems organized along agroecological lines (e.g., closed systems of local manure use applied to local topsoils), with increasing numbers of small- and medium-sized farmers cooperatively organized and receiving restitutive state income support while selling in democratically regulated local and regional markets, could spread. Once under way, this would allow an exhausted Earth's soils, water, air, and nonhuman ecologies to begin to regenerate, even as climate change poses new dangers. In this outcome, the real utopias of cross-over activists and their enterprises could support sustainable farming and food security for all, as activists confront new problems under vastly changed conditions.

It is unclear whether or how soon these countermovements will consolidate or what the social forces and classes composing each would be. Such reactions against disembedded markets have occurred at least twice previously in modernity (Polanyi 1957) but never before at the global scale. US society and its food system have, with the rest of the world, now entered an era of profound uncertainty.

Under these conditions, the enormous gift of local food systems and their activists to human flourishing deserve wider recognition, understanding, and celebration. While such awareness is clearly necessary, it is not sufficient in order for more sustainable and just food systems to emerge, materially expand, and scale up to meet the needs of the future. For that to happen, people must act collectively on their awareness by showing that the politics of the food they eat matters to them—by moving into the fields and gardens and into the streets.

ACKNOWLEDGMENTS

We would first like to acknowledge the hundreds of food activists and other consultants in each of our four sites in North Carolina with whom our research team did interviews and from whose insights and perspectives we learned so much that is to be found in this book. To them all, we are profoundly grateful for the time they provided us but, beyond that, for the gifts they have given the four local food systems we studied.

We would like to thank our research associates Sarah Johnson, Patrick Linder, Kevin McDonough, Jen Walker, and Willie Jamaal Wright; our three community resident researchers, Marilyn Marks, Jasper Lynch, and Hollis Wild; and our research assistant, Carol Lewald. All members of this group, especially the research associates who served as site ethnographers, have contributed to the collection and analysis of the data discussed in this book, and we are enormously thankful to them for their efforts on behalf of the research project.

Don Nonini thanks Malena Rousseau, Molly Greene, and Julio Gutierrez for their help as research assistants. Don is also deeply appreciative of the thoughtful readings and comments of Charles Odom and Harry Phillips at key points in the writing. Finally, Don is grateful beyond words to his partner, Sandy Smith-Nonini, who has put up with so much from him in the course of writing this book.

Jennifer Hammer, Alexia Traganas, and Andrew Katz at New York University Press were instrumental in helping bring this book into existence, and we are deeply grateful to them.

The research project Research on Food and Farming for All (ROFFA) was funded by the Cultural Anthropology Program of the National Science Foundation, Grant #0922229, with Don Nonini as principal investigator and Dorothy Holland as co-principal investigator.

The authors gratefully acknowledge the permission of the publisher for using materials from the following previously published articles:

Dorothy Holland and Diana Gómez Correal, "Assessing the Transformative Significance of Movements & Activism: Lessons from *A Postcapitalist Politics*," *Outlines—Critical Practice Studies* 14, no. 2 (2013): 130–159.

Donald Nonini, "The 'Local Food Movement' and the Anthropology of Global Systems." *American Ethnologist* 40, no. 2 (2013): 267–275.

APPENDIX

"Progress in the Art, Not Only of Robbing the Worker, but of Robbing the Soil"

In this appendix, we review two sets of effects of the operation of the global corporate industrial food economy that contrast with the activities of the small- and medium-scale farming sector.[1] These are the effects of industrial agriculture on Earth ecosystems and its effects on climate change.

HOW IS ECOLOGICAL DEVASTATION CAUSED BY INDUSTRIAL
AGRICULTURE AND LIVESTOCK?

The global food industry and compliant governments work actively to promote large-scale industrial agriculture but, in so doing, ignore or play down its grave negative effects on the environment. Corporate industrial agriculture erodes the Earth's precious topsoils through monocropping cultivation combined with deforestation (Matson et al. 1997, 506, 508). It depletes groundwater reservoirs through wasteful irrigation methods and overpumping (Foley et al. 2005, 571; Matson et al. 1997, 506).[2] It reduces species biodiversity, threatens endangered species, and disrupts pollinators' ecologies through deforestation, monocropped farming's land-use simplifications, and application of synthetic pesticides that destroy species habitats, including that of soil invertebrates (Rockström et al. 2009, 472–474; Center for Food Safety 2022; Gunstone et al. 2021). It pollutes water and air with fertilizers, pesticides, and herbicides to the point that they endanger human health, as in the cases of widely applied glyphosate, the active component of Monsanto/Bayer Crop Science's Roundup pesticide, associated with its GM crops but now demonstrated to be a human carcinogen (IARC 2016), and of synthetic pesticides called "endocrine disruptors" that alter fundamental animal reproductive systems, possibly including human ones (Matson et al. 1997, 508; Cone 2011).

Among these effects, what may be of greatest importance to the themes of this book is that industrial agriculture has over the past century and a half (a geological instant) had the aggregate effect of destroying topsoils in one of the most fertile agricultural regions of the United States: the Midwest Corn Belt. Over the past 150 years, US industrial agriculture has depleted what was originally sixteen inches of loam topsoil by at least one half (Philpott 2020, 6), and it continues to be depleted at a rate far in excess of its natural replenishment (Philpott 2020, 130).[3] The application of industrial methods of monocropping of corn and soybeans; intensive tillage, leaving the farmland without cover crops during the midwestern winters (Philpott 2020, 124–147); and intensive use of chemical pesticides that destroy soil invertebrates, thus compacting soil and reducing its fertility (Gunstone et al. 2021), have led to widescale erosion as heavy rains fall on decomposed topsoils, which are unable to retain water and are washed away, with synthetic fertilizers carried in solution.

One of the most serious by-products of industrial agriculture for the environment and for the survival of species is the use of the Earth's ecosystems as waste sinkholes that take in enormous volumes of industrial synthetic fertilizers, pharmaceuticals, and animal waste (e.g., hog waste) that are toxic to humans and to other organisms on which human beings rely for their existence. Perhaps the most deleterious such waste process is the runoff of nitrogen- and phosphorus-based fertilizers from the huge factory fields of large-scale agriculture, resulting in eutrophication or anoxia of ocean and freshwaters that has deprived, for example, vast areas of the Gulf of Mexico (six thousand to seven thousand square miles) of all oxygen, thus killing marine life in these "dead zones," but also extending to major lake, riverine, and other coastal aquatic systems experiencing episodic eutrophication (Matson et al. 1997, 507; Bennett et al. 2001, 227, 232; Rockström et al. 2009, 474). Eutrophication of coastal waters near the Great Barrier Reef off Queensland from its plantation runoff contributed to the 2016 mass die-off of coral in the reef, which was already under stress from warming ocean waters caused by climate change; some 20 percent of all coral on the reef had died as of mid-2016 (Slezak 2016).

While the practices of small-scale and medium-scale farmers in the US, elsewhere in the global North, and in the global South have material effects on their ecologies, as long as they employ nonindustrial methods of cultivation by multicropping, use locally produced fertilizers and manures, limit the raising of meat animals, and minimize use of pesticides by other methods, these effects do not remotely scale up to the massive ecological degradations associated with industrial agriculture. The methods of agroecology, rehabilitative agriculture, and permaculture are, moreover, designed to avoid these effects (Ferguson and Lovell 2014; Rhodes 2017). No one can say that these methods have not been tried or have not shown significant success (Aznar-Sanchez et al. 2019; Foley et al. 2011). Perhaps it is time to set aside ethnocentric and modernist biases to examine them more carefully, given what is at stake.

WHAT IS THE RELATIONSHIP BETWEEN INDUSTRIAL AGRICULTURE AND CLIMATE CHANGE?

For increasing numbers of people, the most damning indictment of the alliance of the corporate industrial food sector and its state and interstate allies, and how it uses its power, is how all phases of its operations contribute to global climate change. One study puts it bluntly: "Climate change, food and farming are closely interlinked. Food production is a key driver of global warming, accounting for around a quarter of total greenhouse gas emissions—and the majority of the emissions that aren't caused by fossil fuels" (Berners-Lee and Clark 2013, 153). Other studies that make different assumptions about time frame than the one just cited estimate that agriculture makes an even higher contribution, about one-third of all greenhouse-gas emissions (Poore and Nemecek 2018; Oosterveer and Sonnefeld 2012, 89) that have in the past century inexorably led to global warming and to the violent and increasingly unstable changes in climate that it directly causes. In what follows, it is necessary to distinguish between the greenhouse gases that industrial agriculture emits during its various processes (carbon dioxide, methane, and nitrous oxide) from carbon dioxide emissions alone—the largest component by far of all greenhouse-gas emissions due to fossil-fuel burning.

Agriculture's single largest contribution to carbon dioxide emissions is deforestation of tropical forests, including tropical peatlands; as of 2008, it contributed between 4 and 9 percent of all carbon dioxide emissions, placing it second after fossil-fuel use as responsible for CO_2 emissions (although as a percentage of all CO_2 emissions, since 2008, deforestation may have declined in proportion due to the rise in fossil-fuel use) (calculated from Berners-Lee and Clark 2013, 154–155, 236–237n1; van der Werf et al. 2009, 738).[4] It is, however, inaccurate to ascribe deforestation primarily to "agriculture" as such, because specifically corporate agroindustrial agriculture drives deforestation because of business imperatives to continuously expand production and thus expand the amount of land converted from forests to such production.[5] Deforestation of large areas of tropical forests in Amazonia and elsewhere in order occurs to use the land cleared for pasture for farm animals, especially beef cattle, as well as for monocropped production of soybeans, corn, and palm oil used for livestock feed and for oils in processed food manufacture and more recently for biofuels production. Such deforestation connects the agroindustrial corporate logic of continuous increase in resources captured and profits accumulated directly to global climate change (Gerber et al. 2013). According to the Food and Agriculture Organization, "Between 2015 and 2020, the rate of deforestation was estimated at 10 million hectares per year. . . . The area of primary forest worldwide has decreased by over 80 million hectares since 1990. . . . Agricultural expansion continues to be the main driver of deforestation and forest degradation and the associated loss of forest biodiversity" (Food and Agricultural Organization 2020, xvi).

It is tragic and again far more than an issue of mere individual consumer choice that the raising of beef cattle and other meat animals—so beloved in the American diet—is playing a major destructive role in contributing to climate change. Although people appear to make individual consumption choices, these choices "scale up" with effects that can be devastating to the material atmospheric and ecological systems that all humans all depend on, not only for food production but for the other necessities of life. In other words, the perhaps-distinctive American penchant for beef consumption needs to be seriously rethought, irrespective of its implications for human health, because it brings about massive greenhouse-gas emissions, via both deforestation and the methane-emission processes described later in this appendix.

Industrial agriculture also contributes substantially to greenhouse-gas emissions through the production, consumption, and transport of fossil-fuel-based synthetic fertilizers (the "N fertilizer supply chain"), which accounts for 10.6 percent of agricultural greenhouse-gas emissions and

2.1 percent of global greenhouse-gas emissions (Menegat et al. 2022), with the production of fuels for farm machinery adding a much-smaller relative emissions amount to the total.

Although industrial agriculture does directly factor as the second-largest producer of carbon dioxide emissions, it is also the second-largest source of emissions of methane (after leaks from fracking) and the largest source of nitrous oxide—which are greenhouse gases far more potent than carbon dioxide in bringing about atmospheric warming, although their dissipation rate is much greater than CO_2's. It is estimated that over the time frame of a century, agriculturally related methane (primarily emitted from the digestive systems of beef cattle and other ruminants) will contribute 5.8 percent of all greenhouse-gas emissions, while nitrous oxide, which is generated from fossil-fuel-based fertilizers through their chemical decomposition, will make up 6.5 percent of all greenhouse-gas emissions over the same period. Thus, emissions of these two gases due to agriculture make up about 13 percent of all greenhouse-gas emissions over the period of a century (calculated from Berners-Lee 2013, 236n1).

These estimates have important implications for the human population's time frame for reducing greenhouse-gas emissions and seeking to forestall the worst effects of climate change. Over the short to medium term of ten to twenty years—which is the time frame that humanity now needs to work within, given the dire evidence of rapid climate change already occurring and projected in the near future by the Intergovernmental Panel on Climate Change (IPCC 2014, 2018)—methane and nitrous oxide emissions are contributing far more rapidly to planetary warming than emission of slower-acting and less potent carbon dioxide created by fossil-fuel consumption, although the long-term buildup of carbon dioxide in the atmosphere by the global industrial food economy cannot be ignored either.

What is crucial here is that the relatively faster decay rates of methane and nitrous oxide means that rapid reduction in their use and extraction in the food system and the fossil-fuel industry could provide the chance needed to effectively reduce short- to medium-term atmospheric warming by rapidly reducing emissions of these two gases: "According to one study, efforts to cut soot and methane alone could hold down the temperature in 2050 by almost half a degree—far more than would be possible in that timeframe by cutting slow-acting carbon dioxide. Given that this gain could be achieved at relatively little cost and with signifi-cant positive side effects, this is an opportunity that the world can't af-ford not to take" (Berners-Lee and Clark 2013, 149).[6]

Unfortunately, neither the global food industry nor the fracking industry have gotten the news, or rather, they have chosen to ignore and deny it. Increasingly, the survival of much of the planet's human population, including US residents, as well as that of innumerable other species on which humans rely, are being threatened by increased atmo-spheric temperatures and the turbulent weather conditions (hurricanes, droughts, floods) brought on by them. Ironically, with respect to food, as these climatic conditions grow more severe, they will increasingly threaten agricultural production of staple food crops, with one study predicting that by 2050, there will be declines of 13 percent in the yields for irrigated wheat and 15 percent for irrigated rice and a reduction in corn yields in Africa ranging from 10 to 20 percent (Gilbert 2012, 2).

The large-scale plantation-based industrial cultivation methods that constitute the modus operandi of large-scale industrial agriculture have yet other serious consequences in furthering global climate change. First, the combination of overcultivation of topsoils with monocrops, leaving land without cover crops in winter, tilling soils annually, and the intensive use of synthetic pesticides has led to the serious loss of soils' capacity to sequester carbon dioxide. The only reason that so little can be said about this is that the sheer effects of such soil destruction cannot currently be estimated, but they are of a large order of magnitude: "The global food system currently puts a lot of pressure on the world's soils. Intensive farming and overgrazing can both lead to reductions in soil's organic content with the result that more carbon dioxide ends up in the air. The climate impact of this is hard to measure precisely but is poten-tially hugely significant" (Berners-Lee and Clark 2013, 154).

Second, food production is now tied to global climate change through the plantation-based production of biofuels. Since 2007, when the George W. Bush administration pushed through Congress the Renew-able Fuel Standard law, which required biofuels as a fuel additive in the US fuel supply through refining corn, palm oil, and sugar cane, the fate of the world's food and fuel systems have very rapidly become inextri-cably interconnected; the major grain brokers ADM, Bunge, and Cargill were the first to benefit from treating corn as a fuel stock (McMichael

2009, 290). Huge areas of Indonesia, Brazil, and elsewhere in the global South are now being deforested in order to cultivate palm oil and sugar cane in order to harvest their crops to produce biofuels. The energetic costs of producing the biofuels under these industrial conditions (e.g., deforestation and the huge input of fertilizers required) far exceed the energetic benefits received by adding biofuels to the conventional fossil-fuel supply pipeline (Berners-Lee and Clark 2013, 156).

NOTES

INTRODUCTION

1 This new field has its own journal, *Agriculture and Human Values*, and has specific sites of theoretical specialization, such as the University of California Santa Cruz Agro-Food Studies Research Group.

2 Definitions of "sustainable agriculture" are overlapping and revised over time. One that is widely accepted comes from the California-based Funders Agricultural Working Group (2001, quoted in Allen 2004, 82):

   A sustainable food and agriculture system . . . [is] one that:
   - Protects the environment, human health, and welfare of farm animals
   - Supports all parts of an economically viable agricultural sector, and provides just conditions and fair compensations for farmers and workers
   - Provides all people with locally produced, affordable and healthy food
   - Contributes to the vitality of rural and urban communities and the links between them.

3 "Food deserts" are neighborhoods where supermarket chains previously located their stores but, due to the impoverishment of most residents and reduction of their purchasing power, had closed down operations, leaving residents without nearby or convenient food outlets.

4 For instance, "Emissions within farm gate and from agricultural land expansion contributing to the global food system represent 16–27% of total anthropogenic emissions (*medium confidence*). Emissions outside the farm gate represent 5–10% of total anthropogenic emissions (*medium confidence*)" (IPCC 2019, 13).

5 By "surplus" within the food system, we are referring to Gibson-Graham's (2006) and Marx's (1867/1976) theory in which enterprises, in this case food-related enterprises, produce not only commodities (food) but also surplus labor (i.e., not needed for the workers' social reproduction), which when realized as money becomes social wealth to be disposed of, and this is where the question of ethical decision-making about how surplus is to be used comes in.

6 A word of self-criticism, which may more widely reflect on most local food activism but for which we take responsibility: We found no mention among food activists of the need to ensure more just wages and working conditions for farmworkers employed by local (direct-sales) farmers, however they might otherwise be committed to the goals of sustainability, e.g., through organic farming methods. But we (the authors) should have worked with our research associates to ask questions about the working conditions of farmworkers employed by local

farmers. There is no excuse. As Margaret Gray, who carried out a long-term ethnographic study of labor relations between farmworkers and direct-sales farmers in the Hudson Valley of New York, writes, "at least so far, the locavore version of the commons does not include those who raise, tend, and harvest what we put in our mouths" (2014, 131), and yet the food movement "is an influential group with considerable sway over public opinion, and so any amount of scrutiny and awareness would go a long way toward promoting farmworker interests" (145–146).

CHAPTER 1. OUR FOOD AND THE ALLIANCE

1 This circulation of elites between high corporate positions in the food sector and upper management positions in the USDA is what the anthropologist Janine Wedel (2009, 17–18, 105–106, 132–133) has reported elsewhere in the US federal government.

2 One of them, Daniel Amstutz, previously a vice president of Cargill in the 1960s–1970s, later served as US chief agriculture negotiator for the Uruguay Agreement on Agriculture. After his government service, he returned to work in the industry as a lobbyist and as a consultant for Cargill (Clapp 2016, 128). There are many other examples.

3 WTO agreements including the Agreement on Agriculture, also effectively allowed state subsidies by the US and EU to be delinked from the tariff systems that heretofore protected the minimal prices offered to global Southern farmers by allotting these subsidies (which include the massive ones provided US corporate farmers) to a "non-trade-distorting" "Green Box" allowable to the US and EU, unlike the "trade-distorting" "Red Box" tariffs for global Southern countries, which were prohibited (McMichael 2013, 52–53). Casuistry and sophism armored by power won the day for global Northern farm interests. Philip McMichael (2013, 54) cites Food and Agriculture Organization estimates that between twenty and thirty million farmers lost their land from agricultural trade "liberalization."

4 The quotation marks around "emergency food" emphasize its peculiar nature: this food has to be provided *routinely*—not as a sudden state emergency—for large numbers of people who have to experience a more or less continuous emergency of being food insecure.

5 This inference refutes Dickenson's (2020, 137) argument that there is a "pervasive but faulty notion" that "food from pantries comes from private donations," one that "scorns federal entitlements like SNAP that ensure Americans have access to food." While Dickenson's attack on neoliberalism's belittlement of "federal entitlements like SNAP" is justified, her argument that corporate donations were not crucial for the quantities of food provided by the private charitable sector to recipients is not valid. Instead, corporate and business donations have a different logic of connection to the corporate state and neoliberalism, which we trace out here.

6 When farm commodity prices are high, TEFAP aid to food banks falls. So, too, do corporate donations to food banks, because when farm commodities are scarce,

consumers are inclined to buy more food, not less, from retail outlets, and thus less remains unsold and available to donate. This appears to be what happened during Dickenson's (2020) fieldwork and may have led to her incorrect claim (2020, 137).

7 Workers in the food industry—especially in fast foods—are paid so poorly that many qualify for food aid and other government benefits (Allegretto et al. 2013, tables 3, 7).

8 Unfortunately for everyone in the world, the massive amounts of food purchased by households in the US are only partially consumed. As much as 40 percent of food is thrown away by consumers in the US (Berners-Lee 2019, 37, estimated from fig. 1.8) and ends up in landfills, where it generates enormous amounts of methane, most of which is uncaptured and goes into the atmosphere.

CHAPTER 2. THREE MYTHS THAT PROP UP INDUSTRIAL FOODS

1 This is one of the most widely cited studies in nutrition, with 778 citations since its publication in 2004.

2 In 2003, annual medical expenditures attributed to obesity treatments amounted to $75 billion, and health expenses related to obesity cost every US citizen on average $150 annually (Carolan 2011, 75; see the sources he cites for specific studies).

CHAPTER 3. ACTIVISM FOR A MORE SUSTAINABLE AND JUST FOOD SYSTEM

1 We exclude the Carolina Farm Stewardship Association because it consisted primarily of farmers, whom we were unable to interview in this project, except for a very few who found time for local food activism. Its annual convention was a crucial opportunity for small-scale organic farmers throughout North and South Carolina to meet with food activists.

2 From a Marxist orientation, we could examine moral logics as elements within more comprehensive ideologies that organize classes and class coalitions. Thus, moral logics can be seen as elements of what Antonio Gramsci (1971, 376–377) called "organic ideologies," which bring together "historical blocs" of different classes in coalitions opposed to other coalitions of classes within a dialectics shaped by antagonistic or radically different material and social conditions. Gramsci (1971, 376–377) referred to "organic ideologies" that are "historically necessary," in that "they have a validity which is 'psychological,' . . . 'organize' human masses, and create the terrain on which men move, acquire consciousness of their position, struggle, etc." In Gramsci's sense, the moral logics of the food activists we studied operated to mobilize a historic bloc of classes that consists of artisanal small-scale property owners, workers (e.g., cooks, employees of small food enterprises), and professionals / "anti-system radicals" (Huber 2022, 123–124), in opposition to a class coalition around the corporate industrial food economy composed of corporate owners, investors, professionals (e.g., managers and technicians, academics, think-tank pundits), and parts of the petit bourgeoisie, a coalition committed to the accumulation of privately owned capital.

3 Unfortunately, neither moral logic has paid sufficient attention to social justice (in the form of a living wage, benefits, and safe working conditions) for farmworkers and other food workers employed by local farmers and food-processing enterprises (Gray 2014)—a matter we return to in the conclusion.

4 In the interest of full disclosure, Marilyn Marks served as our community resident researcher (CRR) for the Charlotte Mecklenburg site. Our project hired part-time community resident researchers in each site to provide our research associates with specific insider research insights into the local food situation, while also working to further the objectives of the project they were allied with or interested in, in Marilyn's case, the Charlotte Mecklenburg Food Policy Council. The contributions that each CRR made to our understanding of food activism in their locale were invaluable, and we were privileged to be able to work with them.

5 WWOOF stands for "Worldwide Organization of Organic Farmers, a global network of organic farmers who welcome interns to learn about organic farming through homestays and provide room and board but require physical labor from the interns as part of the learning experience. Most hosting organic farms are in Australia, New Zealand, the United States, and the United Kingdom. A "Woofer" is such an intern.

CHAPTER 4. SUSTAINABLE FOOD ACTIVISTS ENGAGE "THE ECONOMY"

1 By "local goods market managers," we mean managers of markets that include local produce and poultry but often also extensively market locally produced processed foods (e.g., baked goods), handicrafts, and the like. For most local food activists, such markets are not purely "farmers markets."

2 While Noah mentioned "efficiency" or "efficient" nine times in his interview, he mentioned "scale" or "scaling" forty-three times.

3 The certification of a farm as organic is overseen by the USDA and its contracting private certifiers. The certification process can take up to four years, and for even a small farmer, there will be an estimated outlay of $700–$1,400 for the certification fee, some of which can be recovered by a USDA cost-sharing refund after certification is completed (USDA 2022).

4 "Mouth feel" is "the powerful sensory force" that is "the way a product interacts with the mouth, as defined more specifically by a host of related sensations, from dryness to gumminess to moisture release. . . . The mouth feel of soda and many other food items, especially those high in fat, is second only to the bliss point in its ability to predict how much craving a product will induce" (Moss 2013, 34).

CHAPTER 5. FOOD SECURITY ACTIVISM IN LEAN AND MEAN TIMES

1 Another federal law provides the corporation indemnity from damages due to illness from consuming the food item as long as the donor acted unintentionally and without negligence (Feeding America 2013).

2 These were the Food Bank of Central and Eastern North Carolina, Second Harvest Food Bank of Metrolina, and the Second Harvest Food Bank of Northwest North Carolina.

CHAPTER 6. SLOW FOOD CHARLOTTE, FARMERS, AND FOOD INSECURITY

1 They had been members of Slow Food Charlotte from at least 2006, when their farm hosted the visit of Alice Waters, celebrity chef and proprietor of Chez Panisse, then a prominent leader of Slow Food USA, to Slow Food Charlotte for a farm-to-fork dinner.

2 One of the six people who mentioned the loss of farmland was a food activist who operated a CSA and repeatedly mentioned the loss of local and regional farmland (seven of the seventeen mentions); the two organizations whose literatures mentioned the loss of farmland were both located outside Charlotte—one was the food policy council of an outlying rural county; the other was a land conservation trust to preserve farmland located in another rural county (four of the seventeen mentions).

CHAPTER 7. F.A.R.M. CAFÉ AND CLASSED LIVES AT ONE'S TABLE

1 Riches (2018) observes that the charitable food system, "food bank nation," now extends throughout the OECD countries that have been incorporated into the US "new world order," which is largely associated with neoliberal political policies and visions.

2 According to the F.A.R.M. Café's mission statement, the food served was to consist of "high quality & delicious meals produced from local sources" (F.A.R.M. Café 2022d). Jen notes that "words like 'sustainably produced,' 'organic,' and 'hormone-free' were left out of the mission statement, but in talking about the type of food that fit the definition of 'healthy' and 'quality' during their [the F.A.R.M. Café founders'] meetings, these indicators were often used."

3 This is the problem about "efficiency" that we discussed in chapter 4, where we examined the challenges of Noah's new position with Appalachian Growers.

4 There is no question that these connections, which have proven essential to the successful evolution of the F.A.R.M. Café, were facilitated by the charismatic and highly sociable character of Renée. Jen writes, "Renée's inclusive style contributes to monthly meetings that were among the most productive I attended all year. There were typically six to ten people present from a diversity of places, . . . and everyone spoke up. People were enthusiastic, and because the majority of the focus was on building the organization in fundraising, every meeting included at least one brainstorming session about who in the room knew whom in the community."

5 Reflecting a long-standing regional preponderance of whites, the racial composition of Watauga County's population in 2022 was almost entirely (94.3 percent)
white, with 4.1 percent Latinx (Hispanic) and 2.1 percent African American (US
Census Bureau 2023).

6 Krais (1993, 172) defines "symbolic violence" as unconscious intimidation, as
when it is applied by men to women to maintain "a relation of domination.
Symbolic violence is a subtle, euphemized, invisible mode of domination that
prevents domination from being recognized as such and, therefore, as misrec-
ognized domination." For example, she refers to academia, where "women are
regularly overlooked when they wish to make a point; they are interrupted when
they speak; male speakers refer to contributions of other male speakers, but not to
those of women," and so on (Krais 1993, 173).

CHAPTER 8. CORNUCOPIA LANDSCAPING, ABUNDANCE, AND FOOD JUSTICE

1 As defined by the national Community Food Security Coalition (CFSC), community food security is "a condition in which all community residents obtain a safe,
culturally acceptable, nutritionally adequate diet through a sustainable food system
that maximizes community self-reliance and social justice" (Food Security 2022).

2 Both Massive Harvest Farm and Lincoln Street Neighborhood Farm—the two
urban farm/gardens in NECD discussed in this chapter—were located in Census
Tract 10.01.

3 Permaculture has been variously referred to as a method of cultivation, a philosophy, and the basis of a movement (Ferguson and Lovell 2014). As method, it has
been defined by its cofounder David Holmgren this way: "Consciously designed
landscapes which mimic the patterns and relationships found in nature, while
yielding an abundance of food, fibre and energy for provision of local needs"
(2004, xix).

4 On the "nonprofit industrial complex" and its relationship to neoliberalism, see
Munshi and Wills 2017, xiii–xxvii.

5 In the interest of protecting the identity of CL and its worker-owners, this website's URL is not provided here.

CHAPTER 9. CLARKETON COMMUNITY GARDEN AND RACIAL AND CLASS DIVISIONS

1 We write in 2023 as the conservative Republican majority of the North Carolina
General Assembly has continued its decade-long campaign to strip voting power
from African Americans through racially based gerrymandering and attempts to
suppress Black votes.

2 Poverty and racial terror during Jim Crow drove one Black mother and her three
young children to leave Rocky Mount for New York City in 1922, as four people
among millions who were part of the "Great Migrations" of African Americans
out of the South to the North and Midwest. Among the three children was five-

year old Thelonious Monk, whose brilliant jazz bebop piano compositions and performances from the 1940s to the 1970s arguably make him Rocky Mount's most famous native son (Kelley 2009, 14). Only belatedly in 2008 did Rocky Mount accord Monk this tribute, when he was posthumously inducted into the Twin County Museum & Hall of Fame (Twin County Museum & Hall of Fame 2022). His family members on both his mother's and father's sides had previously fled the semienslaved labor of sharecropping to seek out employment as laborers in the shops on the Atlantic Coast Line Rail Road and as domestic workers in the growing railway town of Rocky Mount (Kelley 2009, 11).

3 Unfortunately, choosing to affiliate with a gang as a form of livelihood was not a hypothetical risk for Clarketon teenagers and young adults. In Rocky Mount, our ethnographer Willie Jamaal Wright noted, "Gangs are [a] serious issue in certain sections of the city. Two weeks ago, a young Black male was murdered in broad daylight in a gang-related shooting near City Lake, which is a popular walking area. . . . [One community organizer] shared that the man [who] was killed had members of his gang at the site of his murder throwing up gang signs. One child that was at the scene declaring his allegiance to the gang was as young as eight years old." The Reverend David Hemmings also served as a consultant for the county's school district, where, he noted, "I work with the gang violence."

4 When Willie Jamaal started fieldwork in 2011, residues of the plantation economy and Jim Crow still defined the county's physical landscape. The average size of a farm in Edgecombe County was 466 acres, and of the 390 farmers enumerated in the county, 365 of them were whites and only sixteen were African Americans. After soybeans, the largest crop planted consisted of cotton, the traditional crop associated with plantations, and this placed the county's cotton production sixth in the state (USDA 2012a).

CHAPTER 10. ASSESSING THE TRANSFORMATIVE SIGNIFICANCE OF FOOD ACTIVISM

1 By "reluctant subject," Gibson-Graham refer to the retrenched coal miners, managers, and government officials residing in the closed-down mining towns of the Latrobe Valley in Victoria, Australia. Gibson-Graham describe them as bitter and depressed, showing "a patent lack of desire . . . for new economic identities that can energize different enactments of a regional economy" and yet capable of becoming affectively open to new alternatives (2006, 24–25).

2 We have highlighted processes as necessary for significant change. Another route would be to (also) infer a list of substantive achievements. See, for example, the Community Economies Collective website (www.communityeconomies.org), which moves toward inferring the key features that Gibson-Graham considered necessary for a postcapitalist economy.

3 While the local food activists were trying to bypass the global agroindustry, activists concerned about the food insecure, or at least those engaged in social justice efforts, sought to thwart structures of inequality. The efforts of some of those

concerned about the food-insecure segment of the population become important later in the chapter.

4 For many of the activists concerned with creating a "local" food system, the vision and discourses looked inward toward community. Formal ties linking local to regional or national organizations were few. We did not encounter much in-depth take-up of nationally circulating discourses on "civic agriculture" (Lyson 2005), for example, as a basis for action. There were regional conferences, workshops, and speakers, but many of them were addressed more to plans for a new farmers markets or to techniques for slaughtering chickens, for example, than to critical analysis of structural restraints on the growth of local community economies or to political organizing to change bureaucratic barriers to the use of SNAP and/or WIC vouchers at farmers markets.

5 There were a few exceptions. One group—one that was exceptional in several ways—in the western mountains talked at several points about the need to provide farmers with health insurance. Their suggestions were met by a silent reception from the larger community.

6 The moral dilemma mentioned throughout this book pits the small farmer against the low-income person. Those who want to support small farmers think of supporting the expansion of their markets by bringing in high-end shoppers. Market logic leads them to assume that access to good food by low-income people depends on lower price points that, in turn, depend on lower profits for farmers. Another means that we heard some espouse is the "ugly tomato" solution: sell misshapen, about-to-expire produce to low-end customers at discounted prices.

7 Political maneuvers exacerbate this connection. In politicized national debates over allocations for free and reduced school lunches, for example, conservatives have played on the "elitist" aspect of the foodie image. They have charged the celebrity chefs arguing for higher school-lunch budgets with trying to pass their rarefied tastes off on everyone else (Wilson and Holland 2009).

8 My (Holland's) fellow researchers and I have the impression that it was gendered and associated with women, though we have yet to make a systematic check of the data.

9 Unlike the Occupy movement, with its assemblies and passion for deliberation, local food organizations at the time of our study were not directing attention to critical analysis.

CONCLUSION

1 Nonetheless, we also noted the existence of capitalist wage-labor relations between some local farmers and their farmworkers, which belies the myth of the "self-reliant" small local farmer that is so widely held by some foodies and food activists (Gray 2014).

2 We point out that Rev. Hemmings made extensive attempts at interracial cooperation with white members of the local clergy in various food projects, with limited success, but we had no space earlier to recount them.

3 While one could (and should) quarrel with extending this practice as a general equivalence to fair or living wages for permanent employees, it could be justified for poor patrons working for an hour for a meal when they have been let down by the capitalist labor market and the state's social safety net.

4 For instance, in the Midwest Corn Belt, it has been estimated that 57.6 billion metric tons of topsoil have eroded over the past 160 years since settlers began farming (Thaler et al. 2022, 9), equal to a median historic soil loss of 1.8 mm. per year, or a cumulative depth of 270 mm. (11.3 inches) (calculated from Thaler et al. 2022, 9), or more than 50 percent of the sixteen inches of the maximum layer of topsoil deposited by the end of the precolonial era (Philpott 2020, 6).

5 As much as 35 percent of the southern High Plains aquifer may be depleted within the next thirty years (Scanlon et al. 2012). In the California Central Valley, the largest concentrated production region for industrial vegetables, fruits, and nuts in the US, there are already signs of depletion of its aquifers, with saline water being pumped up from drawn-down aquifers and land subsiding due to the exhaustion of subsurface water tables (Philpott 2020).

6 One fix under industrial conditions for soil depletion is no-till cultivation, which requires the application of glyphosate or another herbicide, and these are toxic to humans.

7 "Highly pathogenic and suddenly human-adapted avian and swine influenzas typically first emerge as newly identifiable infections in intensive operations located closer to major cities in both fully industrialized countries and those in the middle of undergoing economic transitions to more industrialized regimes" (Wallace 2020, 79). "Given the genetics of SARS-2—a recombinant of bat and pangolin strains—the increasingly formalized wild food trade in all likelihood played a foundational role in the emergence of the COVID-19 outbreak. . . . The trade, including now pangolin farming, shares with industrial agriculture sources of capital and economic geographies encroaching on Central China's hinterlands" (Wallace 2020, 80).

8 Inflows of phosphates into the oceans now come to eight and half to nine and a half million tons annually, while a maximum limit of eleven million tons is judged sustainable before a critical threshold is exceeded that could allow such an event to occur (Rockström et al. 2009, 474).

9 UBI is an arrangement whereby "every legal resident in a country receives a monthly living stipend sufficient to live above the 'poverty line.' Let's call this the 'no frills culturally respectable standard of living'" (Wright 2010, 5).

10 Wright (2010) presents three possible trajectories out of capitalism. The first, "ruptural transformation," is that of revolution, and he writes that "ruptural strategies for constructing a democratic egalitarian socialism . . . seem implausible in the world in which we currently live" (Wright 2010, 320). Another, "symbiotic transformation," "occurs when working-class empowerment helps solve real problems for capitalists, and occurs only under exceptional circumstances that do not generally apply" (Wright 2010, 337–365). The third is "interstitial transformation" (Wright 2010, 321–336).

11 As morally repulsive initiatives, we might consider, for example, the "drilling" rights held by oil companies for extreme oil extraction in waters of Alaskan native lands, destroying their fishing and hunting grounds, and the "rights" of fracking corporations for fracking under the soils of thousands of farmers and rural residents in the Midwest and South, while poisoning their water; the huge environmental despoliation of the Gulf Coast fisheries and wildlife reserves by the BP *Deepwater Horizon* oil spill; the violence against peaceful Indigenous protestors and their allies perpetrated by police and the private thugs hired by pipeline companies that have expropriated tribal lands (Klein 2014); the use of hundreds of thousands of incarcerated people of color to work for pennies an hour in industries owned by the private prison complex (Wang 2018); and the hundreds of young people of color killed and injured by police and harassed by judges and attorneys by being jailed when they cannot pay bails and court fees (Wang 2018). The list goes on and is mind-numbing.

APPENDIX

1 The quotation in the subtitle is from Marx (1867/1976, 638): "All progress in capitalist agriculture is a progress in the art, not only of robbing the worker, the robbing the soil. . . . The more a country proceeds from large-scale industry as the background of its development, as in the case of the United States, the more rapid is this process of destruction."

2 Michael Carolan (2011, 125) writes, "Cheap food is also premised upon cheap—and thus, inefficiently utilized—water. Irrigation efficiency refers to the ratio of water evaporated to what saturates the soil," and he notes that the efficiency of gravity irrigation is about 40 percent, that of sprinkler systems between 60 and 70 percent, and that of drip irrigation between 80 and 90 percent.

3 Tom Philpott concludes, after his interviews with Richard Cruse ("the foremost authority on soil erosion in the Corn capital") (Philpott 2020, 124), and other soil scientists that "Iowa—and much of the surrounding Corn Belt land—is losing soil at a rate nearly seventeen times the pace of natural replenishment" (Philpott 2020, 130). At this rate, Corn Belt soil fertility will not last more than a few decades, if that.

4 Mike Berners-Lee and Duncan Clark (2013) estimate deforestation from agriculture as contributing 4.0 percent of all CO_2 emissions, while Guido van der Werf et al. (2009, 738) calculate all deforestation and forest degradation, including tropical peatland, as providing about 15 percent. If we estimate that agriculture causes 60 percent of all deforestation (with logging and timbering the other 40 percent; Berners-Lee and Clark 2013, 237n1), then 9 percent of all CO_2 emissions are caused by agriculture.

5 A cause of deforestation in addition to land clearance for agricultural production is logging land for timber undertaken by large-scale profit-making corporations, but compared to deforestation to cultivate feed crops such as corn or soybeans, it is relatively minor.

6   The Berners-Lee and Clark study was published in 2013, before the rapid rise
    in the contribution of the hydraulic fracturing industry to the world's liquified
    natural gas (methane) markets, making it now the largest contributor of methane
    emissions. The rapid increase in liquified natural gas combustion makes the argu-
    ment about the urgency of curbing methane-gas emissions as soon as possible all
    the more compelling, including a decrease in the large percentage of its emissions
    due to industrial agriculture.

REFERENCES

ActionAid International. (2006). *Under the influence: Exposing undue corporate influence over policy-making at the World Trade Organization.* Johannesburg, South Africa: Action Aid International.

Allegretto, S., M. Broussard, D. Graham-Square, K. Jacobs, D. Thompson, and J. Thompson (2013). *The public cost of low-wage jobs in the past-food industry.* Urbana/Berkeley: University of Illinois at Urbana-Champaign/UC Berkeley Labor Center.

Allen, P. (2004). *Together at the table: Sustainability and sustenance in the American agrifood system.* University Park: Pennsylvania State University Press / Rural Sociological Society.

Allen, P. (2008). "Mining for justice in the food system: perceptions, practices, and possibilities." *Agriculture and Human Values* 25: 157–161.

Allen, P., M. FitzSimmons, M. Goodman, and K. Warner. (2003). "Shifting plates in the agrifood landscape: The tectonics of alternative agrifood initiatives in California." *Journal of Rural Studies* 19(1): 61–75.

Alperovitz, G. (2013). *What then must we do? Straight talk about the next American Revolution.* White River Junction, VT: Chelsea Green.

Alvarez, S. E., E. Dagnino, and A. Escobar. (1998). *Cultures of politics, politics of cultures: Re-visioning Latin American social movements.* Boulder, CO: Westview.

Amenta, E., and F. Polletta. (2019). "The cultural impacts of social movements." *Annual Review of Sociology* 45: 279–299.

Anderson, M. D. (2008). "Rights-based food systems and the goals of food systems reform." *Agriculture and Human Values* 25(4): 593–608.

Andrews, G. (2008). *The slow food story: Politics and pleasure.* Montreal: McGill-Queen's University Press.

Angus, I. (2016). "Book review: 'Anthropocene or Capitalocene?' misses the point." *Climate & Capitalism*, September 26, 2016. https://climateandcapitalism.com.

Arrighi, G. (1994/2010). *The long twentieth century: Money, power, and the origins of our times.* London: Verso.

Auborg, A. (2020) "Twenty food cooperatives building resilient communities." *Food Tank*, October 2020. https://foodtank.com.

Aznar-Sanchez, J. A., M. Piquer-Rodriguez, J. F. Valasco-Munoz, and F. Manzano-Agugliaro. (2019). "Worldwide research trends and sustainable land use in agriculture." *Land Use Policy* 87: 104069.

Badgeley, C., J. Moghtader, E. Quintero, E. Zakem, M. J. Chappell, K. Aviles-Vazquez, A. Samulton, and I. Perfecto. (2007). "Organic agriculture and global food supply." *Renewable Agriculture and Food Systems* 22(2): 86–108.

Baer, H. A. (2018). *Democratic eco-socialism as a real utopia: Transitioning to an alternative world system.* New York: Berghahn Books.

Bakhtin, M. M. (1981). *The dialogic imagination: Four essays by M. M. Bakhtin.* Austin: University of Texas Press.

Barber, S., A. George, D. Petersen, J. Thompson, and J. Tzeng. (2008). "African American community of Southeast Rocky Mount, North Carolina: An action-oriented community diagnosis—final report." Paper prepared for graduate course HBHE 741, School of Public Health, University of North Carolina, Chapel Hill.

Barclay, E. (2015). "Your grandparents spend more of their money on food than you do." *The Salt* (blog), National Public Radio, March 2, 2015. www.npr.org.

Barry, T. (2010). "Synergy in security: The rise of the national security complex." *Dollars & Sense: Real World Economics,* 11–15.

Bennett, E. M., S. R. Carpenter, and N. F. Caroco. (2001). "Human impact on erodible phosphorus and eutrophication: a global perspective." *Bioscience* 51(3): 227–234.

Ben-Shalom, Y., R. A. Moffitt, and J. K. Scholz. (2011). *An assessment of the effectiveness of anti-poverty programs in the United States.* Washington, DC: National Bureau of Economic Research.

Berner, M., and K. O'Brien. (2004). "The shifting pattern of food security support: Food stamp and food bank usage in North Carolina." *Nonprofit and Voluntary Sector Quarterly* 33(4): 655–672.

Berner, M., A. Vasquez, and M. McDougall. (2016). "Documenting poverty in North Carolina." School of Government, University of North Carolina at Chapel Hill.

Berners-Lee, M. (2019). *There is no plan(et) B: A handbook for the make or break years.* Cambridge: Cambridge University Press.

Berners-Lee, M., and D. Clark. (2013). *The burning question: We can't burn half the world's oil, coal, and gas. So how do we quit?* Vancouver: Greystone Books.

Berry, W. (1977/1996). *The unsettling of America: Culture and agriculture.* San Francisco: Sierra Club Books.

Beus, C. E., and R. E. Dunlap (1990). "Conventional Versus Alternative Agriculture—the Paradigmatic Roots of the Debate." *Rural Sociology* 55(4): 590–616.

Biersack, A. (2006). "Reimagining political ecology: Culture/power/history/nature." In *Reimagining political ecology,* edited by A. Biersack and J. B. Greenberg, 3–40. Durham, NC: Duke University Press.

Bossy, S. (2014). "The utopias of political consumerism: The search of alternatives to mass consumption." *Journal of Consumer Culture* 14(2): 179–198.

Bray, G. A., S. J. Nielsen, and B. M. Popkin. (2004). "Consumption of high-fructose corn syrup in beverages may play a role in the epidemic of obesity." *American Journal of Clinical Nutrition* 79: 537–543.

Brook, T. (1998). *The confusions of pleasure: Commerce and culture in Ming China.* Berkeley: University of California Press.

Brown, W. (2015). *Undoing the demos: Neoliberalism's stealth revolution*. New York: Zone Books.

Brown, W. (2019). *In the ruins of neoliberalism: The rise of antidemocratic politics in the West*. New York: Columbia University Press.

Burawoy, M. (2020). "A tale of two Marxisms: Remembering Erik Olin Wright (1947–2019)." *New Left Review* 121: 67–98.

Bureau of Labor Statistics. (2016). "Quarterly census of employment and wages." Accessed January 24, 2016. www.bls.gov.

Burns, H. (2017). "Charlotte drops to third-largest banking hub in US." *Charlotte Business Journal*, May 23, 2017.

Carolan, M. S. (2011). *The real cost of cheap food*. London: Earthscan.

Carolina Demography, University of North Carolina, Chapel Hill. (2018). "Suburban and exurban growth in North Carolina's two major metro areas." January 30, 2018. www.ncdemography.org.

Casas-Cortés, M. I., M. Osterweil, and D. E. Powell. (2008). "Blurring boundaries: Recognizing knowledge-practices in the study of social movements." *Anthropological Quarterly* 81(1): 17–58.

Center for Food Safety. (2022). "The role of pesticides in the extinction crisis." Retrieved December 31, 2022, from www.centerforfoodsafety.org.

Chrzan, J. (2004). "Slow food: What, why, and where?" *Food, Culture & Society* 7(2): 117–132.

Clapp, J. (2016). *Food*. 2nd ed. Cambridge, UK: Polity.

Clapp, J., and D. A. Fuchs. (2009). *Corporate power in global agrifood governance*. Cambridge, MA: MIT Press.

Clapp, J., and S. R. Isakson. (2018). *Speculative harvest: Financialization, food, and agriculture*. Rugby, UK: Practical Action / Fernwood.

Cohen, M. (2012). "Mitt Romney's 47% gaffe makes him 100% unsuitable to be president." *The Guardian*, September 18, 2012. www.theguardian.com.

Coleman-Jensen, A., M. Nord, and A. Singh. (2013). *Household food security in the United States in 2012*. ERR-155. US Department of Agriculture, Economic Research Service, September 2013.

Collins, J. L., and V. Mayer. (2010). *Both hands tied: Welfare reform and the race to the bottom in the low-wage labor market*. Chicago: University of Chicago Press.

Community Food Security Coalition. (2001). *The Healthy Farms, Food, and Communities Act: Policy initiatives for the 2002 Farm Bill and the first decade of the 21st century*. Venice, CA: Community Food Security Coalition.

Community Food Security Coalition. (2008). "Welcome to the Community Food Security Coalition." www.foodsecurity.org.

Cone, M. (2011). "Pesticides may block male hormones: Many agricultural pesticides disrupt male hormones, according to new tests." *Scientific American*, February 15, 2011.

Cosier, S. (2019). "The world needs topsoil to grow 95% of its food—but it's rapidly disappearing." *The Guardian*, May 30, 2019.

Dagnino, E. (2005). "'We all have rights, but . . .': Contesting concepts of citizenship in Brazil." In *Inclusive citizenship: Meanings and expressions*, edited by N. Kabeer, 149–163. London: Zed Books.

Dahlberg, K. A. (1991). "Sustainable agriculture—fad or harbinger?" *Bioscience* 41(5): 337–340.

Dean, M. (1999). *Governmentality: Power and rule in modern society*. Thousand Oaks, CA: Sage.

Dees, J. G. (2007). "Taking social entrepreneurship seriously." *Transactions: Social Science and Modern Society* 44(1): 24–31.

Dees, J. G., J. Emerson, and P. Economy. (2002). *Enterprising nonprofits: A toolkit for social entrepreneurs*. New York: Wiley.

De Marco, A., and H. Hunt. (2018). "Race, inequality, poverty and gentrification in Durham, North Carolina." North Carolina Poverty Research Fund, University of North Carolina School of Law, Chapel Hill.

Desmarais, A. A. (2007). *La Vía Campesina: Globalization and the power of peasants*. Black Point, NS: Fernwood.

Dickenson, M. (2020). *Feeding the crisis: Care and abandonment in America's food safety net*. Oakland: University of California Press.

Domhoff, G. W. (2022). *Who rules America? The corporate rich, white nationalist Republicans, and inclusionary Democrats in the 2020s*. New York: Routledge.

Dorning, M. A., J. Koch, D. A. Shoemaker, and R. K. Meentemeyer. (2015). "Simulating urbanization scenarios reveals trade-offs between conservation planning strategies." *Landscape and Urban Planning* 136: 28–39.

Duong, Y. (2019). "Friendship Trays fills a niche in Charlotte's support system." *NC Health News*, January 11, 2019. www.northcarolinahealthnews.org.

Escobar, A. (2008). *Territories of difference: Place, movements, life,* redes. Durham, NC Duke University Press, 2008.

Escobar, A., and S. E. Alvarez. (1992). *The Making of social movements in Latin America: Identity, strategy, and democracy*. Boulder, CO: Westview.

European Commission. (2009). "Press release: EU and Canada settle WTO case on genetically modified organisms." July 15, 2009. http://europa.eu.

Exum, Riyah. (2013). "A real food revolution: Edible backyards." *Durham Voice*, October 4, 2013. https://durhamvoice.org (archived).

F.A.R.M. Café. (2018a). "Our story." Accessed April 12, 2018. http://farmcafe.org.

F.A.R.M. Café. (2018b). "Where does the food come from?" Accessed April 12, 2018. http://farmcafe.org.

F.A.R.M. Café. (2022a). "Interning." Accessed May 13, 2022. http://farmcafe.org.

F.A.R.M. Café. (2022b). "Our community." Accessed May 13, 2022. http://farmcafe.org.

F.A.R.M. Café. (2022c). "Role descriptions—volunteers." Accessed May 13, 2022. http://farmcafe.org.

F.A.R.M. Café. (2022d). "We are F.A.R.M. Café." Accessed May 13, 2022. http://farmcafe.org.

Feeding America. (2013). *Federal tax incentives for food donations*. Washington, DC: Feeding America.

Ferguson, R. S., and S. T. Lovell. (2014). "Permaculture for agroecology: Design, movement, practice, and worldview. A review." *Agronomy for Sustainable Development* 34: 251–274.

Foley, J. A., R. DeFries, G. P. Asner, C. Barford, G. Bonan, S. R. Carpenter, F. S. Chapin, et al. (2005). "Global consequences of land use." *Science* 309: 570–574.

Foley, J. A., N. Ramankutty, K. A. Brauman, E. S. Cassidy, J. S. Gerber, M. Johnson, N. D. Mueller, et al. (2011). "Solutions for a cultivated planet." *Nature* 478: 337–342.

Food and Agriculture Organization. (2020). *The state of the world's forests: Forests, biodiversity, and people, 2020.* Rome: Food and Agriculture Organization. www.fao.org.

Food and Water Watch. (2010). *Food and agricultural biotechnology industry spends more than half a billion dollars to influence Congress.* Washington DC: Food and Water Watch.

Food Security. (2022). "Community food security." Accessed October 6, 2022. https://foodsecurity.org.

Forbes. (2009). "America's ten most impoverished cities." October 12, 2009. www.forbes.com.

Frunza, M.-C. (2015). *Solving modern crime in financial markets: Analytics and case study.* San Diego: Academic Press.

Gaytan, M. S. (2007). "Globalizing resistance: Slow Food and new local imaginaries." *Food, Culture & Society* 10: 97–116.

Gerber, P., H. Steinfeld, B. Henderson, A. Mottet, C. Opio, J. Dijkman, A. Falcucchi, and G. Tempio. (2013). *Tackling climate change through livestock—A global assessment emissions and mitigation opportunities.* Rome: Food and Agricultural Organization.

Gerstle, G. (2022). *The rise and fall of the neoliberal order: America and the world in the free market era.* New York: Oxford University Press.

Gibson-Graham, J. K. (2006). *A postcapitalist politics.* Minneapolis: University of Minnesota Press.

Gilbert, N. (2012). "One third of our greenhouse gas emissions come from agriculture: Farmers advised to abandon vulnerable crops in face of climate change." *Nature,* October 31, 2012. www.nature.com.

Graeber, D. (2004). *Fragments of an anarchist anthropology.* Chicago: Prickly Paradigm.

Gramsci, A. (1971). *Selections from the Prison Notebooks.* Edited by Q. Hoare and G. N Smith. New York: International.

Gray, M. (2014). *Labor and the locavore: The making of a comprehensive food ethic.* Berkeley: University of California Press.

Griffin, N. (2019). "Charlotte suburbs grow faster as developers seek cheap land." *Plan Charlotte* (UNC Charlotte Urban Institute), October 9, 2019.

Grossberg, L. (2010). *Cultural studies in the future tense.* Durham, NC: Duke University Press.

Gudeman, S. (2001). *The anthropology of economy: Commodity, market, and culture.* Oxford, UK: Blackwell.

Gunstone, T., T. Cornellisse, K. Klein, A. Dubey, and N. Donley. (2021). "Pesticides and soil invertebrates: A hazard assessment." *Frontiers in Environmental Science* 9: 643847. https://doi.org/10.3389/fenvs.2021.643847.

Guthman, J. (2008a). "Bringing good food to others: Investigating the subjects of alternative food practice." *Cultural Geographies* 15: 431–447.

Guthman, J. (2008b). "Thinking inside the neoliberal box: The micro-politics of agro-food philanthropy." *Geoforum* 39(3): 1241–1253.

Guthman, J., and M. DuPuis. (2006). "Embodying neoliberalism: Economy, culture, and the politics of fat." *Environment and Planning D-Society & Space* 24: 427–448.

Gutiérrez Escobar, L., and E. Fitting. (2016). "The Red de Semillas Libres: Contesting biohegemony in Colombia." *Journal of Agrarian Change* 16(4): 711–719.

Hanchett, T. W. (2020). *Sorting out the New South city: Race, class, and urban development in Charlotte, 1875–1975.* 2nd ed. Chapel Hill: University of North Carolina Press.

Hardisty, J. V. (1999). *Mobilizing resentment: Conservative resurgence from the John Birch Society to the Promise Keepers.* Boston: Beacon.

Harris, P. (2013). "Monsanto sued small famers to protect seed patents, report says." *The Guardian*, February 12, 2013.

Harvest Moon Grille. (2020). Facebook page. September 9, 2020. www.facebook.com /HMGCart/.

Harvey, F. (2014). "EU under pressure to allow GM food imports from US and Canada." *The Guardian*, September 5, 2014.

Hasenfeld, Y., and E. E. Garrow. (2012). "Nonprofit human-service organizations, social rights, and advocacy in a neoliberal welfare state." *Social Science Review* 86(2): 295–322.

Hassanein, N. (2003). "Practicing food democracy: A pragmatic politics of transformation." *Journal of Rural Studies* 19(1): 77–86.

Hauter, W. (2012). *Foodopoly: The battle over the future of food and farming in America.* New York: New Press.

Heller, C. (2013). *Food, farms and solidarity: French farmers challenge industrial agriculture and genetically modified crops.* Durham NC: Duke University Press.

Holland, D. C., W. S. Lachicotte, D. Skinner, and C. Cain. (1998). *Identity and agency in cultural worlds.* Cambridge, MA: Harvard University Press.

Holland, D. C., D. M. Nonini, C. Lutz, L. Bartlett, M. Frederick, T. Guldbrandsen, and E. Murillo. (2007). *Local democracy under siege: Activism, public interests and private politics.* New York: New York University Press.

Holmgren, D. (2004). *Permaculture: Principles and pathways beyond sustainability.* Hepburn, Australia: Holmgren Design Services.

Hossfeld, L., M. Legerton, and G. Keuster. (2004). "The economic and social impact of job loss in Robeson County, North Carolina 1993–2003." *Sociation Today* 2(2): 1–18.

Huber, M. T. (2022). *Climate change as class war: Building socialism on a warming planet.* London: Verso Books.

IARC (International Agency on Research on Cancer). (2016). "Q&A on glyphosate." March 1, 2016. www.iarc.fr.

Incite! (2007/2017). *The revolution will not be funded: Beyond the non-profit industrial complex.* Durham, NC: Duke University Press.

IPBES (Intergovernmental Science-Policy Platform on Biodiversity and Ecosystem Services). (2019). *Summary for policymakers of the global assessment report on biodiversity and ecosystem services of the Intergovernmental Science-Policy Platform on Biodiversity and Ecosystem Services.* Paris: United Nations.

IPCC (Intergovernmental Panel on Climate Change). (2013). *IPCC summary for policy makers: Climate change 2013: The physical science basis. Contribution of Working Group I to the Fifth Assessment Report of the Intergovernmental Panel on Climate Change.* Cambridge: Cambridge University Press.

IPCC (Intergovernmental Panel on Climate Change). (2014). *Climate change 2014: Synthesis report—Summary for policy makers.* Geneva: Intergovernmental Panel on Climate Change.

IPCC (Intergovernmental Panel on Climate Change). (2018). *Summary for policymakers: Global Warming of 1.5°C—An IPCC special report on the impacts of global warming of 1.5°C above pre-industrial levels and related global greenhouse gas emission pathways, in the context of strengthening the global response to the threat of climate change, sustainable near development, and efforts to eradicate poverty.* Incheon, Korea: Intergovernmental Panel on Climate Change.

IPCC (Intergovernmental Panel on Climate Change). (2019). *Climate change and land: An IPCC special report on climate change, desertification, land degradation, sustainable land management, food security, and greenhouse gas fluxes in terrestrial ecosystems: Summary for policymakers.* Geneva: IPCC.

Isaacs, Martha. (2015). "Power in Dirt: Two Urban Farms and their Strategies for Neighbourhood Cultivation." The Glass-House, July 22, 2015. www.theglasshouse.org.uk.

Israel, M. (2019). "Farms and sprawl: Conservationists worry they're losing the battle." *Carolinas Urban-Rural Connection* (UNC Charlotte Urban Institute), November 5, 2019.

Jasper, J. M. (1997). *The art of moral protest: Culture, biography, and creativity in social movements.* Chicago: University of Chicago Press.

Johnson, C. (2019). "Trumpism, policing and the problem of surplus population." In *Labor in the time of Trump,* edited by J. Kerrisey, E. Weinbaum, C. Hammonds, T. Juravich, and D. Clawson, 169–190. Ithaca, NY: Cornell University Press.

Johnson, S., and J. Kwak. (2010). *Thirteen bankers: The Wall Street takeover and the next financial meltdown.* New York: Vintage Books.

Kapferer, B. (2005a). "Introduction: Oligarchic corporations and new state formations." *Social Analysis* 49(1): 163–176.

Kapferer, B. (2005b). "New formations of power, the oligarchic-corporate state, and anthropological ideological discourses." *Anthropological Theory* 5(3): 285–299.

Kapferer, B. (2010). "The aporia of power: Crisis and the emergence of the corporate state." *Social Analysis* 54(1): 125–151.

Kauffman, J. (2017) "The rise of the modern food cooperative." *SFGate*, March 16, 2017. www.sfgate.com.

Kelley, R. D. G. 2009. *Thelonious Monk: The life and times of an American original.* New York: Free Press.

Khachatourians, G. G. (1998). "Agricultural use of antibiotics and the evolution and transfer of antibiotic-resistant bacteria." *CMAJ* 159(9): 1129–1136.

Klein, N. (2014). *This changes everything: Capitalism vs. the climate.* New York: Simon and Schuster.

Knupfer, A. M. (2013). *Food co-ops in America: Communities, consumption, and economic democracy.* Ithaca, NY: Cornell University Press.

Kornegay, B. (2022). "A vision of community and dignity: The leading women of F.A.R.M. Cafe." *Watauga Democrat* (Boone, NC), April 1, 2022.

Krais, B. (1993). "Gender and symbolic violence: Female oppression in the light of Pierre Bourdieu's theory of social practice." In *Bourdieu: Critical perspectives*, edited by C. Calhoun, E. LiPuma, and M. Postone, 156–178. Chicago: University of Chicago Press.

Kurzman, C. (2004). "The poststructuralist consensus in social movement theory." In *Rethinking social movements: Structure, meaning and emotion*, edited by J. Goodwin and J. M. Jasper, 111–120. Lanham, MD: Rowman and Littlefield.

Kurzman, C. (2008). "Meaning-making in social movements." *Anthropological Quarterly* 81(1): 5–16.

Langemeyer, I. (2011). "Science and social practice: Activity research and activity theory as socio-critical approaches." *Mind, Culture, and Activity* 18(2): 148–160.

Lappe, A. (2011). *Diet for a hot planet: The climate crisis at the end of your fork and what you can do about it.* New York: Bloomsbury.

Laudan, R. (2004). "Slow food: The French terroir strategy, and culinary modernism." *Food, Culture & Society* 7(2): 133–144.

Lohnes, J. D. (2021). "Regulating surplus: Charity and the legal geographies of food waste enclosure." *Agriculture and Human Values* 38: 351–363.

Long, M. A., L. Gonçalves, P. B. Stretesky, and M. A. Defeyter. (2020). "Food insecurity in advanced capitalist nations: A review." *Sustainability* 12: 3654–3673.

Lyson, T. (2005). "Civic agriculture and community problem solving." *Culture and Agriculture* 27(2): 92–98.

Malm, A. (2016). *Fossil capital: The rise of steam-power and the roots of global warming.* London: Verso Books.

Mann, G., and J. Wainwright. (2018). *Climate Leviathan: A political theory of our planetary future.* London: Verso.

Mares, T. M., and D. Peña. (2011). "Environmental and food justice: Toward local, slow, and deep food systems." In *Cultivating food justice: Race, class, and sustainability*, edited by A. H. Alkon and J. Agyeman, 197–220. Cambridge, MA: MIT Press.

Marx, K. (1867/1976). *Capital, volume I.* Translated by B. Fowkes. London: Penguin Books.

Matson, P. A., B. J. Parton, A. G. Power, and M. J. Swift. (1997). "Agricultural intensification and ecosystem properties." *Science* 277: 504–509.

Mattera, P. (2004). *USDA Inc.: How agribusiness has hijacked regulatory policy at the U.S. Department of Agriculture*. Corporate Research Project of Good Jobs First. Washington, DC: Good Jobs First.

McAdam, D. (1982). *Political process and the development of Black insurgency, 1930–1970*. Chicago: University of Chicago Press.

McKittrick, K. (2011). "On plantations, prisons, and a black sense of place." *Social and Cultural Geography* 12: 947–963.

McKittrick, K. (2013). "Plantation futures." *Small Axe* 17: 1–15.

McKittrick, K., and C. A. Woods. (2007). *Black geographies and the politics of place*. Toronto / Cambridge, MA: Between the Lines / South End.

McMichael, P. (2009). "A food regime analysis of the 'world food crisis.'" *Agriculture and Human Values* 26: 281–295.

McMichael, P. (2013). *Food regimes and agrarian questions: Agrarian change and peasant studies*. Halifax, NS: Fernwood.

McMichael, P. (2023). "Covid-19 and the future of food." In *Covid-19 and the global political economy: Crises in the 21st century*, edited by T. DiMuzio and M. Dow, 204–219. London: Routledge.

Menegat, S., A. Ledo, and R. Tirado. (2022). "Greenhouse gas emissions from global production and use of nitrogen synthetic fertilizers in agriculture." *Scientific Reports* 12. https://doi.org/10.1038/s41598-022-18773-w.

Moffitt, R. A. (2015). "The deserving poor, the family, and the U.S. welfare system." *Demography* 52: 729–749.

Morales, A. (2011). "Growing food and justice: Dismantling racism through sustainable food systems." In *Cultivating for justice: Race, class, and sustainability*, edited by A. H. Alkon and J. Agyeman, 149–176. Cambridge, MA: MIT Press.

Morello, P. (2021). "Blog: How food banks and food pantries get their food." *Hunger Blog* (Feeding America), December 29, 2021. www.feedingamerica.org.

Moss, M. (2013). "(Salt+ fat2/satisfying crunch) × pleasing mouth feel = a food designed to addict." *New York Time Magazine*, February 20, 2013.

Munshi, S., and C. Wills. (2017). Foreword to *The revolution will not be funded: Beyond the nonprofit industrial sector*, by Incite!, xiii–xxvii. Durham, NC: Duke University Press.

Nash, J. C. (2005). *Social movements: An anthropological reader*. Malden, MA: Blackwell.

National Agricultural Statistics Service. (2015). *Direct farm sales of food: Results from the 2015 local food marketing practices survey*. Washington DC: National Agricultural Statistics Service, US Department of Agriculture.

National Council for Research on Women and D. L. Schultz. (1993). *To reclaim a legacy of diversity: Analyzing the "political correctness" debates in higher education*. New York: National Council for Research on Women.

National Research Council. (2013). *Supplemental Nutrition Assistance Program: Examining the evidence to define benefit adequacy*. Washington, DC: National Academies Press. https://doi.org/10.17226/13485.

Nestle, M. (2013). *Food politics how the food industry influences nutrition and health.* California Studies in Food and Culture 3. Berkeley: University of California Press.

Nonini, D. (2007). *The global idea of "the commons."* Critical Interventions Series. New York: Berghahn Books.

Nonini, D. (2013). "The 'local food movement' and the anthropology of global systems." *American Ethnologist* 40(2): 267–275.

Nonini, D. (2021). "The triplesidedness of 'I can't breathe': The Covid-19 pandemic, enslavement, and agro-industrial capitalism." *Focaal—Journal of Global and Historical Anthropology* 89: 114–129.

Nonini, D., H. Phillips, S. Weir, and A. Fiske. (2015). *"No one . . . saw what was coming down the road": Twenty years of job loss in North Carolina's First Congressional District.* Durham, NC: CommunEcos.

O'Brien, K. L., and R. M. Leichenko. (2000). "Double exposure: Assessing the impacts of climate change within the context of economic globalization." *Global Environmental Change* 10: 221–232.

One World Everybody Eats. (2022). Home page. Accessed May 4, 2022. www.oneworldeverybodyeats.org.

Oosterveer, P., and D. A. Sonnenfeld. (2012). *Food, globalization and sustainability.* New York: Taylor and Francis.

Osborne, D., and T. Gaebler. (1992). *Reinventing government: How the entrepreneurial spirit is transforming the public sector.* Reading, MA: Addison-Wesley.

Ottinger, G. (2013). *Refining expertise: How responsible engineers subvert environmental justice challenges.* New York: New York University Press.

Outz, E. (2007). *Losing our natural heritage: Development and open-space loss in North Carolina.* Raleigh, NC: Environment North Carolina Research & Policy Center.

Patel, R. (2011). "Survival pending revolution: What the Black Panthers can teach the US food movement." In *Food movements unite! Strategies to transform our food systems*, edited by E. Holt-Gimenez, 115–136. Oakland, CA: Food First Books.

Paul, M. (2017). "Community-supported agriculture in the United States: Social, ecological and economic benefits to farming." *Journal of Agrarian Change* 19: 162–180.

Payne, C. M. (1995). *I've got the light of freedom: The organizing tradition and the Mississippi freedom struggle.* Berkeley: University of California Press.

Peck, J., and A. Tickell. (2002). "Neoliberalizing space." *Antipode* 34(3): 380–404.

Pellizzari, T. (2020). "The radical history of the free breakfast program." *Groundviews Blog* (Solid Ground), September 14, 2020. www.solid-ground.org.

Pesticide Action Network North America. (2022). "Agroecology: The solution to highly hazardous pesticides." Accessed August 16, 2022. www.panna.org.

Petrini, C. (2007). *Taking back life: The earth, the moon and abundance.* Bra (Cuneo), Italy: Slow Food. https://slowfoodlondon.blogs.com.

Philpott, T. (2020). *Perilous bounty: The looming collapse of American farming and how we can prevent it.* New York: Bloomsbury.

Pien, D. (2010). "Black Panther Party's Free Breakfast Program (1969–1980)." Black Past, February 11, 2010. www.blackpast.org.

Polanyi, K. (1957). *The great transformation.* Boston: Beacon.

Polletta, F. (2002). *Freedom is an endless meeting: Democracy in American social movements.* Chicago: University of Chicago Press.

Poore, J., and T. Nemecek. (2018). "Reducing foods' environmental impacts through producers and consumers." *Science* 360: 987–982.

Poppendieck, J. (1998). *Sweet charity? Emergency food and the end of entitlement.* New York: Viking.

Poppendieck, J. (2010). *Free for all: Fixing school food in America.* Berkeley: University of California Press.

Poppendieck, J. (2014). "Food assistance, hunger and the end of welfare in the USA." In *First world hunger revisited: Food charity or the right to food?*, edited by G. Riches and T. Silvasti, 176–190. London: Macmillan.

Pothukuchi, K., H. Joseph, H. Burton, and A. Fisher. (2002). *What's cooking in your food system?* Venice, CA: Community Food Security Coalition.

Povitz, L. D. (2019). *Stirrings: How activist New Yorkers ignited a movement for food justice.* Chapel Hill: University of North Carolina Press.

Price, C., D. Nonini, and E. Fox Tree. (2008). "Grounded utopian movements: Subjects of neglect." *Anthropological Quarterly* 81(1): 127–160.

Public Citizen. (2003). "Backgrounder: The US threats against Europe's GMO policy and the WTO SPS agreement." Washington, DC: Public Citizen.

Public Citizen. (2015a). "Case studies: Investor-state attacks on public interest policies." March 6, 2015. www.citizen.org.

Public Citizen. (2015b). "List of 605 corporate insiders—have confidential access to TPP." Accessed May 20, 2015 (no longer accessible).

Public Citizen. (2015c). "WTO Authorizes over $1 billion in sanctions unless U.S. guts popular country-of-origin meat labels, disproving Obama claim that trade pacts can't undermine public interest policies." December 7, 2015. www.citizen.org.

Public Citizen. (2016). "North Carolina job loss during the NAFTA-WTO period." Accessed February 5, 2016. www.citizen.org.

Ramirez, M. M. (2015). "The elusive inclusive: Black food geographies and racialized food spaces." *Antipode* 47(3): 748–769.

Reagan, R. (1981). "First Inaugural Address of Ronald Reagan." Lillian Goldman Law Library, Yale Law School. https://avalon.law.yale.edu.

Rhodes, C. J. (2017). "The imperative for regenerative agriculture." *Science Progress* 100(1): 80–129.

Riches, G. (2018). *Food bank nations: Poverty, corporate charity and the right to food.* London: Routledge.

Riches, G., and T. Silvasti. (2014). "Hunger in the rich world: Food aid and right to food perspectives." In *First world hunger revisited: Food charity or the right to food?*, edited by G. Riches and T. Silvasti, 1–14. Basingstoke, UK: Palgrave Macmillan.

Roberts, P. (2008). *The end of food.* Boston: Houghton Mifflin.

Robinson, C. J. (1983/2000). *Black Marxism: The making of the Black radical tradition.* Chapel Hill: University of North Carolina Press.

Robinson, J. M., and J. R. Farmer. (2017). *Selling local: Why local food movements matter*. Bloomington: Indiana University Press.

Rockström, J., W. Steffen, K. Noone, A. Persson, F. S. I. Chapin III, E. F. Lambin, T. M. Lenton, et al. (2009). "A safe operating space for humanity." *Nature* 461(24): 472–475.

Rodale, R. (1983). "Breaking new ground: The search for a sustainable agriculture." *The Futurist* 1(1): 15–20.

Roff, R. J. (2008). "Preempting to nothing: Neoliberalism in the fight to de/re-regulate agricultural biotechnology." *Geoforum* 39: 1423–1438.

Rose, N. (1999). *Governing the soul: The shaping of the private self*. London: Free Association Books.

Salmon, M. (2012). "Food fight: A case study of the Community Food Security Coalition's campaign for a fair farm bill." Capstone Collection, paper 2560. SIT Graduate Institute, Brattleboro, VT.

Satterfield, T. (2002). *Anatomy of a conflict: Identity, knowledge, and emotion in old-growth forests*. Vancouver: UBC Press.

Scahill, J. (2008). *Blackwater: The rise of the world's most powerful mercenary army*. New York: Nation Books.

Scanlon, B. R., C. C. Faunt, L. Longuevergne, R. C. Reedy, W. M. Alley, V. L. McGuire, and P. B. McMahon. (2012). "Groundwater depletion and sustainability of irrigation in the US High Plains and Central Valley." *PNAS* 109(24): 9230–9235.

Schor, J. B. (1992). *The overworked American: The unexpected decline of leisure*. New York: Basic Books.

Schor, J. B. (1998). *The overspent American: Upscaling, downshifting, and the new consumer*. New York: HarperPerennial.

Schor, J. B. (2011). *True wealth: How and why millions of Americans are creating a time-rich, ecologically light, small-scale, high-satisfaction economy*. New York: Penguin Books.

Schumpeter, J. (1942). *Capitalism, socialism, and democracy*. New York: Harper and Brothers.

Semuels, A. (2017). "Why it's so hard to get ahead in the South." *The Atlantic*, April 4, 2017.

Singer, P. W. (2003). *Corporate warriors: The rise of the privatized military industry*. Ithaca, NY: Cornell University Press.

Sklair, L. (2001). *The transnational capitalist class*. Oxford, UK: Blackwell.

Slezak, M. (2016). "The Great Barrier Reef: A catastrophe laid bare." *The Guardian*, June 6, 2016.

Slocum, R. (2007). "Whiteness, space and alternative food practice." *Geoforum* 38(3): 520–533.

Slocum, R. (2010). "Race in the study of food." *Progress in Human Geography* 35(3): 303–327.

Slow Food. (1989). "Slow Food manifesto." Accessed July 8, 2022. https://slowfood.com.

Slow Food. (2016). "The central role of food." Accessed July 28, 2022. https://slowfood.com.

Slow Food. (2022a). "About us." Accessed July 28, 2022. www.slowfood.com.

Slow Food. (2022b). "Funded projects." Accessed July 28, 2022. www.slowfood.com.

Slow Food. (2022c). "Slow Food Foundation for Biodiversity." Accessed July 28, 2022. www.fondazioneslowfood.com.

Slow Food. (2022d). "Slow Food Presidia." Accessed July 28, 2022. www.fondazioneslowfood.com.

Slow Food Carolina Piedmont. (2009). "Slow Food Carolina Piedmont Region." Accessed July 13, 2022. https://charlottecultureguide.com.

Smith, H., and W. Graves. (2005). "Gentrification as corporate growth strategy: The strange case of Charlotte, North Carolina and the Bank of America." *Journal of Urban Affairs* 27(4): 403–418.

Smith, N. (2008). *Uneven development: Nature, capital, and the production of space.* Athens: University of Georgia Press.

Smythe, E. (2009). "In whose interests? Transparency and accountability in the global governance of food: agribusiness, the Codex Alimentarius, and the World Trade Organization." In *Corporate power in global agrifood governance*, edited by J. Clapp and E. Fuchs, 93–123. Cambridge, MA: MIT Press.

Sousa Santos, B. D. (2006). *The rise of the global left: The World Social Forum and beyond.* London: Zed Books.

Spinosa, C., F. Flores, and H. Dreyfus. (1997). *Disclosing new worlds: Entrepreneurship, democratic action, and the cultivation of solidarity.* Cambridge, MA: MIT Press.

Stetsenko, A. (2008). "From relational ontology to transformative activist stance on development and learning: Expanding Vygotsky's (CHAT) project." *Cultural Studies of Science Education* 3: 471–491.

Taparia, H., and P. Koch. (2015). "Real food challenges the food industry." *New York Times*, November 8, 2015.

Thaler, E., J. S. Kwang, B. J. Quirk, C. L. Quarrier, and I. J. Larsen. (2022). "Rates of historical anthropogenic soil erosion in the midwestern United States." *Earth's Future* 10(3): 1–16. https://doi.org/10.1029/2021EF002396.

Tollefson, J. (2019). "1 million species face extinction." *Nature* 569: 171.

Twin County Museum & Hall of Fame. (2022). "Thelonious Monk." Accessed October 6, 2022. https://tchof.org.

US Census Bureau. (2010a). "Census urban and rural classification and urban area criteria." Accessed June 20, 2022. www.census.gov.

US Census Bureau. (2010b). "U.S. census tables." Accessed June 20, 2022. https://data.census.gov.

US Census Bureau. (2023). "Quick facts: Watauga County, North Carolina." Accessed June 27, 2023. www.census.gov.

USDA (US Department of Agriculture). (2012a). "County profile: Edgecombe County." National Agricultural Statistics Service. https://agcensus.library.cornell.edu.

USDA (US Department of Agriculture). (2012b). "Special Supplemental Nutrition Program for Women, Infants and Children (WIC): Income eligibility guidelines." Food and Nutrition Service. *Federal Register* 77: 57 (March 23, 2012).

USDA (US Department of Agriculture). (2019a). "A short history of SNAP." Food and Nutrition Service. www.fns.usda.gov.

USDA (US Department of Agriculture). (2019b). "SNAP work requirements." Food and Nutrition Service. www.fns.usda.gov.

USDA (US Department of Agriculture). (2021a). "Federal incentives for businesses to donate food." www.usda.gov.

USDA (US Department of Agriculture). (2021b). *Household food security in the United States in 2020.* Economic Research Service. www.ers.usda.gov.

USDA (US Department of Agriculture). (2022). "Organic certification made simple." Agricultural Marketing Service. www.ams.usda.gov.

Vallianatos, M., R. Gottlieb, and M. A. Haase. (2004). "Farm-to-school—Strategies for urban health, combating sprawl, and establishing a community food systems approach." *Journal of Planning Education and Research* 23(4): 414–423.

van der Werf, G. R., D. C. Morton, R. S. DeFries, J. G. J. Olivier, P. S. Kashibhatla, R. B. Jackson, G. J. Collatz, and J. T. Randerson. (2009). "CO2 emissions from forest loss." *Nature Geoscience* 2: 737–738.

Wallace, R. (2020). "Update: Agriculture, capital, and infectious diseases." In *Transformation of our food systems: The making of a paradigm shift—data/updates/reports—reflections since IAASTD—10 year on,* edited by H. R. Herren, B. Haerlin and IAASTD Group. Berlin: Foundation on Future Farming (Zukunftsstiftung Landwirtschaft).

Wallace, R., A. Liebman, L. F. Chaves, and R. Wallace. (2020). "COVID-19 and circuits of capital." *Monthly Review* 72(1): 1–13.

Wang, J. (2018). *Carceral capitalism.* Pasadena, CA: Sociotext.

Watauga Farmers Market. (2011). Local Harvest, March 19, 2015. www.localharvest.org.

Webb, S., and B. Webb. (1907). *The history of trade unionism.* London: Longmans, Green.

Webb, S., and B. Webb. (1921). *The consumers' co-operative movement.* London: Longmans, Green.

Wedel, J. R. (2009). *Shadow elite: How the world's new power brokers undermine democracy, government, and the free market.* New York: Basic Books.

Weis, A. J. (2007). *The global food economy: The battle for the future of farming.* London: Zed Books.

Wekerle, G. R. (2004). "Food justice movements: Policy, planning and networks." *Journal of Planning Education and Research* 23: 378–386.

Wilson, A. B. (2010). "Case study: Local food in low-wealth communities: Orange and Chatham Counties (Executive summary draft)." Department of Anthropology, University of North Carolina, Chapel Hill.

Wilson, A. B., and D. Holland. (2009). "Movements & moralities: Evolving moral logics of food in the United States." Paper presented for the invited session "Evolving moralities: When the good and the right need to be reconfigured," 108th annual meeting of the American Anthropological Association, Philadelphia, December 5, 2009.

Winne, M. (2008). *Closing the food gap: Resetting the table in the land of plenty*. Boston: Beacon.

Wolford, W. (2009). *This land is ours now: Social mobilization and the meaning(s) of land in northeastern Brazil*. Durham, NC: Duke University Press.

Woods, C. A. (1998). *Development arrested: The blues and plantation power in the Mississippi Delta*. London: Verso.

WRAL News. (2009). "Curiosity takes root as Durham students grow garden." November 19, 2009. www.wral.com.

Wright, E. O. (2010). *Envisioning real utopias*. London: Verso.

Wright, W. J. (2020). "The morphology of marronage." *Annals of the American Association of Geographers* 110(4): 1134–1149.

Yonto, D., and J.-C. Thill. (2020). "Gentrification in the US New South: Evidence from two types of African-American communities in Charlotte." *Cities* 97: 1–10.

Zepezauer, M. (2004). *Take the rich off welfare*. Cambridge, MA: South End.

Zuraw, L. (2015). "House votes to repeal country-of-origin labeling for meat." *Food Safety News*, June 15, 2015. www.foodsafetynews.com.

ABOUT THE AUTHORS

DONALD M. NONINI is Professor Emeritus of Anthropology, University of North Carolina, Chapel Hill. He is the author of *"Getting By": Class and State Formation among Chinese in Malaysia*; coauthor of *Local Democracy under Siege: Activism, Public Interests, and Private Politics*, which won the Delmos Jones and Jagna Sharff Memorial Prize for the Best Book in the Critical Study of North America; author of one other book; and editor or coeditor of three other books.

DOROTHY C. HOLLAND was Boshamer Professor Emeritus of Anthropology, University of North Carolina, Chapel Hill. She was a coauthor of *Local Democracy under Siege: Activism, Public Interests, and Private Politics*, which won the Delmos Jones and Jagna Sharff Memorial Prize for the Best Book in the Critical Study of North America; coauthor of *Identity and Agency in Cultural Worlds*; and editor or coeditor of three other books.

www.ingramcontent.com/pod-product-compliance
Lightning Source LLC
Chambersburg PA
CBHW031137020426
42333CB00013B/418